CATIONIC POLYMERIZATION OF OLEFINS:
A CRITICAL INVENTORY

CATIONIC POLYMERIZATION OF OLEFINS: A CRITICAL INVENTORY

Joseph P. Kennedy

Institute of Polymer Science
The University of Akron

A WILEY-INTERSCIENCE PUBLICATION

John Wiley & Sons   New York · London · Sydney · Toronto

Library of Congress Cataloging in Publication Data:

Kennedy, Joseph Paul, 1928-
    Cationic polymerization of olefins.

    "A Wiley-Interscience publication."
    Includes bibliographical references and index.
    1. Olefines. 2. Polymers and polymerization.
I. Title.

QD305.H7K38   547'.8432'234   74-30124
ISBN 0-471-46909-2

Printed in the United States of America

10 9 8 7 6 5 4 3 2 1

Pour ma Honeybee.
. . . pour tous les cadeaux oubliés.

FOREWORD

Cationic polymerization has achieved not only scientific stature, but also substantial practical importance in the last decades. It is thus very appropriate in reviewing this significant field to have both the academic and industrial aspects put into proper perspective. Professor Kennedy is eminently qualified to do both, combining his years of industrial and academic experience and his personal evaluations to the benefit of the readers. Further, he has the rather unique flair to write not only a scientifically excellent book, but also to make it very readable. He has also succeeded in putting into perspective his field as an integral part of the broader area of electrophilic reactions. His book thus provides a wealth of useful data and information not only for the polymer chemist, but for the organic chemist interested in electrophilic and carbocationic processes.

It is obvious that such a book could have not been written even a short while ago. Rapid progress in our understanding of both the basic and practical aspects of cationic polymerization processes has recently advanced the field to the point where for the first time a thorough discussion of the fundamentals can be combined with explanation, and to some degree prediction of the practical aspects.

I feel that Professor Kennedy has succeeded in all of his goals, and his most enjoyable book will be of great value to everyone interested in the field.

GEORGE A. OLAH

September 1974
Cleveland, Ohio

vii

PREFACE

Cationic polymerization of olefins is a rapidly growing, vigorously expanding discipline that utilizes relatively inexpensive raw materials and leads to useful products. It was founded on solid theoretical concepts, so that research in this field can be scientifically and technologically extremely satisfying. Thus cationic polymerizations occupies a large number of workers both in industrial-application laboratories and in universities.

I have been actively engaged in cationic polymerization research for more than 15 years in both industrial and academic institutions. While in industry, I was often accused of being too academic in outlook, whereas lately in academa, I have heard rumors that I am too practical minded. I am happy and satisfied with these "criticisms." Indeed, I hope that both criticisms are true and correct because I believe that one cannot perform effective industrial research without sustained fundamental inquiry and, similarly, one cannot perform good academic research without paying some attention to usefulness and industrial relevance. I have stated: My research philosophy calls for generation of useful knowledge; my strategy calls for the highest quality fundamental research; my tactic focuses on synthetic-polymer and physical-organic chemistry relevant to modern industry and technology; my hope is that society finds my work applicable (J. P. Kennedy, *Chemtech,* p. 156, March 1974).

In this book I have compiled and organized parts of my working assets that were of great help in formulating many exciting industrial and academic research proposals.

During a research discussion among polymer chemists the question often arises as to whether this or that monomer has been polymerized cationically and if so under what conditions, and what happened in terms of rates, molecular weights, structure, and so on. To find the answers to these questions, the researcher must do a great deal of library work; there is no short cut. Over the years I have assembled a large inventory of this kind of information and have critically examined much of it. It did not take much work to compile this information and render it comprehensive relative to the cationic polymerization of olefins.

Thus this book includes a comprehensive inventory of essentially all cationically polymerizable or oligomerizable olefins described up to 1973 in the scientific literature. Because of its prohibitive size the patent literature has not been searched comprehensively, however, numerous important selections have been made. In addition, this book includes a system of cationically polymerizable olefins, their polymerization behavior, a description, if available, of the physical properties of the polymers or oligomers, and a critical evaluation of much of the available information. The organization and criticism of this material has led in many instances to new insight and better understanding of the field.

JOSEPH P. KENNEDY

September 1974
Akron, Ohio

CONTENTS—————————————————

CATIONIC POLYMERIZATION OF OLEFINS:
A CRITICAL INVENTORY

CHAPTER 1

INTRODUCTION

Cationic polymerizations are polyadditions in which the growing species is electrophilic. The electrophile may be a carbenium ion, oxonium ion, carboxonium ion, sulfonium ion, immonium ion, selenonium ion, siliconium ion, and so on. Electroneutrality is maintained by the presence of a counteranion whose nature is determined by the initiating system employed. In high energy radiation induced cationic polymerizations, the initiation mechanism is poorly understood and the nature of the counteranion is obscure; it is theorized that a solvated, immobilized electron provides electroneutrality.

Why are cationic polymerizations of import to practicing polymer chemists? What are the most important distinguishing earmarks of this field of polymer chemistry? What features of cationic polymerizations make this field of synthetic polymer chemistry unique *vis-à-vis* other chain growth or polyaddition reactions, that is, radical, anionic, and coordinative polymerizations?

Perhaps the first answer that comes to mind is that cationic polymerizations are extremely *useful*. Cationic processes are used to produce a large number of technologically useful, commercially available, inexpensive high polymers and oligomers, for example, polyisobutylene (Vistanex of the Exxon Chemical Company in the United States and Oppanol B of the Badische Anilin & Soda-Fabrik A.G. in Germany), butyl rubber, a random copolymer of isobutylene and isoprene (of the Exxon Chemical Company), polyvinyl isobutyl ether (Oppanol C of the Badische Anilin & Soda-Fabrik A.G.), a random copolymer of trioxane and ethylene oxide (Celcon of the Celanese Corporation of America), and so on. Indeed, these products can be made *only* by cationic techniques.

Another answer is that cationic polymerizations offer a very *great variety* of possibilities to the chemist. Thus, there are many hundreds of cationically responsive monomers and the structural variety of these monomers is also extraordinarily large. Cationically polymerizable monomers include olefins

1

of all kinds, vinyl ethers, cyclic ethers, -sulfides, -imines, aldehydes, oxalates, lactones, lactams, and so on, close to 500 compounds. This number is much larger than the number of monomers polymerizable by anionic or coordinative mechanisms (radicals, of course, initiate the polymerization of the largest numbers of monomers).

Further, many high polymers with unique structures can only be obtained by cationic polymerization processes, for example,

$$-CH_2-\overset{CH_3}{\underset{CH_3}{\diagup\!\!\!\diagdown\;C-}}$$ β-pinene

$$-CH_2-\overset{CH_3}{\underset{CH_3}{C}}-CH_2-$$ 3-Methyl-1-butene

Norbornadiene

$$-\langle\bigcirc\rangle-CH_2-$$ Benzyl chloride

$-O-CH_2-CH_2-CH_2-CH_2-$ Tetrahydrofuran (perfectly alternating ethylene oxide–ethylene copolymer)

$$-CH_2-\overset{CH_3}{\underset{H}{C}}-\overset{CH_3}{\underset{CH_3}{C}}-$$ 3,3-Dimethyl-1-butene

$-O-CH_2-CH_2-O-CH_2-$ Dioxolane (perfectly alternating ethylene–formaldehyde oxide copolymer)

Other than for the production of high polymers, cationic polymerization techniques are used exclusively for the synthesis of numerous important oligomers. These relatively low molecular weight, usually oily products (up to 2000 $\bar{M}_n$) or viscous liquids ($\bar{M}_n \approx 3000–8000$) provide us with a wide range of diverse materials used as lubricating oils, additives, resins, sealants, adhesives, coatings, thickening agents, detergents, processing aids, and so on, products that contribute significantly to our daily comfort, health, and safety.

CHAPTER 2

NOTE ON TERMINOLOGY

CARBENIUM IONS

At the International Colloquium on Cationic Polymerization, 1972 (1), this author urged polymer scientists to adopt Olah's recent proposal (2) to use the term "carbenium ions" to designate trivalent trigonal sp^2-hybridized carbocations, the parent of which is $CH_3^{\oplus}$. The now common term "carbonium ion" should describe pentavalent carbocations, the parent of which is $CH_5^{\oplus}$. This proposition is in line with accepted chemical nomenclature according to which the suffix *-onium* raises the valence state or coordination number of the central atom by one unit, for example, oxonium, sulfonium, ammonium, and phosphonium. German literature still occasionally uses "carbenium ion" to denote $R_3C^{\oplus}$. Carbenium ions and carbonium ions, as well as the species $CH_4^{\oplus}$ which has been postulated to exist by mass spectroscopists, are carbocations:

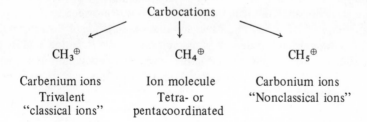

	Carbocations	
$CH_3^{\oplus}$	$CH_4^{\oplus}$	$CH_5^{\oplus}$
Carbenium ions	Ion molecule	Carbonium ions
Trivalent	Tetra- or	"Nonclassical ions"
"classical ions"	pentacoordinated	

The propagating species commonly encountered in cationic polymerizations are thus carbenium ions.

The actual naming of the propagating carbenium ion has not yet been settled and workers in the field use various kinds of nomenclatures to denote the propagating ion; for example,

$$\sim\!\!\sim\!\!CH_2-\overset{\oplus}{\underset{\underset{C_6H_5}{|}}{C}H}$$

is the propagating styryl ion and or $\sim\!\!\sim\!\!CH_2-\overset{\oplus}{C}(CH_3)_2$ is the growing isobutylene cation.

INITIATORS, COINITIATORS, AND PROMOTERS

The chemicals that start the propagation of cationically polymerizable monomers, be these olefins, vinyl ethers, cyclic ethers, and so on, are usually consumed in the course of the initiation and/or propagation steps. Consequently, according to the classical concept of catalysis, these agents are not catalysts but initiators. It is urged that the term "catalyst" be discontinued and the term "initiator" be used exclusively to describe chemical agents that initiate cationic polymerization.

As discussed in Chapter 3, in 1946 Evans, Polanyi, and coworkers discovered that BF_3 requires the presence of traces of water to initiate the polymerization of isobutylene and thus introduced the concept of "coinitiation" into the science of cationic polymerization. Since that time it has been increasingly recognized that purest Friedel-Crafts halides are, more often than not, unable to initiate cationic polymerizations and that they usually (but not always) require for initiation the presence of suitable Brønsted acids (protogens) or other cation sources (cationogens). Indeed many Brønsted acid/Friedel-Crafts acid combinations are among the strongest acids known (BF_3/HF, $SnCl_4/HCl$, $TiCl_4/H_2O$, $AlBr_3/HBr$) and are most effective and useful as cationic initiator systems. In this historical context, then, it is understandable that most authors who are or have been active in this field designated the Friedel-Crafts acid moiety of these combinations the "catalyst" or "initiator," and the Brønsted acid moiety as the "cocatalyst" or "coinitiator." This misleading terminology survived in the face of solid evidence that the true initiating species was provided by the Brønsted acid and not by the Friedel-Crafts acid and that the function of the Friedel-Crafts acid was merely to aid, probably by complexation, the Brønsted acid or other cation source in generating the initiating entity. The early authors recognized that the role of the Friedel-Crafts acid is, in the first approximation, to remove and coordinatively stabilize the halide anion:

$$BF_3 + HF \quad \rightarrow \quad H^{\oplus} \ BF_4^{\ominus}$$

or

$$BF_3 + RF \quad \rightarrow \quad R^{\oplus} \ BF_4^{\ominus}$$

and that *de facto* initiation, that is, the creation of the first propagating species, is accomplished by the Brønsted acid or cation source:

$$H^\oplus + C{=}C \quad \rightarrow \quad H{-}C{-}C^\oplus$$

or

$$R^\oplus + C{=}C \quad \rightarrow \quad R{-}C{-}C^\oplus$$

Since initiation is accomplished by the protogen or other cation provided by the Brønsted acid or other cationogenic species, respectively, these materials and not the Friedel-Crafts acids should be called "initiators." Thus it is that the misleading historical terminology be abandoned and a better, more descriptive nomenclature be adopted, according to which the Friedel-Crafts acid is denoted the "coinitiator" and the Brønsted acid or cationogen the "initiator."

Other terminologies (syncatalysts, duel acids, etc.) that have been used occasionally to designate initiator/coinitiator systems should also be abandoned.

A word of caution on the use of the term "promoter." This term is often used, particularly in the American patent literature, to designate protogenic or cationogenic materials required for initiation in conjunction with Friedel-Crafts acid. This is an obviously misleading terminology and should be abandoned. The term "promoter," however, is useful in describing true promoting, that is, accelerating action. For example, the overall rate of polymerization of THF with oxonium salts $\overset{\oplus}{O}Et_3BF_4^\ominus$ is rather slow but can be accelerated by introducing small amounts of three- or four-membered cyclic ethers. The latter are true promoters—they do not initiate but accelerate a slow reaction. Slow cationic reactions can often be accelerated by using a polar solvent; however, in most cases the role of the polar medium is merely to stabilize the polar transition states by nonspecific solvation. Obviously one should not call these nonspecific solvents "promoters." In certain instances, however, polar solvents may have specific functions and may enter the initiating or propagating event in stoichiometric quantities, for example, the opening of the Al_2Me_6 dimer by CH_3Cl. In these instances the term "promoter" may be appropriate.

The definition of a Friedel-Crafts reaction or Friedel-Crafts acid is rather elusive. In spite of valiant efforts by eminent scientists such as C. K. Ingold* or G. A. Olah†, no satisfactory rigorous definition of a Friedel-Crafts reaction or Friedel-Crafts acid or halide (i.e., an acid or halide that induces a Friedel-Crafts reaction) is available. It appears that experts in this field know what a Friedel-Crafts reaction is—acid, or halide—but cannot define it.

Unfortunately, Natta et al. (6) further diluted the rather vague definition of Friedel-Crafts reactions by coining the term "modified Friedel-Crafts catalysts" to denote halides of multivalent metals in their highest valence state in which the halogens are partly substituted by organic groups. As this terminology does not aid the definition of Friedel-Crafts catalysts, and a modified vague term becomes even vaguer, it should be abandoned.

*"Friedel-Crafts reactions embrace all electrophilic reactions catalyzed by electron-deficient compounds—Lewis acids—whether these are molecules or cations, and include such reactions as are likewise catalyzed by those proton acids which are strong enough to act somewhat like Lewis acids, perhaps like the proton they donate, if it were free, when it would be a Lewis acid (as proton acids themselves are not)" (3).

†". . . Friedel-Crafts type reactions [are] those processes which proceed under the general conditions laid down by the pioneering investigators, and which can also be carried out by the later realization of the general acid-catalyzed nature of the reactions" (4) and ". . . we consider Friedel-Crafts type reactions to be any substitution, isomerization, elimination, cracking, polymerization, or addition reactions taking place under the catalytic effect of Lewis acid type acidic halides (with or without cocatalyst) or proton acid" (5).

REFERENCES

1. J. P. Kennedy, *J. Macromol. Sci., Chem.,* **A6,** 949 (1972).
2. G. A. Olah, *J. Am. Chem. Soc.,* **94,** 808 (1972).
3. C. K. Ingold in *Friedel-Crafts and Related Reactions,* Vol. I, G. A. Olah, Ed., Interscience, New York, 1963, p. vii.
4. G. A. Olah in *Friedel-Crafts and Related Reactions,* Vol. I, G. A. Olah, Ed., Interscience, New York, 1963, p. xi.
5. G. A. Olah, in *Friedel-Crafts and Related Reactions,* Vol. I, G. A. Olah, Ed., Interscience, New York, 1963, p. 29.
6. G. Natta, G. Dell'Asta, G. Mazzantini, U. Giannini, and S. Cesca, *Angew. Chem.,* **71,** 205 (1959).

CHAPTER 3

CATIONIC POLYMERIZATIONS: PAST, PRESENT, AND FUTURE

"PREHISTORIC TIMES" UNTIL FRIEDEL AND CRAFTS

Receding 50 years in the history of the science and technology of cationic polymerizations brings us to "prehistoric times."

The first recorded activity regarding acid catalyzed polymerizations might be Bishop R. Watson's, who described the resinification of turpentine in 1789 as follows (1):

> There is a very curious experiment which illustrates the relation which these four bitumens bear to each other. The most transparent oil of turpentine, resembling naptha, may be changed into an oil resembling petroleum, by mixing it with a small portion of the acid of vitriol; with a larger proportion of the acid, the mixture becomes black and tenacious, like Barbadoes tar; and the proportions of the ingredients may be so adjusted, that the mixture will acquire a solid consistence, like asphaltum.

About 50 years later, in 1839, Deville (2) carried out the first olefin (styrene) polymerization with a Friedel-Crafts metal halide, $SnCl_4$, a system that for many years thereafter attracted the attention of scores of polymer scientists. For instance, in 1938 Williams (3) selected the styrene/$SnCl_4$ system to carry out perhaps the first quantitative kinetic study in the field of cationic polymerization, and many years later over the period 1953-1962 Overberger and his students (4, 5) conducted their now classical work on cationic copolymerization using the same monomer/coinitiator combination.

The first cationic polycondensation was also conducted during these prehistoric times. In 1854 Canizzaro (6) reacted benzyl alcohol with BF_3, $ZnCl_2$, H_2SO_4, and H_3PO_4 and obtained a resinous hydrocarbon, probably a polybenzyl

7

$-C_6H_4-CH_2-$ (see below).

The effectiveness of sulfuric acid and iodine as initiators of the polymerization of styrene and vinyl ethers was first described by Berthelot in 1866 (7) and Vislicenius in 1878 (8), respectively.

A very significant "first" occurred in 1873 when Butlerov and Gorianov reported that isobutylene (9) and propylene (10) oligomerize upon treatment with BF_3 at room temperature. Although this work was not further pursued at that time, these papers concerning the Lewis acid initiated polymerization of aliphatic olefins may be regarded as the harbingers of subsequent huge industries (e.g., lubricating oils, polymer alkylate gasoline, and butyl rubber).

The watershed between these prehistoric times and the next "era of qualitative studies" may be the discovery by Friedel and Crafts that many metal halides are good catalysts for a great variety of reactions (11) and, specifically, that benzyl chloride gives polybenzyl with $AlCl_3$ (12). Much work by others from 1950 to 1970 showed that these polymers were heavily branched, however, under milder conditions (i.e., low temperatures) linear, crystalline products could also be prepared (13):

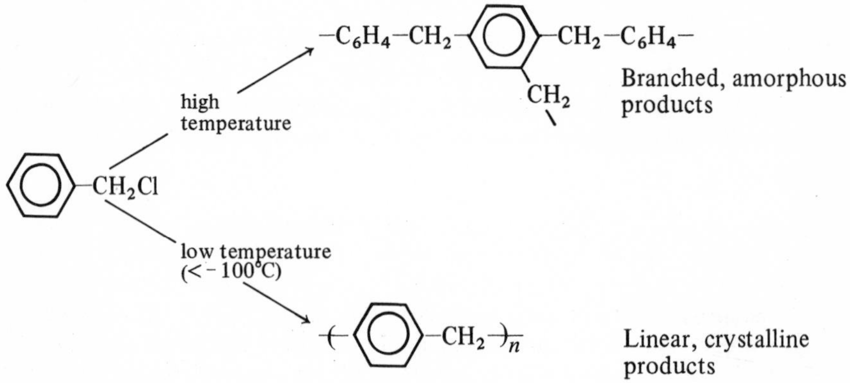

THE ERA OF QUALITATIVE STUDIES UNTIL THE DISCOVERY OF "COINITIATION" IN 1946

The science of polymer chemistry started with Staudinger's pioneering work in the period 1920–1930 when he and his coworkers developed the early chain theory of free radical polymerizations and studied the acid and base initiated polymerization of formaldehyde. Generations of polymer scientists have drawn upon their researches and findings which remain the fountainhead of numerous important commercial developments today (14).

While Staudinger's contributions are all-encompassing and have helped to lay the foundation of synthetic polymer science in general, the mechanistic foundation of the science of cationic polymerization was laid by Whitmore, an organic chemist, in a very brief paper published in 1934 (15). In this paper Whitmore interpreted the reactions of strong acids with olefins for the first time explicitly in terms of carbenium ions, and in less than a thousand words (!) he outlined the modern concept of the mechanism of cationic olefin polymerizations. Forty years later this author can add only very little to the basic views formulated by this pioneer.

Whitmore visualized initiation to proceed by the addition of a proton to the double bond system of an olefin:

$$\ddot{\underset{..}{C}}::\ddot{\underset{..}{C}} + H^{\oplus} = \ddot{\underset{..}{C}}:\ddot{\underset{..}{C}}:H$$

with the simultaneous formation of a carbon with only six electrons. Today we would call this carbon a "propagating carbenium ion." At this point this writer can do little but to repeat Whitmore's statements of 40 years ago and attempt to rewrite parts of his text in 1974 terminology. If Whitmore were a contemporary polymer scientist, he would perhaps put it like this: "The carbenium ion formed in the above process can undergo the following changes: (*1.*) termination with the counteranion, (*2.*) chain transfer by the loss of a proton and simultaneous olefin formation, and (*3.*) propagation by the incorporation of a new monomer. The polymerization mechanism is illustrated with isobutylene. Protonation of isobutylene yields a *t*-butyl cation that can add to another molecule of isobutylene:

The large carbenium ion may eliminate a proton yielding either one or two larger olefins:

The proton may again add to the olefins present and reinitiate the cycle of changes by chain transfer. Thus the process is catalyzed by any substance that gives protons."

Indeed, very little can be added to amplify Whitmore's writing (15). One major point must be made, though. The word "polymerization" as Whitmore used it, today would mean dimerization or, at best oligomerization. While Whitmore's mechanism may be regarded as the basis of today's cationic polymerization mechanisms leading to high polymers, he was interested in the formation of only liquid oligomers, that is, dimers and trimers, and did not even speculate on the formation of high polymers. The important missing links for the preparation of high polymers, that is, the maximization of the propagation rate, and the recognition that low temperatures "freeze out" chain transfer and that the coordinative bonding of simple halogen anions by Lewis acids, for example, $Cl^{\ominus} + AlCl_3 \rightarrow AlCl_4^{\ominus}$, depresses termination, are much later developments of others.

Concurrently, but independently of Whitmore, Chalmers studied the polymerization of vinyl ethers and found that a variety of Friedel-Crafts metal halides are very active coinitiators (16). Significantly also for the history of cationic polymerizations, he correctly recognized the presence of reaction chains but formulated them in terms of a radical process.

About the time these fundamental studies were in progress in university laboratories, studies which not much later culminated in the development of a general theory of carbenium ion intermediates, equally important empirical work was being carried out in various industrial laboratories, notably in Germany and in the United States. These industrial projects, aimed at finding convenient fast processes for the synthesis of lubricating oils and gasolines from low molecular weight olefins, led to the discovery of important polymer chemical concepts, polymeric materials, and polymerization processes which are presented and analyzed below.

The year 1931 is marked in the history of cationic polymerizations by the issuance of German Patent 641,284 to Otto and Müller-Cunradi (17) who assigned it to the I. G. Farbenindustrie AG. In this patent the discoverers described the low temperature (below $-10°C$) polymerization of isobutylene by the use of BF_3 or its complexes to high molecular weight solids. This discovery provided the basis for the entire polyisobutylene industry, an industry of great contemporary industrial significance (polyisobutylenes, butyl rubber, chlorobutyl, etc., see later). This landmark patent was subsequently followed by a series of other important discoveries by Otto and Müller-Cunradi, for example, the polymerization of isobutylene to rubbery solids using aluminum chloride or halides of groups III, V, VI, and VIII of the Periodic Table. The highest molecular weight polymer was obtained by carrying out the polymerization with highly purified monomer at liquid ethylene temperature ($-100°C$) with BF_3 dissolved in liquid ethylene diluent (18). These workers thus recognized the reciprocal effect of temperature on the molecular weights. Only much later was this effect treated in quantitative terms by Flory (see later) and it proved to be the key for the cationic synthesis of high molecular weight polyolefins.

The most important commercial product in the cationic polymerization scene is, undoubtedly, butyl rubber, a random copolymer of isobutylene and isoprene produced by aluminum chloride in methyl chloride diluent. Butyl rubber was discovered by R. M. Thomas and W. J. Sparks in 1937 at the Standard Oil Development Co. (now the Exxon Research and Engineering Co.) and it is of interest to recall some of the major events that led to this discovery.

Around 1930 F. A. Howard, the president of Standard Oil Development Company, was on a business trip to Germany and became interested in Otto's discoveries. Howard was intrigued with the possibility that the new hydrocarbons might find application in the petroleum industry, for example, as thickeners for motor oils. Arrangements were made for Otto to come to the United States and he arrived in 1933 at the Bayway-Linden, New Jersey, refinery laboratories of the Standard Oil Development Company. A small pilot plant was built for making polyisobutylenes of low molecular weights. Not much later Thomas was assigned to work with Otto to make polyisobutylenes of much higher molecular weight and they were able to work out conditions (adding BF_3 to isobutylene/ethylene mixtures) to produce consistently polyisobutylenes with molecular weights ($\overline{M}_w$) approaching 3×10^6. The next step was to produce rubbery polyisobutylenes on a larger (pilot plant) scale for market research. Ultimately it was decided that high polymers of isobutylene were good only for specialty purposes of limited volume because of their inability to vulcanize. Around 1935 Otto returned to Germany leaving Thomas to continue the work in Linden.

The following is a quote from a letter by Thomas in which he describes the crucial "butyl" experiment (19):

> The stage is set for the appearance of butyl rubber. On the last weekend of July 1937, Thomas and his Group Head William J. Sparks were in the laboratory trying ideas. Thomas proposed that they try to copolymerize isobutylene with butadiene using the new aluminum chloride–alkyl halide polymerization catalyst. Butadiene was available in a heavy cylinder obtained from Germany by Otto. Sparks and Thomas prepared 100 cc of isobutylene containing 10% of butadiene (by volume), added powdered dry ice for cooling, and a few cubic centimeters of aluminum chloride dissolved in ethyl chloride (5% concentration). They obtained a white mass of rubber-like polymer which they were unable to dissolve in carbon tetrachloride or other solvents. This was fortunate. It showed definitely that there had been a true copolymerization . . ."

The basic patent for butyl rubber was not issued until 1944 (20).

Further experimentation led to the definition of conditions for the production of soluble, vulcanizable products. Subsequently, Thomas and Sparks found that isoprene was a better comonomer than butadiene because of its higher copolymerization reactivity and because it yielded a faster curing rubber. Moreover, the use of isoprene eliminated cumbersome diene recycling. During World War II when the United States was cut off from natural rubber plantations in the Far East, the federal government helped to finance the development of various synthetic rubbers in this country, including butyl rubber which at that time was primarily used in making inner tubes. Plants were constructed in

Baton Rouge, Louisiana, and Baytown, Texas, to supply the butyl rubber needed during the war. The first commercial butyl rubber plant was on stream in 1943 in Baton Rouge.

Today butyl rubber is produced commercially by the Exxon Chemical Company and its licensee's Columbian Carbon Company in the United States, Esso Petroleum Ltd. in the United Kingdom, Polymer Corporation in Canada, SOCABU in France, and Japan Butyl Co. in Japan. A plant is in the engineering-construction stage by SIR in Italy.

In the commercial process isobutylene, 1.5–4.5 vol. % isoprene, and about three times as much methyl chloride diluent are mixed and cooled to $\sim -100°C$. This monomer feed is injected into a stirred reactor cooled to $\sim -100°C$. Simultaneously with the monomer feed, a stream of catalyst, $AlCl_3$ in methyl chloride solution (~ 0.2 wt. %), is injected. The rate of polymerization is almost instantaneous after mixing the ingredients, and the copolymer, which is insoluble in the polymerization medium, precipitates as it is formed ("slurry" process). The copolymerization is highly exothermic and temperature control is of utmost significance for the synthesis of good quality product. Additional details and a flow plan of the butyl process have been published (21).

The last, most recent chapter of the butyl story is the successful commercialization of chlorobutyl rubber (chlorinated butyl) (22) and Butyl LM (23) by the Exxon Chemical Co. Chlorobutyl is mainly used in inner liner applications, whereas Butyl LM (for low molecular weight) finds its use in sealants, coatings, potting compounds, encapsulation, and so on.

From the vantage point of rubber science, the discovery of butyl rubber is important not only as a discovery of a new general purpose synthetic rubber and some specialty rubber derivatives, but, perhaps even more importantly, as the discovery of the *principle of low functionality*. During the 1930s, rubber scientists believed that elastomeric characteristics could arise only in the presence of predominantly large proportions of C=C double bonds, as in natural rubber, SBR, and Buna. After the synthesis of very high molecular weight rubbery polyisobutylenes by Otto and Thomas at the old Standard Oil Development Company, scientists in this laboratory realized that the presence of double bonds was unnecessary for elastomer properties. Thus, after Otto returned to Germany, Thomas and coworkers set out to synthesize elastomers from isobutylene containing only a limited amount of double bonds, which was visualized to be advantageous not because of the rubberyness obtained but rather because it would bring about sulfur vulcanizability!

The concept of low functionality for vulcanization is one of the most important working concepts of contemporary rubber chemistry and technology. The recently developed sulfur vulcanizable ethylene–propylene terpolymers (EPDM) are a further embodiment of this important principle.

In 1940 Thomas and coworkers published their quantitative studies on isobutylene homopolymerization with BF_3 and $AlCl_3$ (24). This paper was written several years after the discovery of butyl rubber and brought to the attention of polymer researchers the tremendous complexities of this deceptively simple looking polymerization system. The authors described the synthesis of very high molecular weight $(2-5 \times 10^6 \bar{M}_\nu)$ elastomeric polyisobutylenes and the effect of reagent concentrations and temperature on the molecular weights, discussed the effect of "poisons," and last but not least elucidated and proposed the overall structure of polyisobutylene. This research has subsequently been built upon by others; for example, Flory in 1953 replotted Thomas' molecular weight versus temperature data in the log $\bar{M}_\nu$ versus $1/T$ (Arrhenius) fashion and obtained a straight line (25), a procedure that is extensively used by contemporary researchers interested in elucidating the mechanism of polymerizations in general.

Looking at this scene from a perspective of some 40 years, the contemporary science of cationic polymerization was born during the thirties. The two important, largely independent ingredients that were necessary for this birth were the theoretical–laboratory studies by Whitmore and his school in Philadelphia and the technological–commercial success of a major industry conceived by Thomas and coworkers at the Old Standard Oil Development Company in Linden, New Jersey. The theoretical foundation was provided by the new concept of carbenium ion intermediates, while sustained commercial interest was created by the butyl rubber industry.

Another important milestone during this period of the late 1930s and early 1940s is the work by Meerwein and his school in Germany who carried out their now classical investigations on the polymerization of cyclic ethers, particularly tetrahydrofuran, and laid the foundation of this branch of polymer science. These workers discovered efficient initiators for the polymerization of THF in the form of various oxonium and carboxonium salts (26), studied the effect of promoters on these polymerizations, and postulated the now accepted S_N2 propagation mechanism for cyclic ethers. Unfortunately the scope and depth of this work became known to the world only long after World War II (27).

MATURITY AND CONSOLIDATION: CLASSICAL BRITISH INVESTIGATORS AND THE DISCOVERY OF STEREOCONTROL

The war effort in Britain spawned a large amount of olefin research. Sustained British work that started in the thirties concerned and intensified the fundamental investigation of many carbenium ion mechanisms by Ingold and his coworkers at the University of London. However, the important development in regard to cationic polymerizations occurred at the University of Manchester in Polanyi's laboratories. Thus in the years 1946 and 1947 Polanyi, Plesch, Evans, and coworkers reported in a series of papers that exhaustively purified, dry isobutylene fails to polymerize when put in contact with BF_3 and/or $TiCl_4$; however, immediate polymerization ensues upon the introduction of traces of water (28–31). Since initiation requires the presence of moisture in the system, they suggested that the true initiating species is the proton derived from H_2O and proposed the now widely accepted initiation mechanism:

$$BF_3 + HY \rightarrow F_3BYH$$

$$F_3BYH + CH_2 = C(CH_3)_2 \rightarrow F_3BY^\ominus + CH_3-C^\oplus(CH_3)_2$$

These authors call HY (H_2O, HCl, etc), the speices that provides the initiating proton, the "cocatalyst." Unfortunately, this misleading nomenclature is now widely accepted; it would be more descriptive to use the term "initiator" for the protogen and "coinitiator" for the Friedel-Crafts metal halide.

This concept of initiation as conceived and formulated by Polanyi, Plesch, and coworkers has proved of great value for the present generation of polymer scientists.

Many polymerization systems have subsequently been subjected to exhaustive drying, and similar phenomena have been found with styrene, vinyl ethers, and so on. The great merit of these British workers was the demonstration of the effect of trace amounts of impurities on the kinetics of these polymerization systems, a field which up to this recognition was closed to quantitative scientific inquiry. Subanalytical quantities of ubiquitous moisture on the walls of experimental equipment and in routinely "dried" chemicals resulted in irreproducible rate and molecular weight data and frustrated a generation of polymer chemists. By recognizing the role and controlling the amount of protogenic substances in the polymerization systems, reproducible rates and molecular weights could be obtained.

The concept of coinitiation by trace impurities appeared of general validity and, indeed, even in 1963 it appeared proper for Plesch to state that "Isobutylene is not polymerized by metal halides unless an ionogenic substance, the cocatalyst, is present" (32). Only very recent work (33) has indicated that in many systems initiation may proceed in the absence of a conventional protogenic initiating species (see later).

The second landmark event in the 1940s was the discovery of stereocontrol of vinyl polymerizations by Schildknecht and his coworkers (34). The roots of this event can be traced back to developments during World War II in Germany where it was found that vinyl ethers can readily be polymerized to viscous or rubbery, amorphous polymers by the use of Lewis acids such as BF_3. After the war this information became known to American scientists, and a group under Schildknecht at the General Aniline and Film Corporation in Easton, Pennsylvania, became interested in the exploration of the effect of polymerization conditions on the adhesive properties of vinyl isobutyl ether. In the course of this work vinyl isobutyl ether was polymerized with $BF_3 \cdot OEt_2$ in propane diluent at Dry Ice temperature and a solid, horny material was obtained. The physical properties of this polymer, for example, crystallinity, solubility, milling behavior, tack, and hardness, were strikingly different from known polyvinyl isobutyl ethers and were explained in terms of structural regularity of the product. Schildknecht thought this polymer to be syndiotactic, but much later Natta et al. (35) showed that in fact the product was isotactic. However, long before the concept of polymer stereoregularity was formulated by Natta and his school, Schildknecht et al. recognized the tactic nature of their polymer. Unfortunately, Schildknecht's work soon petered out and stopped as it became evident that crystalline polyvinylalkyl ethers are unsuitable for general, large volume engineering applications.

What would have happened to this field had the melting point of polyvinyl isobutyl ether been ~135°C instead of ~80°C?

This era of "maturity and consolidation" may be considered as concluding with three important publishing events: In 1949 Pepper organized the first informal Conference on Cationic Polymerization at Trinity College in Dublin, Ireland, and edited the papers which were presented in the form of a series of publications (36); soon thereafter Plesch organized the second Conference on Cationic Polymerization at the University College of North Staffordshire, Keele, in 1952 and edited and published the proceedings (37); and finally in 1954 Pepper published the first major review of the field (38).

MODERN ERA

By the mid 50's the chemistry and technology of cationic polymerizations had consolidated to the point that the field became a visible, coherent entity with solid scientific foundations and a promising future.

The modern era in the history of cationic polymerizations is best characterized by the great variety of emerging research avenues and the concurrent development of a variety of industrial ventures to be highlighted below.

Japanese researchers acquired international visibility in this field of polymer chemistry in the mid-fifties and since that time their contributions remain significant both in terms of quality and quantity. One of the best known groups led by Okamura in cooperation with Hayashi and Higashimura in Kyoto tackled many fundamental, mainly kinetic, problems, such as the polymerization of vinyl ethers, isobutylene, and styrene and its derivatives induced by Lewis acids and by high energy irradiation. While the main thrust of this group was directed toward the kinetic–mechanistic elucidation of the polymerization behavior of vinyl compounds, another group in Kyoto, under the leadership of Furukawa and Saegusa, was doing pioneering work on cyclic ethers, aldehydes, and acetals. Because of the variety and scope of the contributions by these groups, it is impossible to give a representative reference and the interested reader is advised to consult *Chemical Abstracts* to assess the significance of these investigations. It is obvious though, that the field of cationic polymerizations as we know it today would not be as rich as it is without the contribution of these Kyoto researchers.

An important scientific publication in 1962 describes the cationic oxidative polycondensation of benzene by Kovacic and Kyriakis (39). The polymerization of benzene to poly(p-phenylene) was accomplished under mild conditions by the use of $AlCl_3 \cdot CuCl_2$:

$$\langle \bigcirc \rangle + 2CuCl_2 \xrightarrow[30°, \, 30 \, min]{AlCl_3} -\left(\langle \bigcirc \rangle\right)_n + 2nCuCl + 2nHCl$$

This was the first well characterized homopolymerization of an aromatic ring system. Except for the oxidation, the proposed mechanism is very similar to that of cationic olefin polymerizations (40). Poly(p-phenylene) is a brown oligomer (DP = 4–8), that is highly crystalline, high melting, and insoluble. The molecular weights are not sufficiently high for the development of optimum mechanical properties. Homo- and copolymerization experiments with other aromatic nuclei, such as toluene and chlorobenzene, have been conducted to improve the physical properties by introducing a measure of irregularity in the chain (meta and/or ortho linkages). These materials are semiconducting and show excellent thermal stability. An excellent review including the chemistry, properties, and technology of polyphenylene has recently been published (40).

Another noteworthy paper in 1962 was that of Kennedy and Thomas (41) on the synthesis of the first crystalline aliphatic poly-α-olefin by an intramolecular hydride shift (isomerization) polymerization, that is, the polymerization of 3-methyl-1-butene by the use of a vareity of Lewis acids, in particular with $AlCl_3$ at very low temperatures. Significantly, the polymer was crystalline and it exhibited an unexpected, highly regular structure:

$$-\left(CH_2-C(CH_3)_2-CH_2\right)_n$$

It was proposed that the polymerization involved an intramolecular hydride migration step prior to propagation:

$$H_2C=CH \xrightarrow{R^{\oplus}} R\curvearrowright CH_2-\overset{\oplus}{CH} \xrightarrow{}$$

with the CH / CH_3 \ CH_3 substituents and H_3C / CH_3 \ substituents on the CH group.

$$R\curvearrowright CH_2-CH_2-\overset{CH_3}{\underset{CH_3}{\overset{|}{\underset{|}{C}}}}\oplus \xrightarrow{\text{monomer}} \text{polymer}$$

A large amount of similar systems have subsequently been thoroughly studied and today there exists a large body of information of all types of isomerization polymerizations. For a review see reference 42.

The year 1962 also saw the commercialization of Celcon, an important new trioxane—ethylene oxide copolymer resin by the Celanese Corporation of America making use of a largely home grown cationic polymerization technique. The origin of this plastic can readily be traced back to Staudinger, who studied the polymerization of formaldehyde and described colorless brittle solids made by Lewis acid systems at low temperatures (14). The thermal instability of polyformaldehyde (polyoxymethylene) was also recognized and attributed to a fast depolymerization to monomer (zipping) at room temperature.

The problem of rendering this otherwise attractive material thermally and alkali stable was achieved by Staudinger (14) and after World War II by McDonald (43), a DuPont chemist, by "capping" the polyformaldehyde chain, for example with acetate groups which prevent zipping-initiation at the termini:

$$CH_3COO-CH_2O-CH_2O-CH_2O-\ -\ -\ -CH_2O-CH_2O-COCH_3$$

Delrin

Such resins have been commercialized in 1959 by DuPont (Delrin). Since polymerization of formaldehyde to Delrin is probably an anionic process, this development is mentioned only because it has an important bearing on the development of Celcon.

Trioxane can readily be polymerized by cationic techniques to high molecular weight polyformaldehydes; however, the products are unstable (zipping). Walling, Brown, and Bartz discovered (44) that the copolymerization of trioxane, like that of formaldehyde, with cyclic ethers, for example, ethylene oxide, gives excellent thermally and alkali stable engineering plastics.

The copolymerization is probably carried out by a boron trifluoride etherate initiator system. Since even in such systems the true initiating entity is most likely the proton:

$$BF_3 \cdot OR_2 + H_2O \rightleftharpoons BF_3 \cdot H_2O + OR_2 \rightleftharpoons H^{\oplus}BF_3OH^{\ominus}$$

the termini of this copolymer probably contain HO^- groups. However, unzipping is interrupted by the randomly distirbuted $-CH_2CH_2O^-$ units in the chain ("zipper jammers"):

$$\sim\!\!\sim\!\!\sim\!\!O\!-\!CH_2\!-\!O\!-\!CH_2CH_2\!-\!\!(\!OCH_2\!)_n\!\!-\!O\!-\!CH_2OH$$

Unstable half-acetal

$$\sim\!\!\sim\!\!\sim\!\!O\!-\!CH_2\!-\!O\!-\!CH_2CH_2OH + (n+1)CH_2O$$

Stable acetal

The slight irregularity due to the presence of ethylene oxide units in these copolymers imparts some processing advantages over that of the highly crystalline homopolymer. At the present trioxane–ethylene oxide copolymers are commercially available from Celanese Corporation of America (Celcon) and Farbwerke Hoechst AG (Hofstaform).

In 1963 Plesch edited a major book on cationic polymerization (45). This publication represents a valuable summary of the status of the science of cationic polymerizations (exclusive of technological matters) up to 1962.

In 1964 Gandini and Plesch (46) proposed the concept of pseudocationic polymerization, that is, in certain systems the propagating entity is not ionic but perhaps an ester. This conclusion was reached during a reexamination of some of Pepper's findings with the styrene/$HClO_4$/methylene chloride system. In addition to confirming Pepper's kinetic scheme, the authors noted that the polymerizing solution is colorless and nonconductive; however, when polymerization is almost complete conductivity and color appear simultaneously. The addition of fresh styrene results in the resumption of polymerization in the absence of color and conductivity which, however, reappear at the end of styrene consumption. In contrast to conventional cationic polymerizations, small amounts of water do not affect the rate. The molecular weights were very low. The authors theorized (46) that the chain carrier is not a carbenium ion and later proposed that propagation occurs by monomer insertion into an ester, for example

$$-\underset{\underset{\displaystyle C_6H_5}{|}}{C}H\!-\!OClO_3 \ .$$

A number of other aromatic systems have since been described to proceed by this mechanism, for example, styrene/H_2SO_4, and p-methoxystyrene/CF_3COOH, and acenaphthylene/I_2.

The concept of pseudocationic polymerizations fits well into the fabric of cationic polymerizations. In pseudocationic systems the interaction between the propagating species and counterion is best visualized as being a covalent link. On the other end of the spectrum of propagating cationic entities are the completely "free" carbenium ions encountered, for example, in high energy irradiation initiated cationic polymerizations. In between these two extremes are the bulk of systems investigated, those in which the chain carriers are ion pairs, solvated ions, contact ion pairs, and so on. Indeed, the exploration and definition of the ionicity of propagating carbenium ions is one of the most urgent outstanding tasks in this field.

In 1965 Vandenberg, a research chemist of the Hercules Powder Co., announced a new class of specialty synthetic rubbers based on epichlorohydrin (47). Two members of this new family have been developed: amorphous polyepichlorohydrin and a 1:1 mole copolymer of epichlorohydrin and ethylene oxide:

$$\mathrm{CH_2\!-\!CH\!-\!CH_2Cl} \xrightarrow{\text{catalyst}} \mathrm{\text{--}(CH_2\!-\!CH\!-\!O)\text{--}}_n \qquad n \sim 5000$$

with $\mathrm{CH_2Cl}$ substituent.

$$\mathrm{CH_2\!-\!CH\!-\!CH_2Cl} + \mathrm{CH_2\!-\!CH_2} \xrightarrow{\text{catalyst}} \mathrm{\text{--}(CH_2\!-\!CH\!-\!O\!-\!CH_2\!-\!CH_2\!-\!O)\text{--}}_n$$

with $\mathrm{CH_2Cl}$ substituent.

$$n \sim 20{,}000$$

The starting point for this development was the discovery that Et_3Al/H_2O systems give high molecular weight polymers at a fast rate. While the catalyst used today for the commercial manufacture of these rubbers is a modification of the original Et_3Al/H_2O system by a chelating agent, for example, acetyl acetone, and propagation is probably by a coordinative mechanism, the basic discovery that provided the first impetus involved a purely cationic system: Et_3Al/H_2O.

These epi rubbers, as they are called now, exhibit a unique combination of good low temperature properties, oil resistance, flame resistance, molding and extrusion characteristics, tack, and air permeability; they are commercially available from B. F. Goodrich Company (since 1965) under the trade name Hydrin, and from Hercules, Inc. as Herclor.

Penton, poly(3,3-bischloromethyloxetane), was first sold by Hercules Inc. this same year; however, commercial production stopped in 1971. The development of Penton can be traced back to Farthing and Reynolds (48) who in 1954 in a short but important publication described the high polymerization of 3,3-bis-chloromethyl oxetane by $BF_3 \cdot H_2O$ using methylene chloride diluent:

$$ClH_2C \diagdown C \diagup CH_2Cl$$
$$CH_2 \diagdown O \diagup CH_2 \quad \xrightarrow[\text{low temperatures}]{BF_3} \quad \text{---}(O-CH_2-\underset{CH_2Cl}{\overset{CH_2Cl}{C}}-CH_2 \text{---})$$

The commercial polymerization process has probably been carried out by the use of BF_3 etherate and sulfur dioxide diluent at low temperatures. In the presence of SO_2 the polymer precipitates as a fine powder and can be readily recovered. This polymer is highly crystalline (m.p. = 180°C) and consequently largely insoluble in common solvents; it exhibits high chemical resistance and, due to its high chlorine content, is self-extinguishing. Penton can readily be fabricated into complicated molded shapes, and it exhibits a unique combination of rigidity under static stress at high temperatures (almost no creep even at 138°C).

Last but not least before we leave the "modern era" and turn to the "present time," the important contributions of Olah of the Case-Western Reserve University, on the physical-organic chemistry of small carbocations should be noted. While the activities of Olah and coworkers lie outside the mainstream of cationic polymerizations *per se*, this work is nonetheless of great direct value not only as an inspiration for new polymerization experiments, particularly in cationic polycondensations [evidence for this already exists (49)], but also as a possible way of characterizing the elusive propagating electrophilic species.

Thus in 1968 (50) Olah and Schlosberg stated that methane, ethane, and so on, can be oligomerized to branched hydrocarbons by the extremely strong ("magic") acid FSO_3H-SbF_5 and proposed the following mechanism to explain their findings:

$$CH_4 \underset{SbF_5}{\overset{FSO_3H}{\rightleftharpoons}} [CH_5^{\oplus}] \longrightarrow CH_3^{\oplus} + H_2$$

$$C_2H_6 \underset{-H^{\oplus}}{\overset{H^{\oplus}}{\rightleftharpoons}} [C_2H_7^{\oplus}] \overset{-H_2}{\longrightarrow} C_2H_5^{\oplus} \underset{H^{\oplus}}{\overset{-H^{\oplus}}{\rightleftharpoons}} CH_2=CH_2$$

$$C_3H_8 \underset{-H^{\oplus}}{\overset{H^{\oplus}}{\rightleftharpoons}} [C_3H_9^{\oplus}] \underset{H^{\oplus}}{\overset{-H_2}{\rightleftharpoons}} C_3H_7^{\oplus}$$

$$\rightleftharpoons [C_4H_{11}^{\oplus}] \rightleftharpoons C_4H_9^{\oplus}$$
$$\downarrow$$
$$\text{etc.}$$

Not since the classical workers of some 40 years ago, such as Whitmore, Ingold, and Meerwein, completed their researches on carbenium ions has this field undergone such revolutionary and rapid progress as when Olah and his group started their studies on the direct observation and characterization of carbocations. This work has changed some of our basic concepts of electrophilic species and its review even in capsule form is beyond the scope of this brief history (for a recent review consult reference 51).

THE PRESENT

The field of cationic polymerizations is growing rapidly. Many groups around the world are deeply interested in this chemistry and are committed to its further development, because of the availability of a large variety of cationically polymerizable monomers, potential commercial interest due to reasonably inexpensive monomers and Friedel-Crafts metal halides, endless possibilities of homo-, co-, block, and graft polymerizations and polymer modifications, and so on. Some of the work by major groups in progress today in various parts of the globe is discussed below.

Plesch, perhaps the foremost pioneer who maintained his interest in this field since the early forties is interested in elucidating the details of polymerization initiation of hydrocarbons (isobutylene) by Friedel-Crafts halides, for example, $AlBr_3$, in the absence of protogenic impurities. His latest theory invokes self-dissociation of dimeric aluminum bromide and initiation by the $AlBr_2^{\oplus}$ moiety:

$$(AlBr_3)_2 \ \rightleftharpoons \ AlBr_2^{\oplus} \ AlBr_4^{\ominus}$$

$$AlBr_2^{\oplus} + CH_2{=}C(CH_3)_2 \ \longrightarrow \ Br_2Al{-}CH_2{-}C^{\oplus}(CH_3)_2 \ \xrightarrow{M} \ etc.$$

Another key proposition is the existence of polymerization inactive π complexes, for example, $AlBr_3 \cdot CH_2{=}C(CH_3)_2$. Other work in his laboratory concerns the kinetics of pseudocationic polymerization and polarography of carbenium ions (52).

Ledwith in Liverpool maintains his interest in carbenium ion salt ($R^{\oplus}SbCl_6^{\ominus}$ or $BF_4^{\ominus}$ where $R^{\oplus}$ = trityl, tropylium, etc.) initiated polymerizations of vinyl ethers, vinyl carbazole, and so on (53). Pepper concentrates on the elucidation of kinetic details of pseudocationic polymerization, particularly the styrene/$HClO_4$ system. Some of his latest work involves the use of the stop-flow technique at low temperatures (54).

In Paris, Sigwalt, who over the years developed elegant techniques to study fundamentals of olefin (isobutylene, cyclopentadiene, styrene, etc.) polymerizations with soluble Friedel-Crafts halides ($TiCl_4$, $SnCl_4$, etc.), is continuing to work in this field (55). One of his earlier coworkers, Marèchal in Rouen, generated a large body of information on polymerization behavior of indene and its derivatives. Recently this very productive researcher became interested in the application of quantum chemistry to define monomer reactivity (55). Pinazzi and his group in LeMans continue their systematic studies on the polymerization behavior of small rings (56).

Italian workers Giusti and Magagnini at the University of Pisa, in cooperation with Cesca, Priola, and others of the SNAM Progetti laboratories, conduct important investigations on the low temperature polymerization of isobutylene and isobutylene/isoprene mixtures using, among other systems, alkylaluminum/alkyl halide initiators (57). Aside from the high quality of the fundamental information generated, this work is of great practical significance as it describes how acceptable quality butyl rubber can be produced using much milder conditions (i.e., higher temperatures) than in the conventional butyl process.

Over the last 10 years Marek in Czechoslovakia has studied the polymerization of isobutylene with Friedel-Crafts halides. This author showed conclusively that this monomer can be polymerized, for example, by $AlBr_3$, in the total absence of protogenic species. Very recently Marek and Toman found that the polymerization rate of isobutylene initiated by VCl_4 and other Friedel-Crafts halides can be accelerated by light (58). Unfortunately, another productive Czechoslovakian group led by Zlamal who worked on the isobutylene/$AlCl_3$ system, has not published in this field since 1969 (59).

The famous Kyoto group that started under Okamura and is now led by Higashimura and Hayashi maintains high interest in catalytic and irradiation initiated cationic polymerizations. For example, Higashimura and Masuda recently indicated the possibility of molecular weight distribution control by the use of two simultaneously propagating styryl ion–counterion species, that is, $R^\oplus$ $ClO_4^\ominus$ and $R^\oplus$ $SnCl_4ClO^\ominus$ (60). Hayashi and coworkers recently showed the coexistence of cationic and anionic polymerizations in a UV illuminated mixture of cyclohexene oxide and nitroethylene (61).

Furukawa, whose interests are extremely broad and wide ranging, continues to actively study cationic polymerizations. One of his most recent contributions deals with the fundamentals of the effects of various Friedel-Crafts halides and triphenyl carbenium ion initiators and solvents on the reactivity of olefins in copolymerizations (62). His former coworker Saegusa, who became one of the leaders in the field of cationic polymerization of heterocyclics, in particular cyclic ethers and oxazolines, recently developed analytical techniques for the detailed kinetic investigation of cyclic ethers and provided great insight into their polymerization behavior (63, 64).

Continued studies on cyclic diene polymerizations are in progress by Aso and by Kunitake at Kyushu University is presently studying the mechanistic aspects of the polymerization of substituted cyclopentadienes for example,

and the cyclopolymerization of divinyl compounds and dialdehydes (66).

Yamashita in Nagoya contributed significantly to our store of knowledge on the polymerization of cyclic ethers, esters, and particularly, dioxolane. Recently he prepared a block copolymer of tetrahydrofuran and 3,3-bischloromethyl-oxetane (BCMO) by reacting a "living" polytetrahydrofuran with BCMO (67). The field of cyclic acetals also proved to be a fertile research area over the years for Penczek in Lodz (68) and Enikolopyan in Moscow (69).

Since Hoechst AG is a producer of polytrioxane acetal resin, it is not surprising that the company is interested in cationic polymerization as evidenced by occasional recent publications in the scientific literature (70). The tragic, untimely death of Jaacks of Mainz in 1971 cut short the career of a promising young researcher who was also interested in polyoxymethylenes, a field to which he contributed more than 40 publications.

Heublein and Jena tried to copolymerize isobutyl vinyl ether with isobutylene and propylene, but the evidence is yet insufficient for the existence of a truly random copolymer (71).

The field of cationic episulfide polymerizations is kept active by the systematic contributions of Goethals in Ghent (72), Sigwalt and coworkers in Paris (73), and Lautenschlager at the Dunlop Co. in Canada (74).

The field of cationic oxazoline polymerizations is receiving continuous detailed interest both from the mechanistic–reactivity point of view by Saegusa (64) and Kagiya in Kyoto (75) and from the structure–property characterization angle by Litt and coworkers at Case Western Reserve University (76).

In the United States Patricia and Peter Dreyfuss in Akron, who have made important contributions to the study of tetrahydrofuran polymerization, recently organized a symposium on Cyclic Ethers and Sulfides and edited and published the proceedings (77).

Vandenberg of Hercules, Inc., who over many years has contributed significantly to the science and technology of vinyl and cyclic ether polymerizations by cationic and coordination catalysis, is still active in this field (78).

Bailey and coworkers in Maryland recently published a series of interesting papers describing the cationic polymerization of certain spiro compounds and formals, compounds that expand on polymerization (79).

And, finally, work in this author's laboratory in Akron concerns the development of new organic chemical principles/concepts adaptable for exploitation in the field of cationic polymerizations. Major recent events include the synthesis of a new class of graft copolymers (EPDM-*g*-polystyrene, PVC-*g*-polyisobutylene, chlorobutyl-*g*-polyindene, etc.), block copolymers (styrene-isobutylene), random copolymers (isobutylene-β-pinene), kinetic and chemical proof for the absence of chain transfer reactions in isobutylene polymerization initiated by certain alkylaluminum systems, and the finding that halonium ions, for example, Cl_2, Br_2, are most effective polymerization initiators (80). The last discovery has also been made by Italian workers (57).

THE FUTURE

What lies ahead? One way to guess the future is to imagine one possesses unlimited funds, a motivated, talented crew, good research facilities, and 10-15 years of reasonably good health. What projects would one tackle, what objectives would one want to obtain? Some possibilities are given below.

Research should be initiated to observe, characterize, and determine quantitatively the concentration of propagating carbenium ions in a polymerization system. There is indication that $[C^\oplus] <<< [C]$, where $C^\oplus$ = active propagating ion and C = added initiator. Why is this so? Olah recently developed convenient methods for the direct observation of a large variety of electrophiles (many carbenium ions, oxonium ions, halonium ions, etc.). This and other methods (polarography, conductivity) should be further developed to study polymerization active species. The study of ionicity in anionic polymerizations is far advanced.

In a similar vein, research should be carried out to define the nature of the propagating species. Under what conditions are these solvent separated ion pairs, contact ion pairs, free ions, and so on:

$$\overset{\delta\oplus\ \ \delta\ominus}{\sim\!\!\sim\!C\!-\!G} \ \rightleftharpoons\ \overset{\oplus\ \ \ominus}{\sim\!\!\sim\!C\cdot G} \ \rightleftharpoons\ \overset{\oplus\ \ \ominus}{\sim\!\!\sim\!C/\!/\,G} \ \rightleftharpoons\ \overset{\oplus\ \ \ \ominus}{\sim\!\!\sim\!C\ +\ G}$$

Are the counteranions monomeric or are they aggregates? A predictive, at least semiquantitative concept of counteranion nucleophilicities is needed. The inorganic chemistry of the counteranions is still in its infancy.

The exploration of effect of solvents on the elementary steps of polymerizations is also completely empirical at the present.

Concurrently with these researches aimed at the characterization of polymerization active species, the absolute kinetics, that is, the absolute rate constants must be worked out. There are only very few simple systems where we know k_p.

The chemistry of chain breaking steps (chain transfer and termination) must be developed. While we have made progress in the understanding of initiation, we know very little about chain breaking. Ultimately progress in this area will lead to our control of molecular weight, molecular weight distribution, and yield—most desirable goals.

Leads developed in these areas should be applied to the search for "living" carbenium ion polymerization. Living oxonium ion polymerizations have already been attained and have led to very interesting new block polymers.

Systematic sequential polymer research has already started in this field and this work should be intensified as it could lead to the synthesis of completely new materials and, consequently, to new concepts.

The recent recognition that in addition to protons and carbenium ions, halonium ions may also initiate cationic polymerizations leads to the expectation that all kinds of electrophilic species may function as initiating species. Thus, initiation by $NO_2^{\oplus}$, $NO_2^{\oplus}$, $R-SO_2^{\oplus}$, $HgR^{\oplus}$, and so on could yield new polymers with well defined, functionalized head-groups.

While the number of cationically polymerizable monomers is very large ($\sim$500), only an insignificant number of these are being exploited commercially. As industry becomes increasingly sophisticated it will need, or will be willing to use, an increasingly larger number of specialty materials, some of which will undoubtedly be produced by cationic polymerization techniques.

The future of cationic polymerizations is at the moment completely dependent on the availability and price of petrochemicals. Efforts should be made to break this dependence. One way may be to develop biological (microorganism, vegetable) sources for raw materials, such as terpenes, which are produced by coniferous plants, of which β-pinene yields an important commercial resin.

REFERENCES

1. Bishop Watson, *Chemical Essays*, 5th ed., Vol. III, printed for T. Evans and sold by J. Evans, London, 1789.

2. M. Deville, *Ann. Chim., (Paris)*, 75, 66 (1839).

3. G. Williams, *J. Chem. Soc.*, 1938, 246, 1938, 1046, and 1940, 775.

4. C. G. Overberger and F. Endres, *J. Am. Chem. Soc.*, 75, 6349 (1953).

5. G. F. Endres, V. G. Kamath, and C. G. Overberger, *J. Am. Chem. Soc.*, 84, 4813 (1962) and previous papers in this series.

6. S. Canizzaro, *Ann. Chem. Pharmacol.*, 90, 252 (1854) and 92, 113 (1854).

7. M. Berthelot, *Bull. Soc. Chim. France*, 6, (2) 294 (1866).

8. J. Wislicenius, *Ann.*, 92, 106 (1878).

9. A. M. Butlerov and V. Gorianov, *Ann.*, 169, 146 (1873).

10. A. M. Butlerov and V. Gorianov, *J. Russ. Chem. Soc.*, 5, 302 (1873).

11. C. Friedel and J. M. Crafts, *Compt. Rend.*, 84, 1450 (1877) and 85, 74 (1877).

12. C. Friedel and J. M. Crafts, *Bull. Soc. Univ.*, **43**, 53 (1885).

13. J. P. Kennedy and R. B. Isaacson, *J. Macromol. Sci.*, **i**, 541 (1966).

14. H. Staudinger, *Die Hochmolecularen Organische Verbindungen, Kautschuk und Cellulose*, Springer, Berlin, 1932.

15. F. C. Whitmore, *Ind. Eng. Chem.*, **26**, 94 (1934).

16. W. Chalmers, *Can. J. Res.*, **7**, 464, 472 (1932).

17. M. Otto and M. Müller-Cunradi, German Patent 641,284 to I. G. Farbenindustrie, July 26, 1931.

18. M. Otto and M. Müller-Cunradi, U.S. Patent 2,203,873 to I. G. Farbenindustrie, July 1, 1937.

19. R. M. Thomas, letter to J. P. Kennedy, Jan. 2, 1968.

20. R. M. Thomas and R. W. Sparks, U.S. Patent 2,356,128 to Standard Oil Development Co., 1944.

21. J. P. Kennedy in *Copolymerization*, G. E. Ham, Ed., Wiley, New York, 1964, Chap. 5.

22. F. P. Baldwin, D. J. Buckley, I. Kuntz, and S. B. Robison, *Rubber Plastics Age*, **42**, 500 (1961).

23. J. Glazman, "Low Molecular Weight Butyl Rubber" presented at 28th Annual Technical Conference, Society of Plastic Engineers, New York City, May 6, 1970.

24. R. M. Thomas, W. J. Sparks, P. K. Frolich, M. Otto, and M. Müller-Cunradi, *J. Am. Chem. Soc.*, **62**, 276 (1940).

25. P. J. Flory, in *Principles of Polymer Chemistry*, Cornell University Press, Ithaca, New York, 1953.

26. H. Meerwein, German Patent 741,478 to I. G. Farbenindustrie, 1939.

27. H. Meerwein, D. Delfs, and H. Morschel, *Angew. Chem.*, **72**, 927 (1960).

28. A. G. Evans, D. Holden, P. H. Plesch, M. Polanyi, H. A. Skinner, and W. A. Weinberger, *Nature*, **157**, 102 (1946).

29. A. G. Evans, G. W. Meadows, and M. Polanyi, *Nature*, **158**, 94 (1946).

30. A. G. Evans and M. Polanyi, *J. Chem. Soc.*, **1947**, 252.

31. P. H. Plesch, M. Polanyi, and H. A. Skinner, *J. Chem. Soc.*, **1947**, 257.

32. P. H. Plesch, in *The Chemistry of Cationic Polymerization*, P. H. Plesch, ed., Macmillan, New York, 1963, p. 149.

33. A review by J. P. Kennedy, *J. Macromol. Sci., Chem.*, **A6**, 329 (1972).

34. C. E. Schildknecht, S. T. Gross, R. H. Davidson, J. M. Lambert, and A. O. Zoss, *Ind. Eng. Chem.*, **40**, 2104 (1948).

35. G. Natta, P. Corrandini, and I. Bass, *Makromol. Chem.*, **18–19**, 455 (1955).

36. D. C. Pepper, *Sci. Proc. Roy. Dublin Soc.*, **25**, 131 (1950).

37. P. H. Plesch, Ed., *Cationic Polymerization and Related Complexes*, Heffer and Sons, Cambridge, 1953.

38. D. C. Pepper, *Quart. Rev.*, **8**, 88 (1954).

39. P. Kovacic and A. Kyriakis, *Tetrahedron Letters*, **11**, 467 (1962).

40. J. G. Speight, P. Kovacic, and F. W. Koch, *Macromol. Sci., Rev. Macromol. Chem.*, **5**, 295 (1971).

41. J. P. Kennedy and R. M. Thomas, *Makromol. Chem.*, **53**, 28 (1962).

42. J. P. Kennedy, *Encycl. Polymer Sci. Technol.*, **7**, 754 (1967).

43. R. N. McDonald, U.S. Patent 2,768,994 to E. I. DuPont deNemours Co., 1956.

44. C. T. Walling, F. Brown, and K. W. Bartz, U.S. Patent 3,027,352 to Celanese Corp., 1960.

45. P. H. Plesch, Ed., *The Chemistry of Cationic Polymerization*, Macmillan, New York, 1963.

46. A. Gandini and P. H. Plesch, *Proc. Chem. Soc.*, **1964**, 240.

47. E. J. Vandenberg, *Rubber Plastics Age*, **46**, 1139 (1965).

48. A. C. Farthing and R. J. W. Reynolds, *J. Polymer Sci.*, **12**, 503 (1954).

49. D. T. Roberts and L. E. Calihan, *J. Macromol. Sci., Chem.*, **A-1**, 1629, 1641 (1974).

50. G. A. Olah and R. H. Schlosberg, *J. Am. Chem. Soc.*, **90**, 2726 (1968).

51. G. A. Olah, *J. Am. Chem. Soc.*, **94**, 808 (1972).

52. P. H. Plesch, Lecture presented at International Symposium on Cationic Polymerization, Rouen, France, Sept. 1973, and *Makromol. Chem.*, **175**, 1065 (1974).

53. A. Ledwith, Lecture at International Symposium on Cationic Polymerization, Rouen, France, 1973, and *Makromol. Chem.*, **175**, 1117 (1974).

54. M. D. Sorgo, D. C. Pepper, and M. Szwarc, *J. Chem. Soc.*, **1973**, 419.

55. E. Maréchal, *J. Macromol. Sci., Chem.*, **A7**, 433 (1973).

56. C. P. Pinazzi, J. Cattiaux, and J. C. Brosse, *Makromol.*, **169**, 45 (1973).

57. M. Baccaredda, M. Bruzzone, S. Cesca, M. DiMaina, G. Ferraris, P. Giusti, P. L. Magagnini, and A. Priola, *Chim. Ind. (Milano)*, **55**, 109 (1973).

58. M. Marek and L. Toman, *J. Polymer Sci.*, Symp. No. 402, 339 (1973).

59. Z. Zlamal, *Kinetics and Mechanisms of Polymerization Series*, Vol. I, Part II, G. E. Ham, Ed., Dekker, New York, 1969, Chap. 6.

60. T. Higashimura and T. Masuda, *Polymer J.*, **5**, 278 (1973).

61. M. Irie, S. Tomimoto, and K. Hayashi, *J. Polymer Sci., Chem.*, **11**, 1859 (1973).

62. J. Furukawa, E. Kobayashi, and S. Taniguchi, Lecture C11 presented at International Symposium on Cationic Polymerization, Rouen, France, Sept. 1973.

63. T. Saegusa, H. Fujii, S. Kobayshi, and H. Ando, *Macromolecules*, **6**, 26 (1973).

64. T. Saegusa, Lecture presented at International Symposium on Cationic Polymerization, Rouen, France, Sept. 1973; *Makromol. Chem.*, **175**, 1199 (1974).

65. O. Ohara, C. Aso, and T. Kunitake, *Polymer J.*, **5**, 49 (1973).

66. C. Aso, T. Kunitake, and S. Tagami, *Progr. Polymer Sci. Japan*, **1**, 149 (1972).

67. Y. Yamashita and K. Chiba, *Polymer J.*, **4**, 200 (1973).

68. S. Penczek, Lecture presented at International Symposium on Cationic Polymerization, Rouen, France, Sept. 1973; *Makromol. Chem.*, **175**, 1217 (1974).

69. N. S. Enikolopyan, Lecture presented at International Symposium on cationic Polymerization, Rouen, France, Sept. 1973.

70. H. Cherdron, *J. Macromol. Sci., Chem.*, **A6**, 1077 (1972).

71. G. Heublein and W. Römer, *Makromol. Chem.*, **163**, 143 (1973).

72. E. J. Goethals, Lecture presented at International Symposium on Cationic Polymerization, Rouen, France, Sept. 1973; *Makromol. Chem.*, **175**, 1309 (1974).

73. P. Dumas, N. Spassky, and P. Sigwalt, *Makromol. Chem.*, **156**, 65 (1972).

74. F. Lautenschlager, *J. Macromol. Sci., Chem.*, **A6**, 1089 (1972).

75. T. Kagiya, T. Matsuda, M. Nakano, and R. Hirata, *J. Macromol. Sci., Chem.*, **A6**, 1631 (1972).

76. M. H. Litt and T. Matsuda, *Polymer Preprints*, **15**, 753 (1974).

77. M. P. Dreyfuss, *J. Macromol. Sci., Chem.,* **A7,** 1359 (1973).

78. E. J. Vandenberg, *Polymer Preprints,* **15,** 208 (1974).

79. W. J. Bailey, H. Katsuki, and T. Endo, *Polymer Preprints,* **15,** 445 (1974).

80. J. P. Kennedy, Lecture presented at International Symposium on Cationic Polymerization, Rouen, France, Sept. 1973 and *Makromol. Chem.,* **175,** 1101 (1974).

CHAPTER 4

PHENOMENOLOGY OF CATIONIC MONOMERS

Conceptually, any chemical that can be cationated and in turn is able to attack another molecule of the original chemical is a cationically polymerizable monomer or, briefly, a "cationic monomer." Thus cationic monomers are bases or, more precisely, nucleophiles.[a]

Cationic monomers include olefins with electron releasing substituents,

$$CH_2=CH\leftarrow R \quad \leftrightarrow \quad \overset{\delta\ominus}{CH_2}-\overset{\delta\oplus}{CH}-R$$

and consequently "electron rich" double bonds. It is a truism, nonetheless it is often overlooked, that to obtain high polymer by cationic propagation the double bond must be the site of the highest nucleophilicity or cationic affinity. Since the attacking cation tends to seek out the most nucleophilic site in the molecule, if this site is not the double bond, nonpropagating complexes may form and propagation may come to an end. This explains why acrylates, vinyl acetate, acrylonitrile, and so on are not cationic monomers and why they cannot be polymerized cationically. For example, acrylonitrile is not a cationic monomer because the nucleophilic $-CN$ group irreversibly complexes the electrophile (the cation or the Lewis acid):

$$CH_2=\underset{CN}{CH} \xrightarrow{+R^\oplus} CH_2=\underset{C\equiv N-R}{CH_\oplus} \leftrightarrow CH_2=\underset{\underset{\oplus}{C=N-R}}{CH} \leftrightarrow \overset{\oplus}{CH_2}-\underset{C=N-R}{CH} \leftrightarrow \text{etc.}$$

[a]The term "nucleophile" in our context denotes organic compounds able to coordinate with carbenium ions; similarly, the term "nucleophilicity" refers to the ability to coordinate with cations. The terms "bases" and "basicity" refer to interactions (coordination) with the proton.

The situation is similar with acrylates or vinyl acetate:

$$
\underset{\substack{\\ \text{CH}_2=\text{C} \\ \vert \\ \text{C}=\text{O} \\ \vert \\ \text{O} \\ \vert \\ \text{CH}_3}}{\overset{\text{CH}_3}{}}
\;\;\xrightarrow{+\text{R}^\oplus}\;\;
\underset{\substack{\\ \text{CH}_2=\text{C} \\ \vert \\ \oplus\text{C}-\text{O}-\text{R} \\ \vert \\ \text{O} \\ \vert \\ \text{CH}_3}}{\overset{\text{CH}_3}{}}
\;\;\longleftrightarrow\;\;
\underset{\substack{\\ \overset{\oplus}{\text{CH}}_2-\text{C} \\ \vert \\ \text{C}-\text{O}-\text{R} \\ \vert \\ \text{O} \\ \vert \\ \text{CH}_3}}{\overset{\text{CH}_3}{}}
\;\;\longleftrightarrow\;\;
\underset{\substack{\\ \text{CH}_2=\text{C} \\ \vert \\ \text{C}-\text{O}-\text{R} \\ \vert \\ \oplus\text{O} \\ \vert \\ \text{CH}_3}}{\overset{\text{CH}_3}{}}
\;\longleftrightarrow \text{etc.}
$$

$$
\underset{\substack{\\ \text{O} \\ \vert \\ \text{C}=\text{O} \\ \vert \\ \text{CH}_3}}{\overset{\text{CH}_2=\text{CH}}{}}
\;\;\xrightarrow{+\text{R}^\oplus}\;\;
\underset{\substack{\\ \text{O} \\ \vert \\ \oplus\text{C}-\text{O}-\text{R} \\ \vert \\ \text{CH}_3}}{\overset{\text{CH}_2=\text{CH}}{}}
\;\;\longleftrightarrow\;\;
\underset{\substack{\\ \oplus\text{O} \\ \Vert \\ \text{C}-\text{O}-\text{R} \\ \vert \\ \text{CH}_3}}{\overset{\text{CH}_2=\text{CH}}{}}
\;\;\longleftrightarrow\;\;
\underset{\substack{\\ \text{O} \\ \vert \\ \overset{\oplus}{\text{C}}=\text{O}-\text{R} \\ \vert \\ \text{CH}_3}}{\overset{\text{CH}_2=\text{CH}}{}}
\;\longleftrightarrow \text{etc.}
$$

These monomers carry electron withdrawing substituents and as such have "electron-poor" double bonds: they are polymerizable by anions or bases ("anionic polymers").

The term "polymerizability" implies the overall tendency of a monomer to give a high molecular weight product. A high degree of polymerizability indicates a high rate of initiation combined with a high rate of uninterrupted propagation to high polymer.

Cationic polymerizability necessarily implies that the cationated species will remain sufficiently reactive and will sustain propagation. For example, in $\text{CH}_2=\text{CH}-\text{NH}_2$ cationic attack occurs readily; however, propagation does not proceed because of the prohibitively high stability of the species formed:

$$
\underset{\substack{\\ \vert \\ \text{NH}_2}}{\overset{\text{CH}_2=\text{CH}}{}}
\;\;\xrightarrow{+\text{H}^\oplus}\;\;
\underset{\substack{\\ \vert \\ \text{NH}_2}}{\overset{\text{CH}_3-\overset{\oplus}{\text{CH}}}{}}
\;\;\longleftrightarrow\;\;
\underset{\substack{\\ \Vert \\ \underset{\oplus}{\text{NH}_2}}}{\overset{\text{CH}_3-\text{CH}}{}}
\;\;\longleftrightarrow\;\;
\underset{\substack{\\ \vert \\ \underset{\oplus}{\text{NH}_3}}}{\overset{\text{CH}_2=\text{CH}}{}}
\;\;\longleftrightarrow\;\; \text{etc.}
$$

Unavailability of the cationic center can, of course, also prevent propagation. For example, monomers such as 1,1-diphenylethylene and 2,4,4-trimethyl-1-pentene can readily be cationated, however, the resulting carbenium ions are severely hindered ("buried") and are unable to sustain propagation because of steric compression in the transition state:

$$CH_2=C\underset{\overset{|}{CH_2}}{\overset{CH_3}{\diagup}} \quad \xrightarrow{+H^{\oplus}} \quad CH_3-\overset{\oplus}{C}\underset{\overset{|}{CH_2}}{\overset{CH_3}{\diagup}} \quad \longrightarrow \quad \text{dimers and trimers}$$

with $H_3C-\overset{|}{\underset{|}{C}}-CH_3$ groups (CH_3).

Similarly, 1,2-disubstitution also reduces polymerizability. Thus β-substituted styrenes are reluctant to polymerize to high molecular weight products probably because of steric compression. In this context, it is of interest that the polymerizability of indene and/or indene derivatives is appreciable and that these structures give respectable molecular weights, indicating that the polymerizability of β-styrene derivatives can be restored by "cyclic restriction":

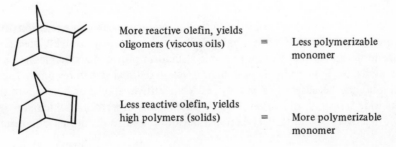

Fast polymerization to low molecular weight product

Fast dimerization

Fast polymerization to high molecular weight product

Fast polymerization to high molecular weight product

The effect of reactivity and steric compression on the polymerizability of olefins can be dramatically demonstrated with methylenecycloalkanes. While competition experiments show that 2-methylenenorbornane is more reactive than norbornene, 2-methylenenorbornane gives only oligomers and norbornene leads to solid polymers of reasonably high molecular weight under essentially identical conditions ($AlEtCl_2$, $\sim -78°C$):

More reactive olefin, yields oligomers (viscous oils) = Less polymerizable monomer

Less reactive olefin, yields high polymers (solids) = More polymerizable monomer

The oligomerization of 2-methylenenorbornane probably involves a tertiary cation transition state similar to that arising in the oligomerization of methylenecyclohexane and 1-methylcyclohexene, olefins that also yield only low molecular weight oils:

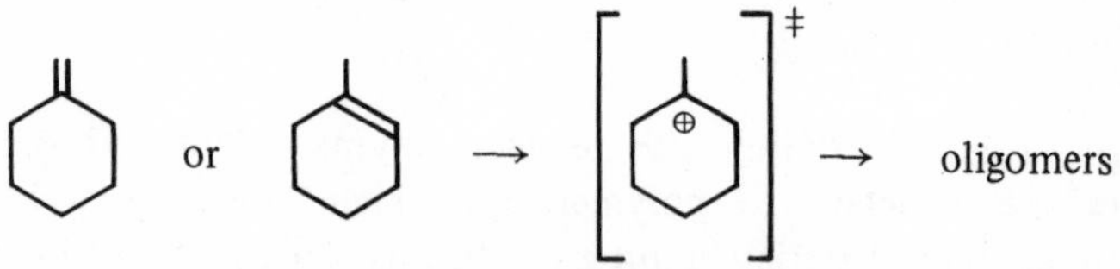

Evidently vinyl polymerizations giving rise to 1,1-disubstituted cycloalkanes are in general prohibited because of steric compression during propagation. 3-Methylenecyclohexene polymerizes to a product with exclusively 1,4-microstructure and no detectable amount of 1,2 or 3,4 addition occurs:

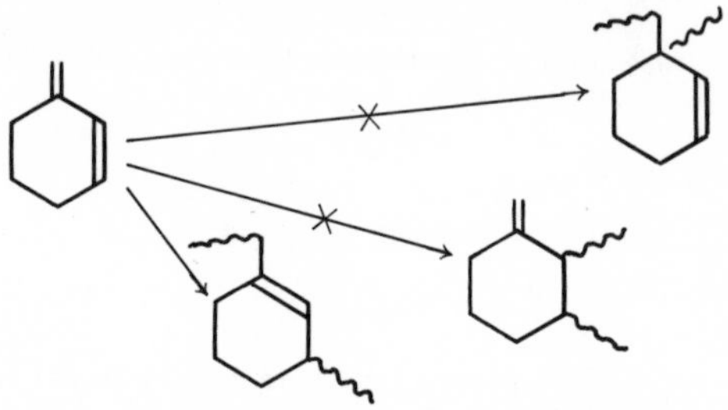

However, if a mechanism is available by which the sterically unfavorable cyclic carbenium ion can be avoided, propagation may proceed to high polymer. This situation obtains in the β-pinene system where the initial attack on the exo-methylene group giving rise to a "buried" cyclic carbenium ion is followed by a rearrangement to a sterically favorable propagating cation, and consequently, high molecular weight products are formed:

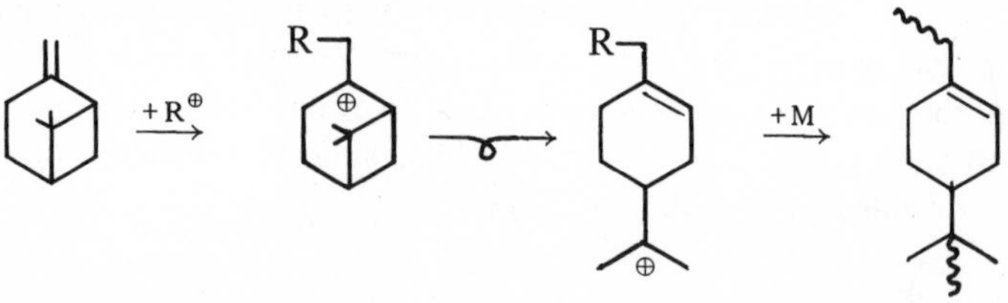

Isomerizations are sometimes undesirable, sometimes desirable reactions during cationic polymerizations. Carbenium ions are inherently unstable species and are prone to rearrange to more stable structures provided a suitable mechanism is available. For example, propylene gives ill defined structures under cationic conditions because of a variety of competitive alkylations, isomerizations, hydride transfers, and so on. Isobutylene, on the other hand, which propagates via relatively stable tertiary cations, does not lead to disturbing side reactions and gives rise to high molecular weight, unbranched structures.

Isomerization, however, can be controlled and may be extremely useful in providing unique high polymer structures. The classical example of a controlled isomerization coupled with polymerization is the case of 3-methyl-1-butene, a monomer that gives crystalline high polymer:

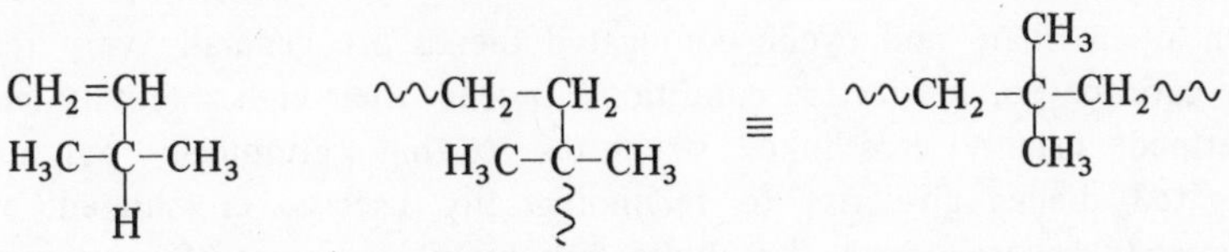

$$
\begin{array}{ccc}
\underset{\underset{\displaystyle H}{|}}{\underset{\displaystyle H_3C-C-CH_3}{|}}{CH_2=CH}
&
\underset{\underset{\displaystyle\wr}{|}}{\underset{\displaystyle H_3C-C-CH_3}{|}}{\sim\!\sim\!CH_2-CH_2}
\equiv
&
\underset{\underset{\displaystyle CH_3}{|}}{\overset{\overset{\displaystyle CH_3}{|}}{\sim\!\sim CH_2-C-CH_2\!\sim\!\sim}}
\end{array}
$$

a,ω-Diolefins may undergo intra-intermolecular polymerization, particularly if the resulting structure is a five- or six-membered ring:

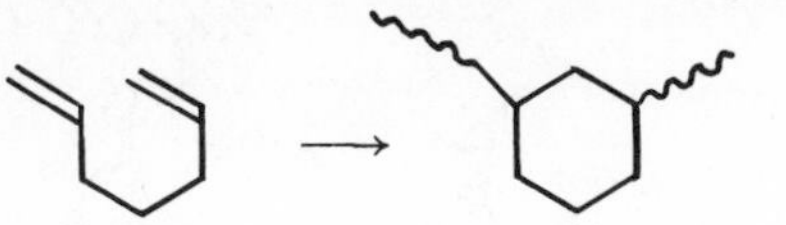

Intra-intermolecular reactions have also been proposed to occur with various dienes and partially conjugated trienes (e.g., myrcene and allocimene), however, because of the large number of side reactions the structures obtained are not "clean" and have not been fully characterized.

Olefins containing two dissimilar bonds may be useful for specific applications. For example, 2-methyl-1,5-heptadiene

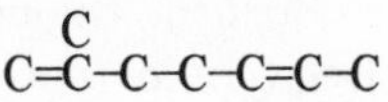

$$
\overset{\displaystyle C}{\underset{\displaystyle }{C=C-C-C-C=C-C}}
$$

could be copolymerized cationically with other olefins through the more reactive terminal double bond and the second, less reactive internal unsaturation can subsequently be used for other derivatization purposes.

Styrene and many of its derivatives can readily be polymerized cationically. The influence of steric compression on β-substituted styrenes has been mentioned above. Inductive and resonance effects importantly influence the reactivity of styrene and derivatives. Electron donating substituents increase the polymerizability of styrene derivatives, whereas electron withdrawing substituents cause a decrease. For example, p-chlorostyrene is much less reactive and p-methoxystyrene much more reactive than the parent compound. It is of interest that p-methoxystyrene is cationically polymerizable at all in spite of the presence of a nucleophilic $-OCH_3$ group. Evidently, effective electron donating by the p-methoxy group produces the highest electron density at the double bond, the site of cationic attack (vinyl ethers $C=C-O-R$ are also extremely reactive under cationic attack, but in this case high polymerizability is explained by enhanced ion stability due to direct resonance interaction:

$$C-\overset{\oplus}{C}-O-R \longleftrightarrow C-C=\overset{\oplus}{O}-R).$$

In this context it is not too surprising that it is difficult to polymerize *m*-methoxystyrene by cations.

Both open chain and cyclic conjugated dienes are generally very reactive under cationic polymerization conditions. Indeed, their high reactivity leads to ill defined, usually crosslinked structures so that cationically polymerized conjugated dienes give rise to technologically useless, crosslinked, chalky products. In general terms, the allylic carbenium ion formed from conjugated diolefins may attack the next diolefin in at least two ways:

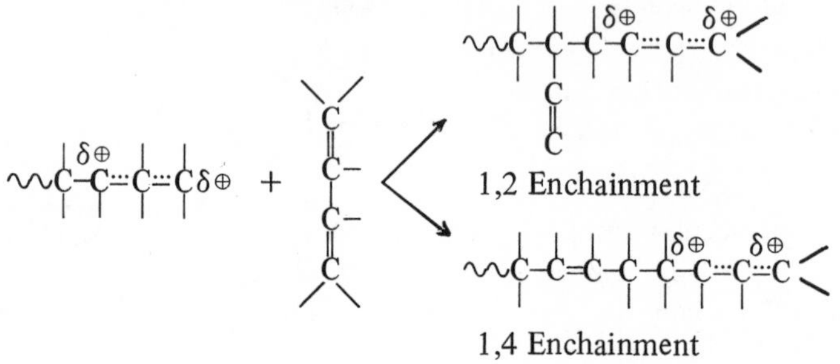

1,2 Enchainment

1,4 Enchainment

leading to two different enchainments. In practice it is difficult to control these competing routes in conventional cationic polymerization systems and a mixture of ill defined products is obtained. The mechanism is further complicated by the stereochemistry involved (cisoid–transoid monomer conformation, allylic carbenium ion conformation, *cis* or *trans*-1,4 enchainment, etc.), penultimate effects, and the reactivity of both mesomeric forms of the allyl cation,

$$\sim\!\!\!\sim\!\!C-C=C-\overset{\oplus}{C} \text{ and } \sim\!\!\!\sim\!\!\!\sim\!\!C-\overset{\oplus}{C}-C=C.$$

Only in exceptional instances can clean structures be obtained from conjugated dienes by cationic techniques. This is the case, for example, with 2,5-dimethyl-1,3-hexadiene $(CH_3)_2C=CH-CH=C(CH_3)_2$, where the secondary allylic ion is probably completely hindered sterically and propagation occurs only by the terminal tertiary cation:

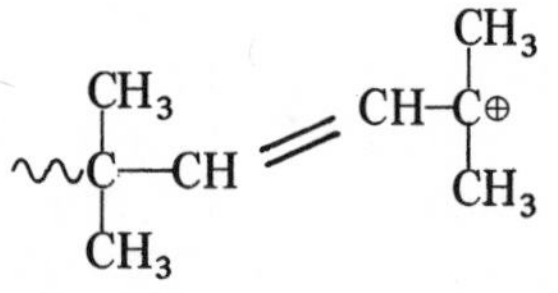

Thus the product has a well defined repeat unit and a crystalline *trans*-1,4-poly-(1,1,4,4-tetramethyl-1,3-butadiene) can be obtained.

Presently the most important industrial application of conjugated dienes in this field is the use of isoprene as a comonomer in butyl rubber manufacture.

Vinyl ethers are readily polymerized by cations and they are very reactive cationic monomers. Their reactivity is due to the high stability of the carbenium ions, resonance stabilized by the unshared pair of electrons on the neighboring oxygen:

$$\overset{\oplus}{C}-C-O-R \ \longleftrightarrow \ C-C\overset{\oplus}{=}O-R.$$

Conjugation of the carbenium ions with oxygen giving rise to extremely stable cationic species explains why vinyl ethers cannot be readily copolymerized to random copolymers with olefins or dienes. The copolymerization of different vinyl ethers is, of course, feasible and has been described.

Vinyl ethers occupy an important historical position in polymer chemistry: poly(vinyl isobutyl ether) was the first stereoregular synthetic polymer so recognized by Schildknecht and coworkers (see Chapter 3). Surprisingly, high molecular weight crystalline α,β-disubstituted ethers have been obtained recently by Higashimura and coworkers. Vinyl ethers possess a certain industrial significance—polyvinyl isobutyl, -ethyl, and so on ethers are used in pressure sensitive tapes and other applications.

Cyclic ethers are an important class of cationic polymers. Most three-, four-, five-, and higher membered cyclic ethers readily propagate by oxonium ions. The laws of polymerizability of cyclic ethers are fairly well understood. The most important parameters are ring strain and basicity. Tetrahydropyran and substituted tetrahydrofurans (THF) are reluctant to polymerize ($\Delta G \approx 0$ or > 0). Poly-THF gives an excellent elastomer, although the current price of THF is too high to justify the commercial development of poly-THF rubbers. Epichlorohydrin, however, is the basis of a promising new line of specialty elastomers: epi rubbers.

Cyclic formals and acetals are also important cationic polymers that propagate via oxonium ions. A large amount of scientific work has been done on dioxolane, the simplest member of this class, and its various derivatives. The most important monomer among cyclic formals is trioxane, the cyclic trimer of formaldehyde. Trioxane polymers containing a few percent of ethylene oxide (or dioxolane) impart heat stability to the product and lead to Celcon, a highly successful engineering plastic.

Aliphatic aldehydes, their derivatives, and dialdehydes are cationically readily polymerizable to interesting high polymers. For example, the polymerization of formaldehyde, acetaldehyde, acrolein, glutaraldehyde, and chloral have been studied in some depth and considerable scientific literature is available. In contrast, aromatic aldehydes do not give high polymers under cationic conditions.

Simple ketenes are reactive cationic polymers.

Among cyclic esters, various groups of lactones, lactides, carbonates, and oxalates have been studied under cationic polymerization conditions, particularly by Japanese investigators.

Limited studies have also been carried out with sulfur and nitrogen containing compounds.

Finally, a few classes of miscellaneous compounds exist, for example, benzene and its derivatives, benzyl halides and their derivatives, acetylenic compounds, azo-compounds, and ferrocenes, that have been polymerized by cationic means and gave very interesting and unexpected products.

CHAPTER 5

THE ORGANIZATION OF THE LIST OF CATIONICALLY POLYMERIZABLE OLEFINS

Table 1 is a compilation of the structures of cationically polymerizable olefins discussed in this volume. This list includes essentially all the olefins described up to 1973 in the scientific literature as polymerizing by cationic techniques. Though much of the material has also been extracted from the patent literature, that part of this survey is not complete.

The organization of this list starts with simple aliphatic olefins and systematically progresses to more complicated structures.

To preserve the coherence of the organization, in addition to pure hydrocarbon olefins a few hetero atom containing olefin derivatives are included in appropriate places. For example, chloroprene (2-chloro-1,3-butadiene) appears among conjugated dienes; halogenated styrene derivatives and other hetero atom substituted (alkoxy, hydroxy, etc.) aromatic olefins are listed together with the parent olefin compounds.

In a few cases the classification of certain compounds was decided arbitrarily. For example, 2-phenyl-1,3-butadiene can be viewed as a conjugated diene derivative or as an aromatic olefin. In these borderline cases the reader should consult all the appropriate subclasses to locate the desired monomer.

Small ring compounds, that is, cyclopropane and cyclobutane derivatives, comprise compounds that polymerize by the opening of the highly strained ring. Thus ethylcyclopropane belongs to this class, but vinylcyclopropane does not because the latter polymerizes by the vinyl function and not by the cyclopropane ring.

Compounds such as 1,2-dimethylenecyclohexane, 1-vinylcyclohexene, and 1-methylene-2-cyclohexene can be found among cyclic conjugated dienes (and not among alicyclic olefins) because they behave as conjugated dienes under

37

polymerization conditions. For the same reason certain indene derivatives have also been listed in this subclass.

The aromatic olefins comprise three subclasses: (*1.*) Styrene and Derivatives contain all styrene derivatives beside the pure hydrocarbons, including some hetero atom containing compounds such as halogenated styrenes and alkoxy substituted compounds. (*2.*) Indene and Derivatives are listed separately because of the large number of these compounds. Most of the publications in this group are by Maréchal and coworkers. (*3.*) Miscellaneous Aromatic Olefins contain essentially all other aromatic olefins that could not be easily listed in the first two subclasses.

In total this survey comprises about 280 compounds.

Table 1 is useful not only as a record and quick reference for monomers that have been investigated in the past, but also as a road map for future research. Serious students of cationic polymerization will easily recognize gaps in this table and will capitalize from the missing information. For instance, this list might be of use for copolymerization research. Compared to free radical copolymerization, cationic copolymerization is a relatively little explored field. Glancing at the large number of cationic monomers available, this dearth of information cannot be due to a limited number of monomers. Monomer pairs that homopolymerize by the same basic mechansim and therefore could copolymerize can easily be identified in this list. For example, according to the scientific literature, the copolymerization of simple monomer pairs such as 3-methyl-1-butene/4-methyl-1-pentene and 3-methyl-1-butene/conjugated dienes has not been investigated.

Or, on the basis of the structures known to be cationically polymerizable, one can invent new, hitherto uninvestigated structures whose cationic polymerizability can be predicted and which might lead to interesting ultimate properties. For example, while *p*-methoxystyrene is readily polymerizable and allylbenzene leads to oligomers, *p*-methoxyallylbenzene has not yet been polymerized by cationic technique. It is conceivable that the following yet unknown structure could be synthesized with this monomer:

Or, the polymerization of allylcyclopropane might yield the repeat:

Similarly, the cationic polymerization of 3,4-dimethyl-1-pentene has not yet been studied. It would be of interest to know if the isomerization polymerization of two successive hydride migrations would give:

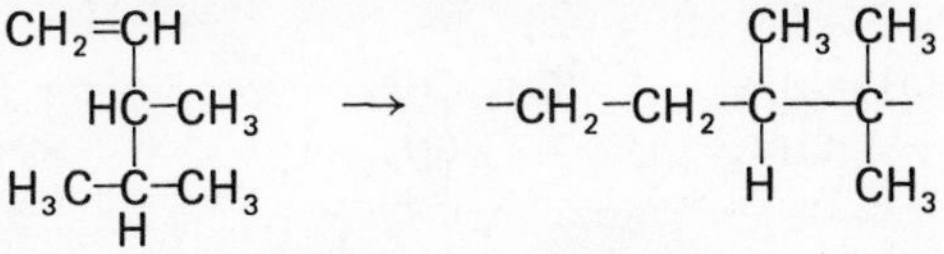

$$CH_2=CH \quad\quad\quad\quad\quad CH_3\ CH_3$$
$$HC\text{–}CH_3 \quad\longrightarrow\quad -CH_2\text{–}CH_2\text{–}C\text{—}C\text{–}$$
$$H_3C\text{–}C\text{–}CH_3 \quad\quad\quad\quad\quad H\quad\ CH_3$$
$$H$$

a structure identical to that obtained from 4,4-dimethyl-1-pentene.

It would be interesting to examine systematically the polymerization behavior of 2-alkoxy substituted butadienes, $CH_2=COR\text{–}CH=CH_2$, compounds that represent a cross between vinyl ethers and conjugated dienes. It would be of value to elucidate the microstructure (1,4 or 1,2 enchainment) of these polymers. Perhaps they would provide a way to vinyl ether–diene copolymers.

TABLE 1 CATIONIC OLEFINS (Cationically Polymerizable or Oligomerizable Olefins)

I. ALIPHATIC OLEFINS

   A. Straight-Chain Olefins and Nonconjugated Diolefins

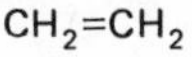

$CH_2=CH_2$	$CH_2=CH$ CH_3	$CH_2=CH$ CH_2 CH_3	CH_3 CH CH CH_3	$CH_2=CH$ CH_2 CH_2 CH_3	$CH_2=CH$ CH_2 CH_2 CH_2 CH_3	$CH_2=CH$ CH_2 CH_2 CH_2 C_nH_{2n+1}
			cis and *trans*			where n = 3, 5, 7, 11
1	**2**	**3**	**4**	**5**	**6**	**7**

   B. Branched α-Olefins

      a. Branches α to the Double Bond

$CH_2=CH$ CH $CH_3\ \ CH_3$	$CH_2=CH$ CH $CH_3\ \ CH_2$ CH_3	$CH_2=CH$ CH $CH_2\ \ CH_2$ $CH_3\ CH_3$	$CH_2=CH$ $H_3C\text{–}C\text{–}CH_3$ CH_3
1	**2**	**3**	**4**

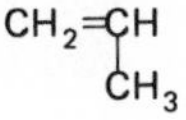

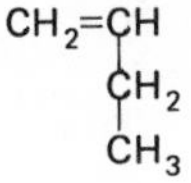

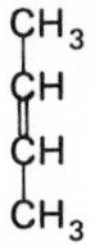

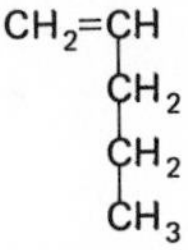

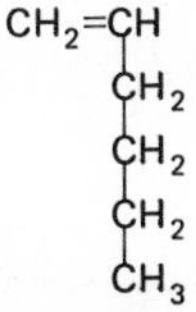

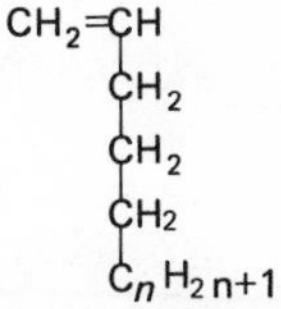

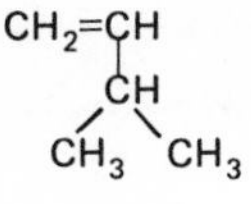

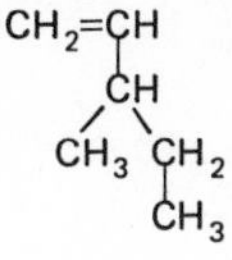

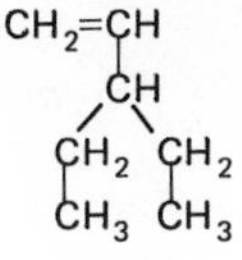

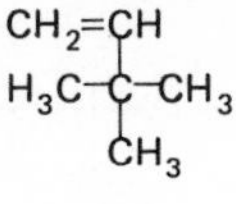

      b. Branches β to the Double Bond

$CH_2=CH$ CH_2 CH $CH_3\ \ CH_3$	$CH_2=CH$ CH_2 CH $CH_3\ \ CH_2$ CH_3	$CH_2=CH$ CH_2 $H_3C\text{–}C\text{–}CH_3$ CH_3
1	**2**	**3**

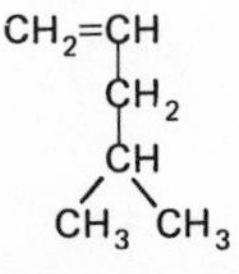

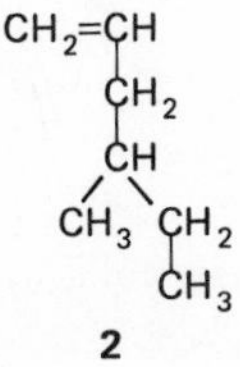

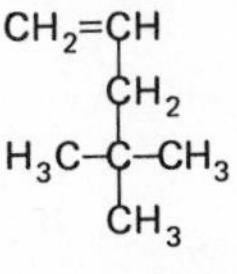

Table 1 (continued)

c. Branches γ, δ etc. to the Double Bond

CH₂=CH
  CH₂
  CH₂
  CH
CH₃ CH₃

1

CH₂=CH
  CH₂
  CH₂
  CH
CH₃ CH₂
      CH₃

2

CH₂=CH
  CH₂
  CH₂
  CH₂
  CH
CH₃ CH₃

3

C. Isoolefins

CH₂=C(CH₃)
     CH₃

1

CH₂=C(CH₃)
     CH₂
     CH₃

2

CH₃–C(CH₃)
      CH
     CH₃

3

CH₂=C(CH₃)
     CH₂
     CH₂
     CH₂
     CH₃

4

CH₂=C
CH₃
CH₂
CH₂
CH₂
CH₃

5

CH₂=C(CH₃)
     CH
   CH₃ CH₃

6

CH₂=C(CH₃)
H₃C–C–CH₃
     CH₃

7

CH₂=C(CH₃)
     CH₂
 H₃C–C–CH₃
      CH₃

8

CH₂=C(CH₃) CH₂=C(CH₃)
     CH₂———CH₂

9

D. Alicyclic Olefins, Diolefins, and Derivatives

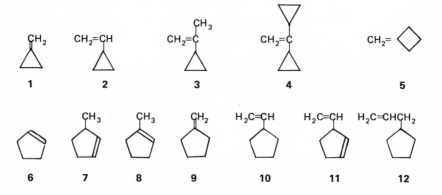

1   **2**   **3**   **4**   **5**

6   **7**   **8**   **9**   **10**   **11**   **12**

40

Table 1 (continued)

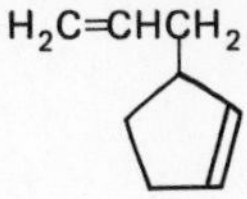

H₂C=CHCH₂ ... structures 13-19

$H_2C=CHCH_2$

| 13 |  14 | 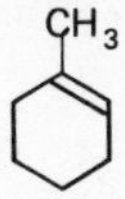 15 | 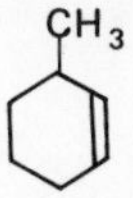 16 | 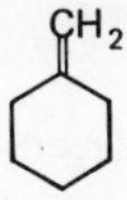 17 | 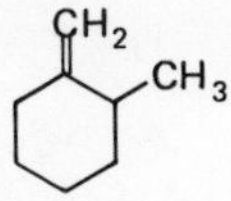 18 | 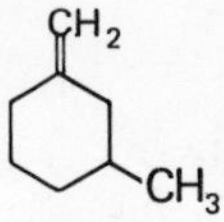 19 |

| 20 | 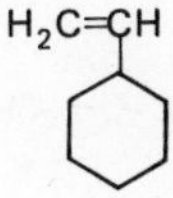 21 | 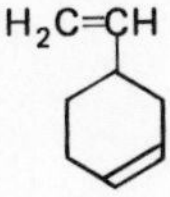 22 | 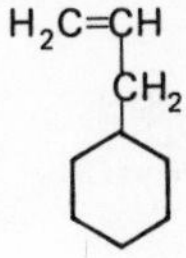 23 | 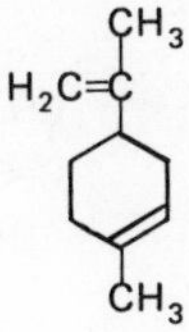 24 | 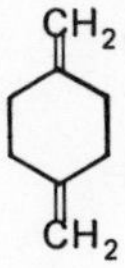 25 | 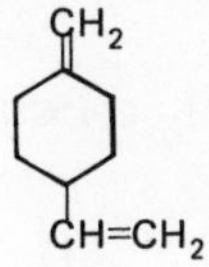 26 |

| 27 | 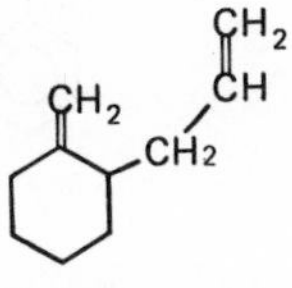 28 | 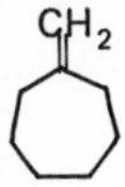 29 | 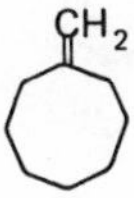 30 | 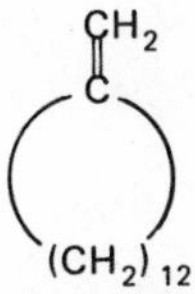 31 |

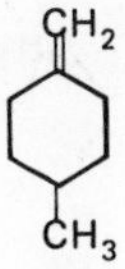

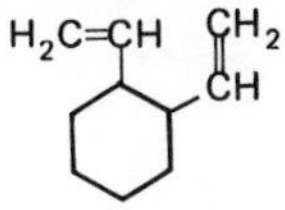

E. Conjugated Olefins

a. Straight and Branched Chain Conjugated Dienes and Trienes

$CH_2=CH-CH=CH_2$ $CH_2=\overset{\underset{\displaystyle CH_3}{|}}{C}-CH=CH_2$ $CH_2=\overset{\underset{\displaystyle CH_3}{|}}{\underset{}{C}}-\overset{\displaystyle CH_3}{\underset{}{C}}=CH_2$ $CH_2=CH-CH=CH-CH_3$

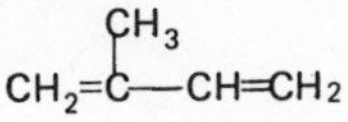

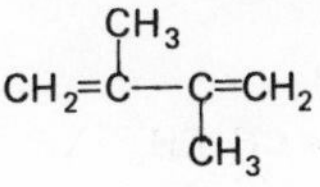

| 1 | 2 | 3 | 4 |

| 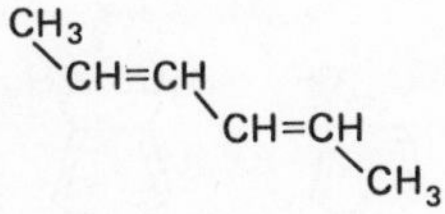 5 | 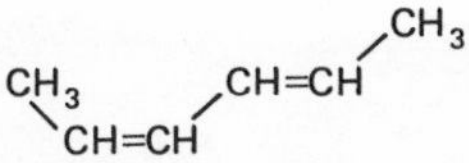 6 | 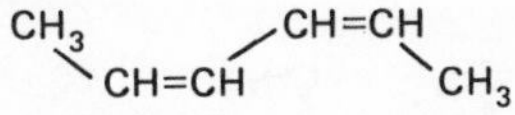 7 |

41

Table 1 (continued)

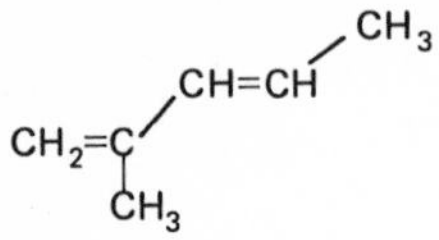

8

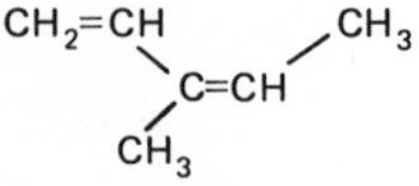

9

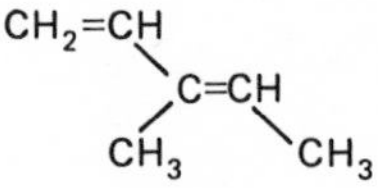

10

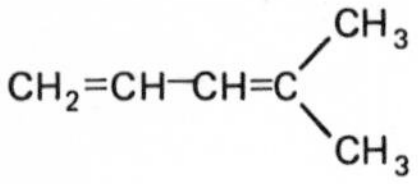

CH$_2$=CH–CH=C(CH$_3$)$_2$

11

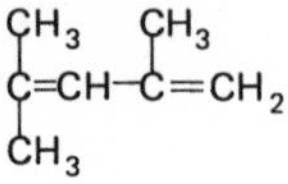

12

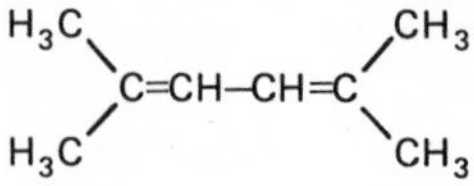

13

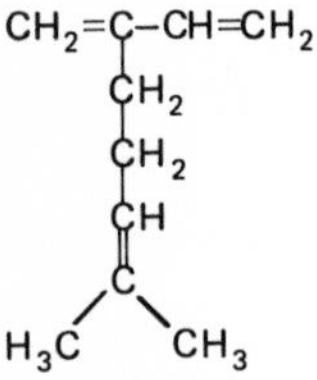

14

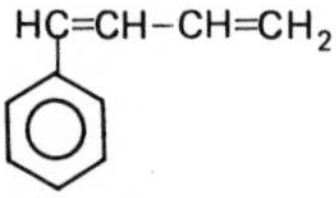

15

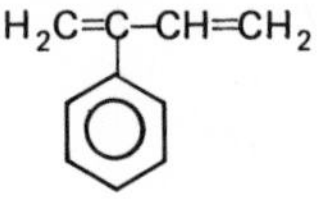

16

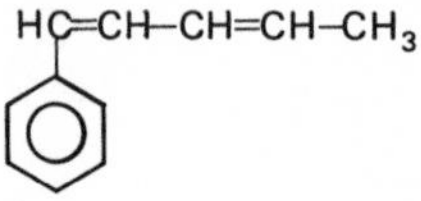

17

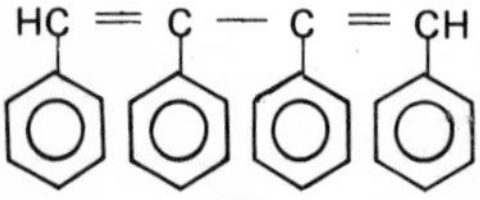

18

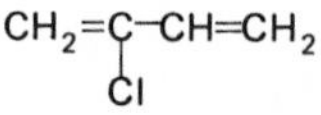

19

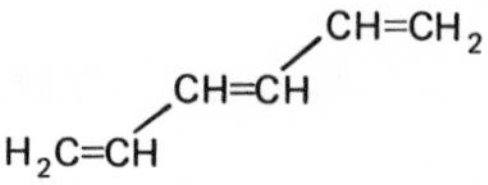

20

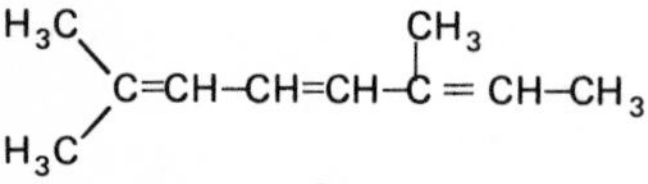

21

b. *Cyclic Conjugated Dienes*

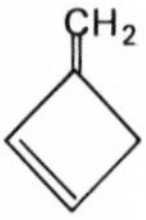

1 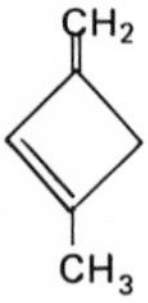   2    3 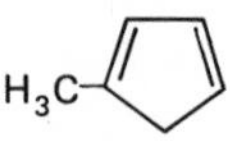   4 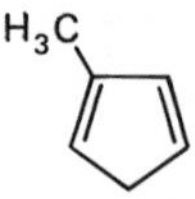   5 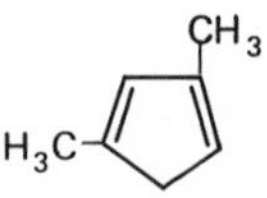   6 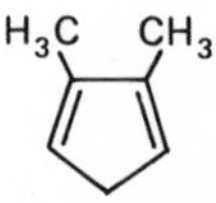   7

42

Table 1 (continued)

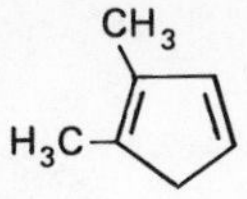

8 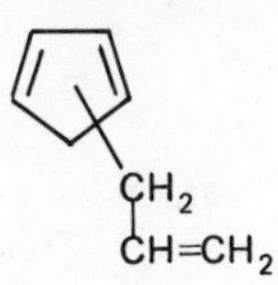   9 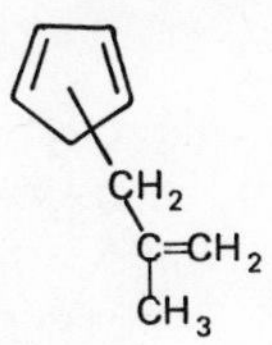   10 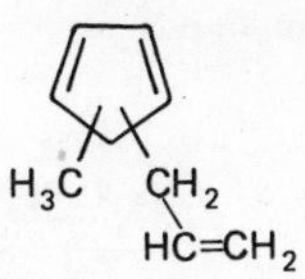   11 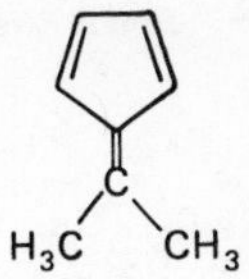   12

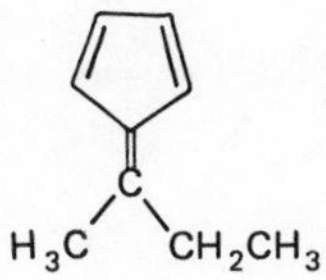

13    14 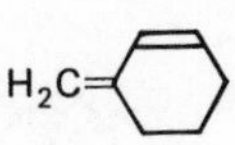   15 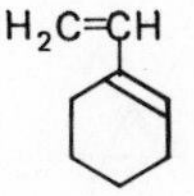   16 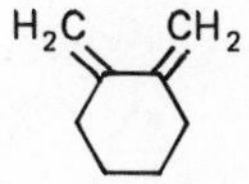   17

18 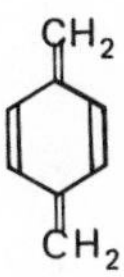   19 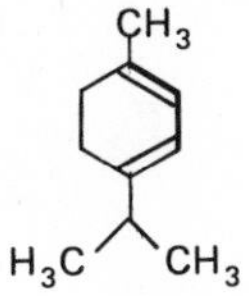   20 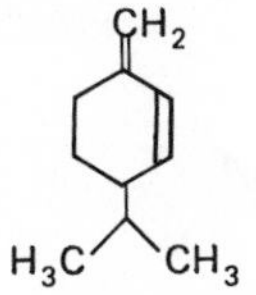   21 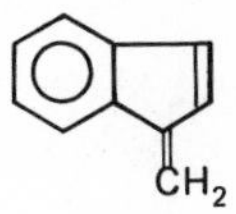   22

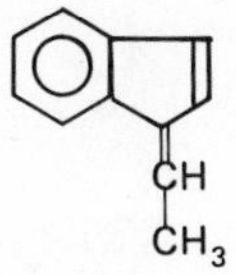

23 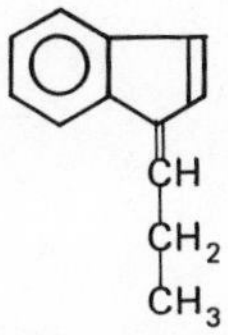   24 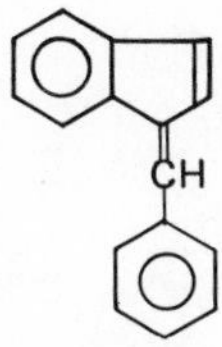   25 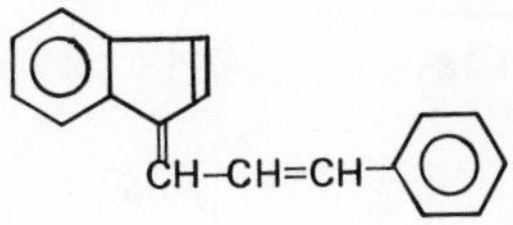   26

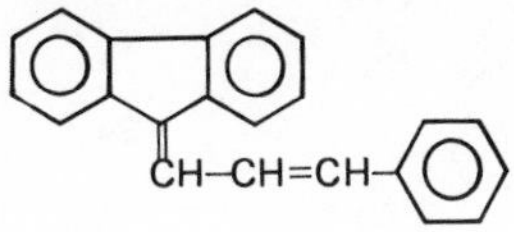

27 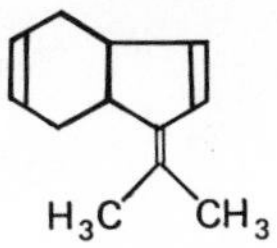   28 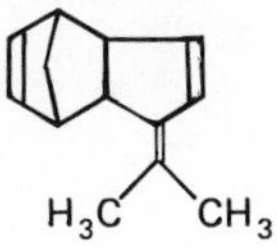   29

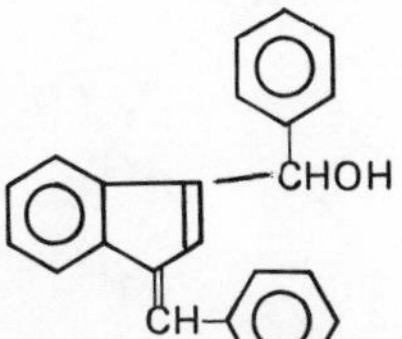

30

Table 1 (continued)

F. Small Ring Compounds

CH_2CH_3

$CH_2-CH_2-CH_3$

CH_3
$CH-CH_3$

H_3C
CH_3

CH_3
CH_3

1 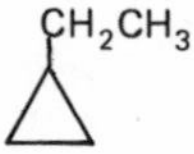   2 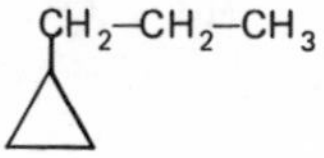   3 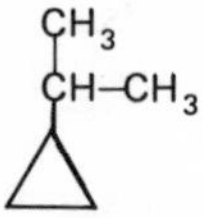   4 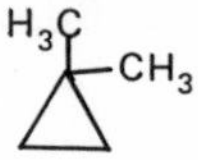   5 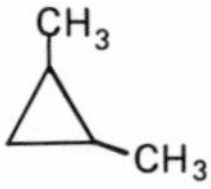   6

7    8    9    10 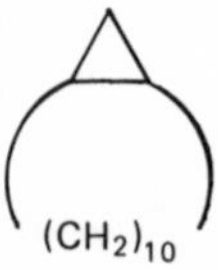   11

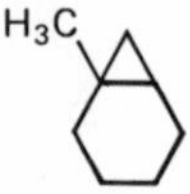

$(CH_2)_{10}$

H_3C H_3C H_3C H_3C H_3C

CH_3

$(CH_2)_{10}$

12 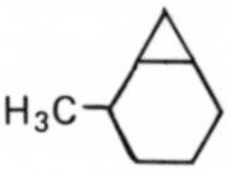   13    14 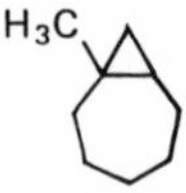   15 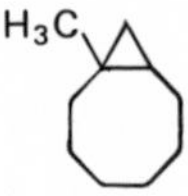   16 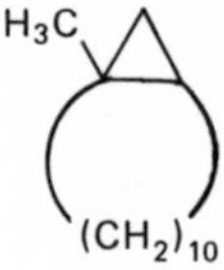   17

$(CH_2)_{12}$

18    19    20    21    22    23    24

H_3C H_3C CH_3 CH_3
H_3C H_3C H_3C
Cl Cl Cl Cl Cl Br

Cl Cl

H_3C

Cl Cl

Cl

Cl Cl

25 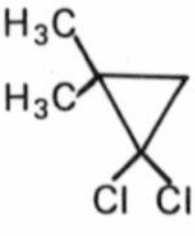   26 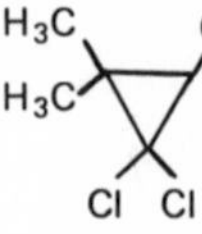   27 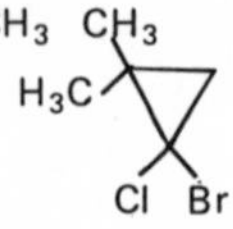   28 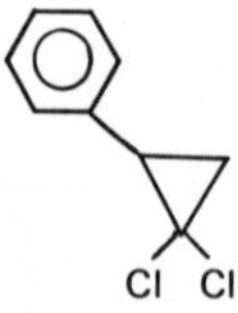   29 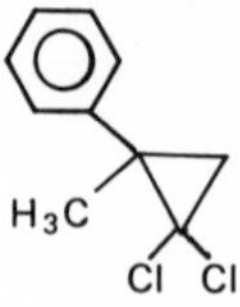   30 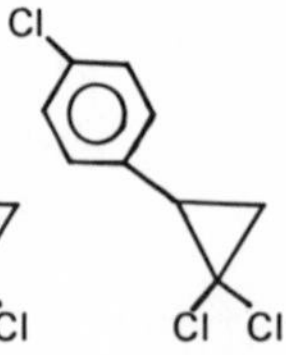   31

Table 1 (continued)

H$_2$C=CH
H$_3$C
Cl Cl

32

Cl Cl
(CH$_2$)$_4$

33

Cl Cl
(CH$_2$)$_5$

34

Cl Cl
(CH$_2$)$_6$

35

H$_3$C Cl Cl
(CH$_2$)$_3$

36

H$_3$C Cl Cl
(CH$_2$)$_4$

37

H$_3$C Cl Cl
(CH$_2$)$_5$

38

H$_3$C Cl Cl
(CH$_2$)$_6$

39

Cl Br
(CH$_2$)$_4$

40

Cl Cl
(CH$_2$)$_{10}$

41

Br Br
(CH$_2$)$_{10}$

42

H$_3$C Cl Cl
(CH$_2$)$_{10}$

43

H$_3$C Br Br
(CH$_2$)$_{10}$

44

45

CH$_3$

46

CH$_2$CH$_3$

47

H$_3$C–CH–CH$_3$

48

G. Bicyclic Olefins

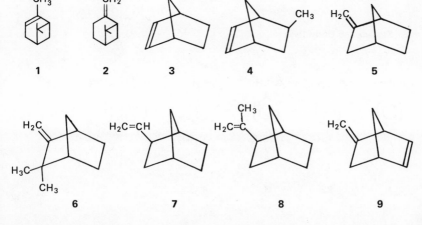

CH$_3$

1

CH$_2$

2

3

CH$_3$

4

H$_2$C

5

H$_2$C
H$_3$C
CH$_3$

6

H$_2$C=CH

7

CH$_3$
H$_2$C=C

8

H$_2$C

9

45

Table 1 (continued)

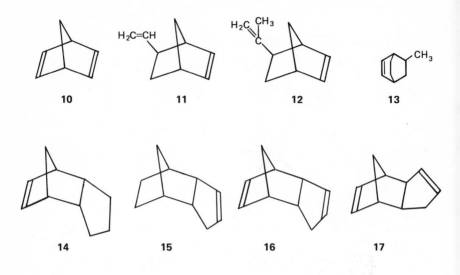

10   11   12   13

14   15   16   17

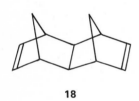

18

A. Styrene and Derivatives

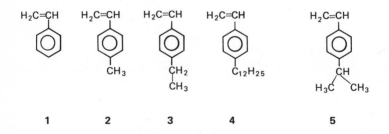

1   2   3   4   5

Table 1 (continued)

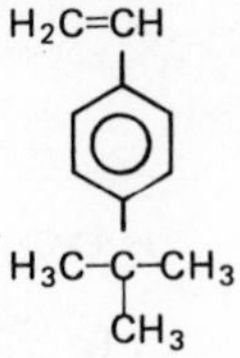

6 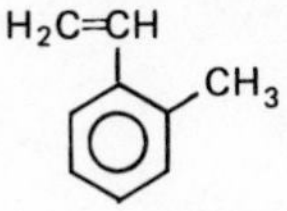   7 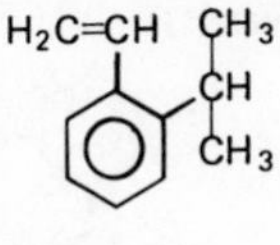   8 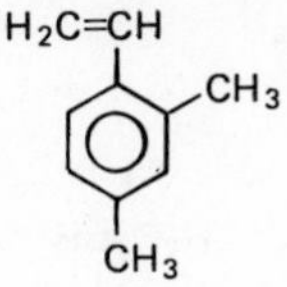   9 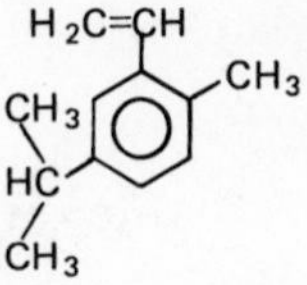   10

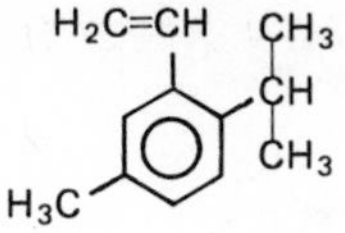

11 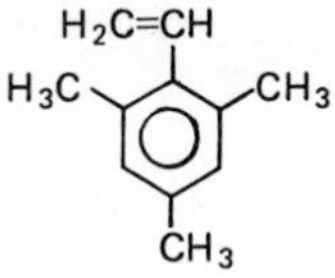   12 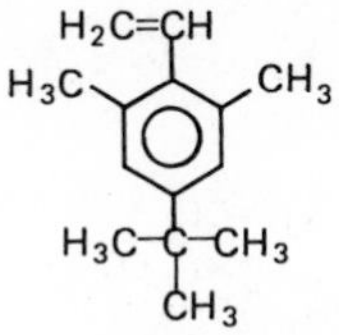   13 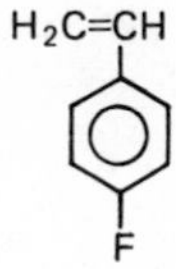   14

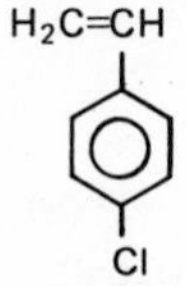

15 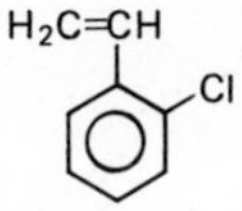   16 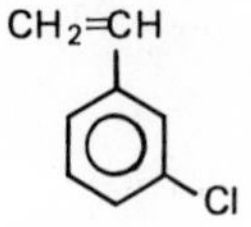   17 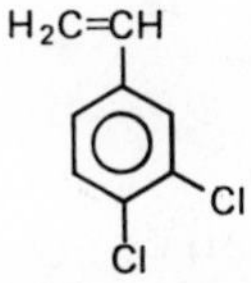   18 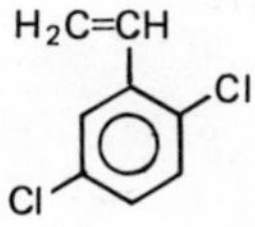   19 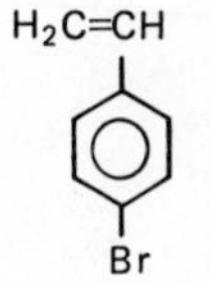   20

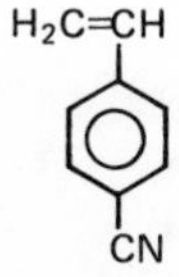

21 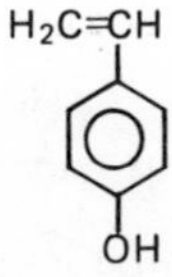   22 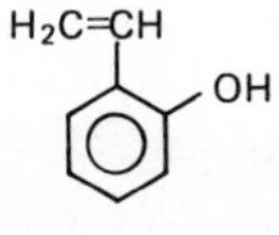   23 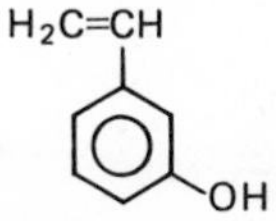   24 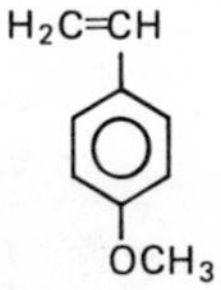   25

47

Table 1 (continued)

26 27 28 29 30

31 32 33 34

35 36 37 38 39

40 41 42 43

Table 1 (continued)

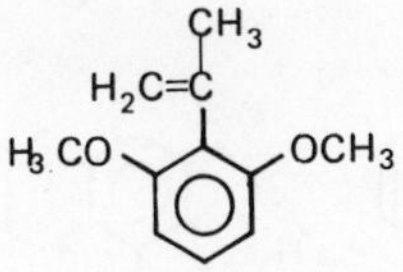

44

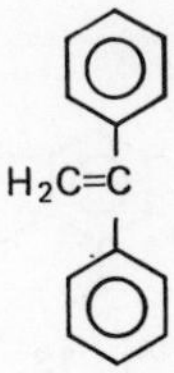

45

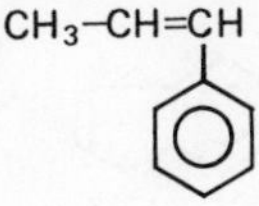

cis and *trans*

46,47

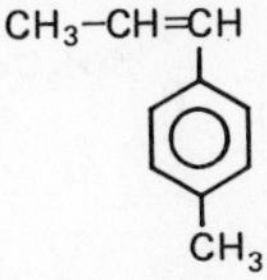

cis and *trans*

48,49

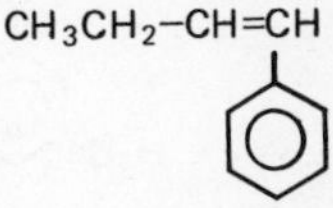

cis and *trans*

50,51

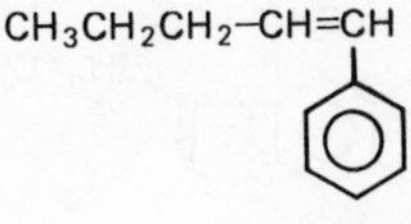

cis and *trans*

52,53

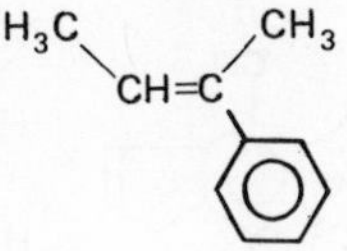

54

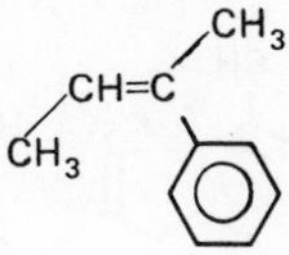

55

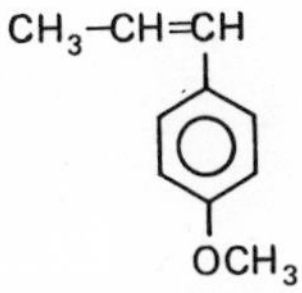

cis and *trans*

56,57

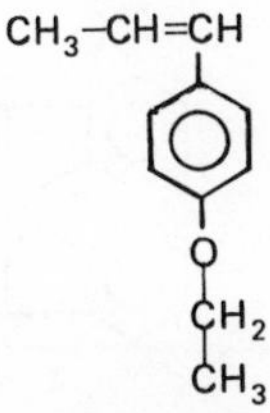

cis and *trans*

58,59

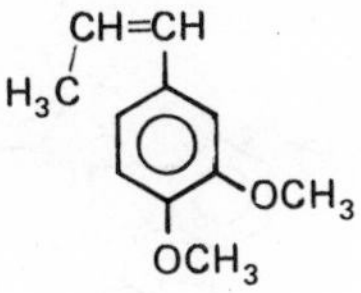

60

B. Indene and Derivatives

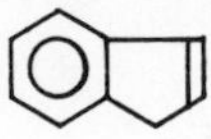

1

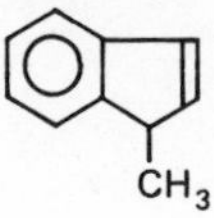

2

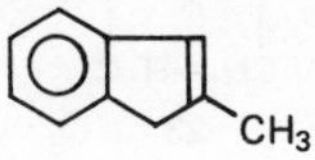

3

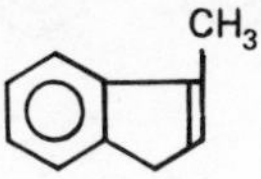

4

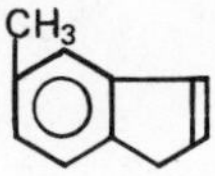

5

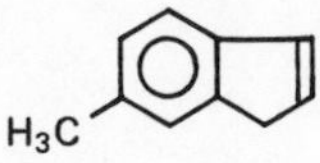

6

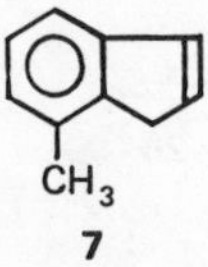

7

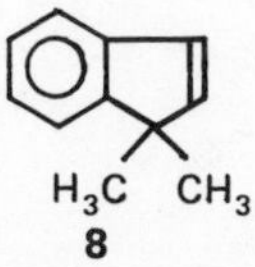

8

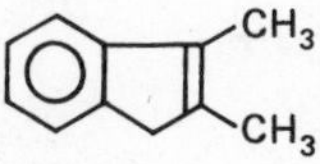

9

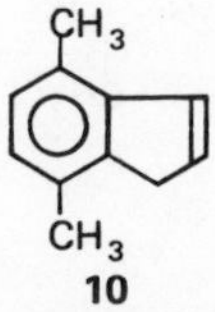

10

Table 1 (continued)

11 **12** **13** **14**

15 **16** **17** **18**

19 **20** **21**

22 **23** **24**

25 **26**

Table 1 (continued)

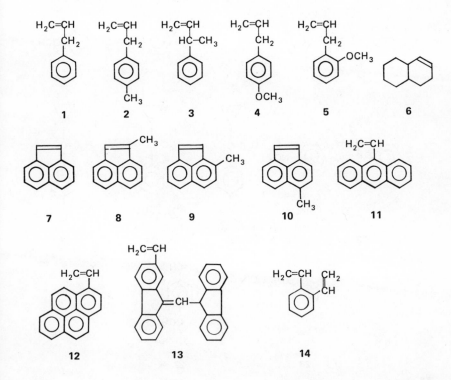

27

28

29

Br–

CH₂–C=C–CH₂

CH₂CH₂CH₂Br

30

31

32

33

CH₂CH₂CH₂CH₂Br

OCH₃

H₃CO

H₃CO

C. Miscellaneous Aromatic Olefins

1

H₂C=CH
CH₂

2

H₂C=CH
CH₂

CH₃

3

H₂C=CH
HC–CH₃

4

H₂C=CH
CH₂

OCH₃

5

H₂C=CH
CH₂

OCH₃

6

7

8

CH₃

9

CH₃

10

CH₃

11

H₂C=CH

12

H₂C=CH

13

H₂C=CH
CH

14

H₂C=CH   CH₂
CH

51

Table 1 (continued)

cis and trans
18,19

15   **16**   **17**   **20**   **21**   **22**   **23**   **24**   **25**   **26**   **27**

Table 1 (continued)

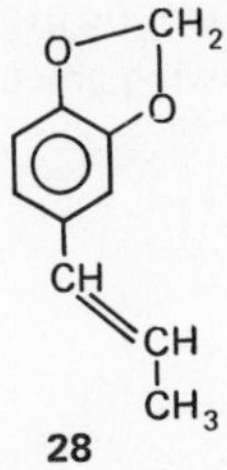

28

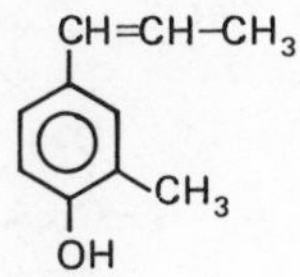

29

I. ALIPHATIC OLEFINS

A. n-OLEFINS AND NONCONJUGATED n-DIOLEFINS

a. Ethylene $CH_2=CH_2$

Ethylene cannot be polymerized to high molecular weight and/or linear polymers by cationic techniques. It is of historical interest that Ipatieff and Grosse found in 1936 that purest $AlCl_3$ will not polymerize ethylene at 50 atm unless traces of protogenic materials, such as moisture and hydrogen chloride, are present (1). Indeed, ethylene is so inert under cationic conditions that it may be used as a diluent in low temperature polymerization of isobutylene with $AlCl_3$ (2).

However, under more heroic conditions with $AlCl_3$ or $BF_3/SO_3/HF$, with phosphoric acid contacts at high temperatures and pressures, or with trifluoromethanesulfonic acid at room temperature ethylene can be oligomerized or cooligomerized to low molecular weight oils or products boiling in the gasoline range (2–7). Synthetic lubricating oils were produced from ethylene on a large scale in Germany during World War II. The process involved the oligomerization of a pure ethylene feed with $AlCl_3$ at 200°C and 880 psi (6).

That ethylene cannot be polymerized to high molecular weight products but can be oligomerized to oils or gasolines is due to the difficulty of protonation and/or cationation of ethylene under mild conditions. Even if these processes did occur [the formation and existence of primary carbenium ions in conventional liquid phase systems have been seriously questioned (8)], rapid isomerization and/or elimination would follow giving rise to ill defined structures. In this context the series of experiments by Schmerling (9,10) should be recalled. This author alkylated ethylene with various tertiary and secondary alkyl halides in conjunction with Friedel-Crafts acids ($AlCl_3$, $FeCl_3$, etc.) in the range −30 to 30°C and analyzed the products. For example, in one of these famous experiments t-butyl chloride was reacted with ethylene (9) and, among other products, 1-chloro-3,3-dimethylbutane was obtained. The following mechanism was proposed to explain the data:

$$
\begin{array}{c}
\overset{\displaystyle C}{\underset{\displaystyle C}{C-C-Cl}} + AlCl_3 \;\rightleftharpoons\; \overset{\displaystyle C}{\underset{\displaystyle C}{C-C}}{}^{\oplus} AlCl_4{}^{\ominus}
\end{array}
$$

$$
\overset{\displaystyle C}{\underset{\displaystyle C}{C-C}}{}^{\oplus} + C{=}C \;\rightleftharpoons\; \overset{\displaystyle C}{\underset{\displaystyle C}{C-C-C-C}}{}^{\oplus} \;\rightleftharpoons\; \overset{\displaystyle C}{\underset{\displaystyle C}{C-C-\underset{\oplus}{C}-C}} \;\rightleftharpoons\; \overset{\displaystyle C\;C}{C-C-\underset{\oplus}{C}-C}
$$

$$
\overset{\displaystyle C}{\underset{\displaystyle C}{C-C-C-C}}{}^{\oplus} + \overset{\displaystyle C}{\underset{\displaystyle C}{C-C-Cl}} \;\longrightarrow\; \overset{\displaystyle C}{\underset{\displaystyle C}{C-C-C-C-Cl}} + \overset{\displaystyle C}{\underset{\displaystyle C}{C-C}}{}^{\oplus}, \text{ etc.}
$$

The series of reactions proposed to explain the products formed in the alkylation of ethylene with isobutane and $AlCl_3/HCl$ (a process used to produce enormous amounts of 2,3-dimethylbutane for high octane gasoline) is very similar to the above mechanism:

$$
\overset{\displaystyle C}{\underset{\displaystyle C}{C-C-C}} + C{=}C + \xrightarrow[HCl]{AlCl_3} \overset{\displaystyle C}{\underset{\displaystyle C}{C-C-C}} + C-C^{\oplus}AlCl_4{}^{\ominus}
$$

$$
\overset{\displaystyle C}{\underset{\displaystyle C}{C-C-C}} + C-C^{\oplus} \;\longrightarrow\; \overset{\displaystyle C}{\underset{\displaystyle \underset{\oplus}{C}}{C-C-C}} + C-C
$$

$$
\overset{\displaystyle C}{\underset{\displaystyle \underset{\oplus}{C}}{C-C-C}} + C{=}C \;\longrightarrow\; \overset{\displaystyle C}{\underset{\displaystyle C}{C-\underset{\oplus}{C}-C-C}} \;\rightleftharpoons\; \overset{\displaystyle C}{\underset{\displaystyle C}{C-C-\underset{\oplus}{C}-C}} \;\rightleftharpoons\; \overset{\displaystyle C\;C}{C-C-\underset{\oplus}{C}-C} \;\longrightarrow
$$

products

While these series of equations account for the products obtained, the formalism used postulates or implies highly endothermic events, highly energetic primary carbenium ions, and an equilibrium between primary, secondary, and tertiary cations, controversial concepts in light of recent developments in this chemistry (for a recent review see reference 11). It has been estimated that tertiary carbenium ions are ~33 kcal/mole and secondary carbenium ions are ~22 kcal/mole more stable than primary ones (12). An alternate mechanism is one that assumes ethylene insertion into polarized –C–Cl bonds and an equilibrium between various covalent chlorides. Another proposition by Karabatsos (8) to avoid the existence of primary cations in liquid phase isomerizations is the assumption of biomolecular reactions and/or the transitory existence of protonated cyclopropanes. For example, the isotope position rearrangement of *t*-butyl chloride under the influence of aluminum chloride

$$
\begin{array}{ccc}
\quad C & & \quad C \\
| & & | \\
(C\text{--}\overset{*}{C}\text{--}Cl & \longrightarrow & \overset{*}{C}\text{--}C\text{--}Cl) \\
| & & | \\
C & & C
\end{array}
$$

can be explained by the following sequence (13, 14):

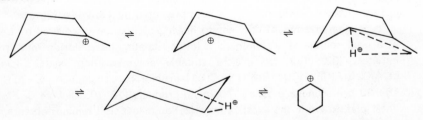

Or, for the ring expansion of methylcyclopentane under the influence of a strong Lewis acid that was commonly assumed to proceed according to:

Karabatsos proposes (8) the intermediary formation of protonated cyclopropane derivatives:

Recently Kramer also suggested protonated cyclopropanes instead of methide shifts to explain isomerization results (15). For example, isobutyl chloride in strong acid media gave isobutane and n-butane in approximately equilibrium concentrations. The following mechanism was proposed to explain the results:

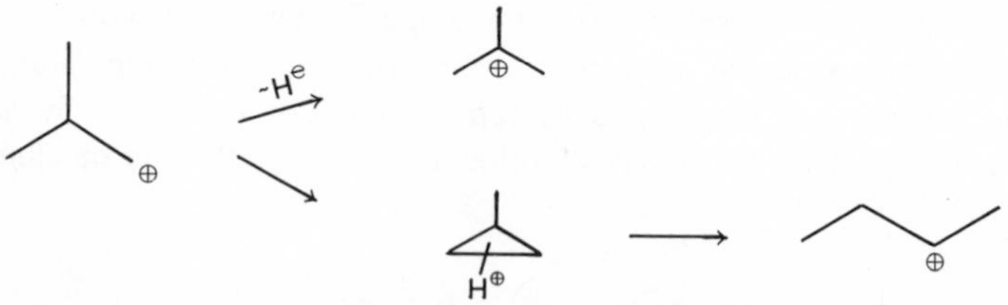

Evidently the oligomerization of ethylene with conventional Friedel-Crafts acids might start with species other than the simple ethylenium ions. Our understanding of the existence and/or behavior of primary carbenium ions in the liquid phase is incomplete.

A discussion of the cationic polymerization of ethylene would not be complete without mentioning the investigations by Nash, Hall, and coworkers, and Fischer. These researchers are of particular interest in the history of Ziegler/Natta catalysis. Significantly, Nash, Hall and coworkers found (16–19) that viscous oils were obtained when ethylene was heated with $AlCl_3$ containing some metallic aluminum at elevated temperatures and pressures (150°C and 50 atm). The intermediate formation of alkylaluminum halides was recognized and in separate experiments it was shown that Et_2AlCl and ethylene produced large quantities of linear α-olefins containing an even number of carbon atoms. Törnquist pointed out (20) that since Et_2AlCl in these experiments may have contained traces of Et_3Al, Hall and Nash may have carried out the first olefin "growth reaction."

Along similar lines, Fischer, while experimenting to produce lubricating oils from ethylene with Friedel-Crafts acids, discovered that solid polyethylene can be obtained by adding some metallic aluminum to catalysts containing $AlCl_3$ and $TiCl_4$ (21). The single example in Fischer's patent is of historical importance. This author premixed $AlCl_3$, $TiCl_4$, and Al powder in n-heptane and pressurized with ethylene in a steel bomb. The polymerization at 155°C and 58 atm produced in good yields a "very light, almost white powder." The interesting history of these developments has been presented by Törnquist (20).

The oligomerization of ethylene to linear α-olefins was discovered by Langer who used soluble alkylaluminum chloride/titanium tetrachloride mixtures as initiators (22). For example, a suitable polymerization charge involved $Et_2AlCl/EtAlCl_2/TiCl_4/t\text{-BuOH}$ (6:6:1:1) in chlorobenzene diluent at −20°C and 15–500 psi ethylene pressure. The main products (70–100%) were C_{12} to C_{20} linear olefins. While the exact polymerization mechanism remains obscure, it is conceivable that the reaction involves a preliminary coordination of the ethylene molecule on the cationic species $EtTiCl_2^{\oplus}$ followed by insertion:

$$R_2AlCl + TiCl_4 \longrightarrow [\text{complex}] \longrightarrow RAlCl_3^{\ominus} \cdot RTiCl_2^{\oplus}$$

$$\overset{\oplus}{RTiCl_2} + CH_2{=}CH_2 \longrightarrow \overset{\delta\oplus}{RTiCl_2} \longrightarrow R{-}CH_2{-}CH_2{-}\overset{\oplus}{TiCl_2}$$

$$\underset{\delta\oplus}{H_2C{\overset{\ddagger}{=}}CH_2}$$

The olefins are formed by the following hydride transfer displacement:

$$R{-}\underset{\underset{H_2C{\overset{\ddagger}{=}}CH_2}{\overset{|}{H}}}{CH}{-}CH_2{-}\overset{\oplus}{TiCl_2} \longrightarrow R{-}CH{=}CH + \underset{H_3C{-}CH_2}{\overset{\oplus}{\underset{|}{TiCl_2}}}$$

Very similar ethylene oligomerization chemistry has been studied by Bestian and Clauss (23) who produced mixtures of α-olefins, branched olefins, and internal olefins with $CH_3AlCl_2 \cdot CH_3TiCl_3$ initiator in chlorinated hydrocarbon solvents at −70°C. The discussion of mixed alkylaluminum/transition metal halide catalysts (Ziegler/Natta catalysts) is beyond the scope of this book.

b. Propylene $CH_2{=}CH{-}CH_3$

Propylene is a fairly reactive cationic monomer and gives low molecular weight products in the lubricating oil range in the presence of certain Friedel-Crafts acids such as BF_3, $AlCl_3$, and $AlBr_3$ and Brønsted acids such as H_3PO_4. High molecular weight and/or linear polypropylenes cannot be produced by present day conventional cationic techniques but are readily obtainable by anionic coordinated Ziegler-Natta polymerization; however, this chemistry is beyond the scope of our discussion. Thus the survey of the cationic polymerization of propylene is, necessarily, a summary of various oligomerization techniques and an inquiry into the reasons why cationic mechanisms produce only low molecular weight products of ill defined structures.

Butlerov and Gorianov were the first to investigate the cationic polymerization of propylene. In 1873 they stated (24) that propylene can be oligomerized to oily products (which did not contain dimers and trimers) in the presence of BF_3 in sealed tubes at room temperature. Russian scientists maintained their high interest in this chemistry to the present day and have made important contributions along the way. Topchiev and Paushkin found that the rate of oligomerization of propylene with BF_3 is strongly enhanced in the presence of "activators" such as H_2O and HF (25). During World War II synthetic oils of this type were produced from propylene.

Besides BF_3 and/or activated BF_3 complexes, a variety of Brønsted and

Friedel/Crafts acids and combinations thereof have been used in propylene oligomerization studies. Among the Brønsted acid systems, various phosphoric acid processes have been developed. This chemistry has been surveyed and criticized by Fontana (26). Interesting studies have been carried out, for example, with $AlCl_3$ using an ethyl chloride solution in the range −20 to −100°C (27), or with $AlBr_3$ promoted by HBr in propane solvent from −78 to 26°C (28). The reader interested in the details of propylene oligomerizations should consult reference 29 and references therein.

An important watershed in the study of cationic polymerization of propylene (and higher α-olefins, for that matter) is the work of Fontana (30–32). This author and his coworkers thoroughly investigated the important polymerization parameters (temperature, pressure, feed rate, concentrations, etc.) in the $AlBr_3$/HBr/propylene/propane system and studied their effect on the yield, molecular weight, and molecular weight distribution of the oligomers. The purpose of these studies was to develop oligomers used as viscosity index (V.I.) improvers for lubricating oils. It was found that the V.I. of the oligomers of propylene (or 1-butene) synthesized by $AlBr_3$ can be greatly increased by the use of HBr in the synthesis. As shown in Figure 1 the effect was particularly strong in the range −20 to −40°C (31).

Significantly, the higher viscosities obtained with $AlBr_3$ in the presence of HBr might be unique to the $AlBr_3$/HBr system since a similar effect has not been found when HCl was added to $AlCl_3$ or BF_3 used to polymerize propylene.

A thorough analysis of the large amount of kinetic data also resulted in important conceptual advances in the understanding of the reaction mechanisms of cationic α-olefin polymerizations. Reproducible kinetic investigations lead to a quantitative interpretation of the data and herein lies the long range merit of Fontana's work.

Rate measurements in a batch reactor showed that the oligomerization of propylene in the presence of $AlBr_3$/HBr (4:1) at −78°C is rapid until it reaches a peak, then it quickly falls to a plateau where it remains over most of the run until it slowly decays to zero. The final products were plastic semisolids. The rate profile of a typical experiment is shown in Figure 2 (30).

To explain the data, the authors proposed the existence of relatively long lived complexes comprising a carbenium ion−tetrabromoaluminate anion pair associated with or coordinated with a monomer. This species might be represented as:

$$R—CH_2 \qquad CH_2$$
$$H\overset{\oplus}{C} \quad AlBr_4^{\ominus} \quad \overset{\parallel}{C}H$$
$$CH_3 \qquad CH_3$$

where R = H or polypropylene, H during initiation or a polymer chain during

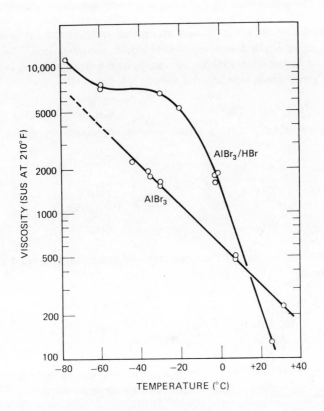

Figure 1. Viscosities of polypropylenes obtained with AlBr$_3$/HBr (31).

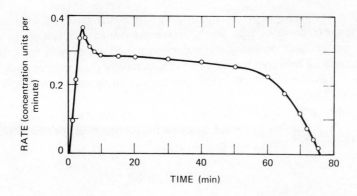

Figure 2. Rate Profile of a propylene Polymerization with AlBr$_3$/HBr at $-78°$C (30).

propagation. The authors in no way attempted to define the above species, the existence of which is based on purely kinetic considerations. The induction period, that is, the rapidly increasing portion of the rate curve, was explained by assuming a mechanism akin to autocatalytic reactions:

$$\underset{\underset{CH_3}{|}}{\overset{\overset{CH_3}{|}}{HC^\oplus}}\ AlB_4{}^\ominus\ \underset{\underset{CH_3}{|}}{\overset{\overset{CH_2}{\parallel}}{CH}}\ +\ HBr\ \xrightarrow{\ k'\ }\ \underset{\underset{CH_3}{|}}{\overset{\overset{CH_3}{|}}{HC^\oplus}}\ AlBr_4{}^\ominus\ +\ \underset{\underset{CH_3}{|}}{\overset{\overset{CH_3}{|}}{CH{-}Br}}$$

$$\underset{\underset{CH_3}{|}}{\overset{\overset{CH_3}{|}}{HC{-}Br}}\ +\ AlBr_3\ \xrightarrow{\ rapidly\ }\ \underset{\underset{CH_3}{|}}{\overset{\overset{CH_3}{|}}{HC^\oplus}}\ AlBr_4{}^\ominus$$

The middle, flat portion of the rate curve was explained by a rapid associative equilibrium,

$$\underset{\underset{CH_3}{|}}{\overset{\overset{CH_3}{|}}{HC^\oplus}}\ AlBr_4{}^\ominus\ +\ \underset{\underset{CH_3}{|}}{\overset{\overset{CH_2}{\parallel}}{CH}}\ \underset{}{\overset{K}{\rightleftharpoons}}\ \underset{\underset{CH_3}{|}}{\overset{\overset{CH_3}{|}}{HC^\oplus}}\ AlBr_4{}^\ominus\ \underset{\underset{CH_3}{|}}{\overset{\overset{CH_2}{\parallel}}{CH}}$$

followed by a rate determining incorporation of the complexed monomer into the ion pair to produce a new, one monomer unit larger ion pair:

$$\underset{\underset{CH_3}{\diagup}}{\overset{\overset{CH_3}{\diagdown}}{HC^\oplus}}\ AlBr_4{}^\ominus\ \underset{\underset{CH_3}{|}}{\overset{\overset{CH_2}{\parallel}}{CH}}\ \longrightarrow\ \underset{CH_3}{\overset{\overset{\displaystyle CH_3}{\underset{|}{HC}}{-}CH_2}{}}\ \underset{\underset{CH_3}{\diagup}}{\overset{\diagdown}{HC^\oplus}}\ AlBr_4{}^\ominus$$

Thus the key intermediate in this mechanism is the relatively long lived ion pair–monomer complex that slowly rearranges during propagation to a one unit larger ion pair, or during the induction period reacts with the promoter HBr to produce isopropyl bromide that in turn rapidly gives another mole of the ion pair. Independent kinetic experiments were in agreement with this basic hypothesis (30).

The $AlBr_3$ is present in the reaction either as a complex ion pair or as a complex monomer–ion pair; thus denoting by [c] the $AlBr_3$ or HBr concentration, whichever is limiting, we obtain:

$$[c]\ =\ [M_nC]\ +\ [M_nCM]$$

where M_nC denotes the initiating or propagating ion pair $\sim C^\oplus AlBr_4{}^\ominus$, and M_nCM is the complex between the ion pair and monomer $\sim C^\oplus AlBr_4{}^\ominus C_3{}^=$.

The above equilibrium and the propagation controlling step, respectively, can be written as:

$$M_nC + M \underset{}{\overset{K}{\rightleftharpoons}} M_nCM$$

and

$$M_nCM \xrightarrow{k} M_{n+1}C$$

which gives the kinetic equation for propagation:

$$-\frac{d[M]}{dt} = \frac{kK[C][M]}{1 + K[M]}$$

This expression was found to quantitatively describe the experimental results.

Depending on the relative values of k and K, the same propagation mechanism may give a reaction order anywhere between zero and unity with respect to monomers.

For the propagation step a cyclic transition state was proposed (33):

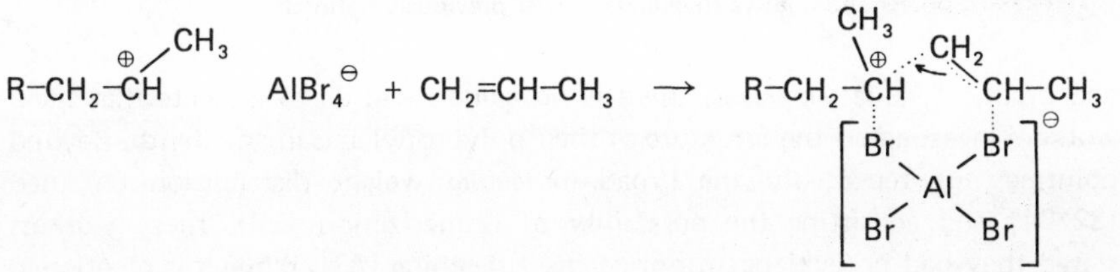

It was recognized that termination must also occur and an ingenious hydride transfer followed by proton transfer was proposed to account for this event (32). This mechanism is one of the few proposals that explains true (permanent) termination of cationic polymerization.

Fontana's last contribution (34) concerned molecular weight distribution studies and the further substantiation of the hydride transfer reaction. Results of the molecular weight distribution studies are very interesting as the product of a fractionation of a polypropylene prepared by a semicontinuous method at $-44°C$ (32) having an average degree of polymerization of ~55 ($\overline{M}_n$ ~3300), contained about 10 wt. % of a fraction with $\overline{M}_n = 830,000$. Indeed, fractionation studies showed that a total of about 30 wt. % of the end-product had a number average molecular weight in excess of 10^5 (see Table 2). This was the first and only time the existence of cationically polymerized polypropylene having molecular weights close to 10^6 has been demonstrated. The investigation of the structure–property relation of this material would have been of great theoretical

TABLE 2   SUMMARY OF EXPERIMENTAL RESULTS ON THE FRACTIONATION OF POLYPROPYLENE (34)

Fraction j	Weight[a] (%)	TP_{210}[b]	RTP[b]	Molecular Weight
1	3.97	0.49	0.441	597
2	6.39	0.92	0.614	1,055
3	3.66	1.19	0.646	1,529
4	5.00	1.36	0.667	1,843
5	5.00	1.60	0.690	2,708
6	3.04	1.78	0.693	3,585
7	4.62	2.04	0.716	17,300
8	7.94	2.24	0.747	19,780
9	4.64	2.93	0.787	32,020
10	4.87	3.58	0.822	35,790
11	12.64	3.76	0.827	46,770
12	5.18	6.72	0.866	101,200
13	9.36	7.60	0.872	136,900
14	9.27	10.33	0.918	154,300
15	4.76	12.6	0.923	300,000
16	9.66	24.6	0.998	829,200

[a]Adjusted to 100% yield. Actual yield 97.4%. Sample weight 455.0 g.
[b]Thickening power and relative thickening power previously defined.

and practical value; however, there is no evidence at all to indicate that these workers investigated the structure of their polypropylenes in any depth. Beyond pointing out repeatedly the broad molecular weight distribution obtained (32–34) and admitting the possibility of isomerization (33), these workers stated they had no evidence for or against branching (33). While it is mentioned that the second order transition temperature of this molecule is $-30°C$ and that it produces gels (?) at higher concentrations in oil, these authors reveal tantalizingly little of the structure and/or properties of these materials. It is unfortunate that Fontana's investigations stopped around 1952 [his last paper (34) was written long after he had left this field] as these results should have been further exploited.

A largely empirical investigation on the cationic oligomerization of propylene with aluminum halides to produce viscosity index improvers (oil additives) was carried out by Kölbel and coworkers (35). These authors studied the effect of the nature of the Friedel-Crafts acid ($AlCl_3$, $AlBr_3$, AlI_3), the solvent (nitrobenzene, carbon disulfide, ethyl bromide, chloroform, etc.), the presence or absence of H_2 or $LiAlH_4$, and the effect of temperature, regulators (t-butyl chloride), inhibitors, and so on. They concluded that it is difficult to control the mechanism of the oligomerization or the structure of the product, although molecular weights can be readily controlled (lowered) by experimental variables (35).

As a consequence of Fontana's publications, two groups, Ketley and Harvey (36) and Immergut and coworkers (37), became interested in the elucidation of the structure of cationically produced polypropylene. These authors set out to characterize by infrared spectroscopy the structure of polypropylenes obtained by Friedel/Crafts acids, a formidable task. Both groups adapted Fontana's semicontinuous technique for a simple batch method and used $AlBr_3$ and other Friedel-Crafts acids to produce highly viscous, liquid polypropylenes (of unknown molecular weight or molecular weight distribution). Ketley and Harvey (36) assigned the bands appearing at 769 and 739 cm^{-1}, respectively, to ethyl and n-propyl groups, whereas Immergut et al. (37) attributed these to "short branches." Ketley and Harvey developed a quantitative infrared method, determined the relative abundance of various methyl, ethyl, and propyl groups, and proceeded to extend Fontana's hydride transfer proposition to explain these findings. For example, the presence of propyl groups was attributed to intermolecular $H^{\ominus}$ transfer:

$$\sim CH_2-\underset{\underset{CH_3}{|}}{CH}-CH_2\sim \;+\; \overset{\oplus}{\underset{\underset{CH_3}{|}}{CH}}-CH_2\sim \;\longrightarrow\; \sim CH_2-\overset{\oplus}{\underset{\underset{CH_3}{|}}{C}}-CH_2\sim \;+\; \underset{\underset{CH_3}{|}}{CH_2}-CH_2\sim$$

or to two intramolecular $H^{\ominus}$ shifts:

$$\sim CH_2-\underset{\underset{CH_3}{|}}{CH}-CH_2-\overset{\oplus}{CH}\cdot \;\xrightarrow{\sigma}\; \sim CH_2-\underset{\underset{CH_3}{|}}{CH}-\overset{\oplus}{CH}-\underset{\underset{CH_3}{|}}{CH_2} \;\xrightarrow{\sigma}\; \sim CH_2-\overset{\oplus}{\underset{\underset{CH_3}{|}}{C}}-CH_2-CH_2-CH_3$$

Ethyl group   Propyl group

On the basis of these and other well established carbenium ion reactions, a highly branched structure for cationically obtained polypropylenes was proposed. Ketley and Harvey (36) found very little infrared evidence for the presence of

$$-CH_2-\underset{\underset{CH_3}{|}}{CH}-CH_2- \;,$$

that is, isolated methyl groups in their products, and consequently concluded that rearrangement must accompany almost every propagation step. Immergut and coworkers (37) came to essentially identical conclusions. However, none of these authors discussed the puzzling problem of chain termination or problems connected with the molecular weights or their distribution. [In Immergut's paper (37) misprints appear in Equations I and II; also the statement "In reaction IIb an original methyl becomes a methylene group" is quite puzzling.]

The kinetics of the polymerization of propylene with BF_3/CH_3OH initiators in CH_2Cl_2 solution in the range -15 to $-80°C$ has been studied by Eastham and coworkers (38), but serious experimental difficulties prevented the proposition of a satisfactory mechanism (39).

Tinyakova et al. stated (40) that diethylaluminum chloride in ethyl chloride diluent rapidly polymerizes propylene at ~0°C to an oily product of molecular weight ~1000. Interestingly, the addition of trialkylaluminum (alkyl group unspecified, probably ethyl) to the system completely inhibited this oligomerization. A very similar observation is that by Kennedy and Langer (41) who noted the rapid oligomerization of propylene when they bubbled hydrogen chloride gas into this olefin at -78° until saturation was reached and then added small amounts of Et_3Al. Polymerization under these conditions is probably due to the *in situ* formation of $AlCl_3$ or $EtAlCl_2$. It is interesting, however, that polymerization was much delayed when the addition sequence of the reagents was: propylene, Et_3Al, HCl. An explanation may be that complex formation between propylene and triethylaluminum ($Et_3Al \cdot CH_2 = CHCH_3$) delayed the reaction between the hydrogen chloride and Et_3Al.

Recently Swiss researchers obtained atactic propylene oligomers by coordination catalysis, that is, mixtures of titanium salts and alkylaluminum halides (42). The structure of the product was determined by IR and NMR spectroscopy. The regular head-to-tail enchainment could be readily seen from the characteristic IR bands at 1155 cm^{-1}. The important finding was the demonstration that head-to-tail enchainment can be obtained by controlling the initiator acidity. Whereas relatively little regular head-to-tail growth occurred with the acidic $TiCl_4/EtAlCl_2$ (1:5) system, head-to-tail propagation became more important as the overall acidity of the initiator was decreased by the use of alkoxides of titanium or Et_2AlCl instead of $EtAlCl_2$. Some of the data are reproduced in Table 3. Evidently it is hopeless to strive for head-to-tail polypropylene by a conventional cationic route, but there is hope of obtaining something useful, perhaps a truly high molecular weight polypropylene, via cationic coordination polymerization.

Certain acidic molecular sieves oligomerize propylene to a mixture of products (43, 44). The logarithim of the first order rate constant (obtained from the slope of pressure versus time plots) gave a satisfactorily linear correlation with the weight percent reactive alumina present in the molecular sieve (45).

Immergut, Kollman, and Malatesta claim to have copolymerized propylene with isoprene (46); however, the evidence for this is not convincing as only ill characterized products (probably mixtures) were obtained. The copolymerization (or more precisely the cooligomerization) of propylene and diolefins with Friedel-Crafts acids has been claimed in the patent literature (47). For example, mixtures of propylene and butadiene have been reacted with BF_3 gas at -60°C and mixtures of oily, soluble and insoluble products of unknown structures were obtained (47).

c. 1-Butene, 2-Butenes, and Higher *n*-Olefins

1-Butene is readily polymerizable by both Brønsted and Lewis acids. Liquid

TABLE 3 ORIENTATION TO REGULAR HEAD-TO-TAIL GROWTH. [Ti] = 20 $\times$ 10^{-3} moles/l; T = 5°C; solvent = benzene (42)

Ti Compound	Al Compound	Al/Ti	$p\,C_3H_6$ (atm)	Propylene Uptake (mole/1 hr)	$\overline{M_n}{}^a$	Solid Polymer (%)	Regular Growth (%)
TiCl$_4$	EtAlCl$_2$	5	1	2.2	516	0	28
(EtO)$_3$TiCl	EtAlCl$_2$	5	1	1.0	340	0	49
(EtO)$_4$Ti	EtAlCl$_2$	5	1	0.12	432	3.5	87
TiCl$_4$	Et$_2$AlCl	2	1	2.2	555	3.6	70
TiCl$_4$	Et$_2$AlCl	2	5	4.5	770	6.2	68
(EtO)TiCl$_3$	Et$_2$AlCl	1.3	5	2.2	1100	17.0	94

^{a}Number average molecular weight of the oligomeric part, excluding the solid polymer mentioned in the seventh column.

oligomers (mostly decenes and dodecenes) are obtained with solid phosphoric acid (48) or with BF_3/HPO_3 systems (49).

The best published cationic polymerization research work to date on 1-butene was carried out by Fontana and coworkers (26, 31–33) who employed the $AlBr_3/HBr$ initiator system and conditions described earlier in conjunction with propylene (see above). The overall findings with 1-butene resembled those obtained with propylene, although some characteristic differences have been found. For example, under quite similar conditions, the maximum viscosity average molecular weight was observed at about $-35°C$ for poly-1-butene (31) as seen in Figure 3, whereas it was about $-50°C$ with polypropylene (32). The maximum in the molecular weight–temperature curve was attributed to a competition between proton (chain) transfer leading to low molecular weight products at higher temperatures and hydride transfer leading to a broadening of the molecular weight distribution. Ketley and Harvey (36) state that under similar conditions the polymerization of 1-butene "follows the same course" as that of propylene, but the reference they quote does not contain information on 1-butene. They go on to state that poly-1-butene contained n-propyl and n-butyl side groups indicating successive H shifts during polymerization.

Olah et al. (50) polymerized 1-butene to liquid oligomers by using a variety of BF_3 complexes at $-80°C$.

The 2-butenes are also reactive cationic monomers but yield only low oligomers with acid systems, for example, H_3PO_4 or Friedel-Crafts halides. Meier (51) studied the polymerization of 2-butene. In the gas phase a very rapid polymerization occurred immediately upon mixing moist 2-butene and BF_3. Remarkably, more C_9 and C_{10} hydrocarbons formed than C_8 hydrocarbons, a large proportion of saturated low molecular weight products were obtained, and infrared studies indicated the presence of neopentyl groups. In the liquid phase, polymerization was not instantaneous upon mixing the reagents and a certain threshold BF_3 concentration was necessary to start the reaction; also catalyst activity rapidly decreased and fresh BF_3 had to be added to reach higher conversions. Meier also claimed that 1-butene and 2-butene copolymerize (51).

Certain clays exhibit polymerization activity for 2-butene isomers (52). Fontana mentioned a few experiments with 2-butenes and $AlBr_3/HBr$ (31), and Allcock and Eastham used $BF_3–CH_3OH$ in some polymerization studies (53); both groups obtained low molecular weight products. The cis ⇌ trans isomerization of 2-butene was studied in depth by Eastham over a period of years (reference 39 and references therein).

Higher 1-olefins, that is, 1-pentene, 1-hexene, and so on, can also be oligomerized by cationic initiators but relatively little published research is available on these monomers. Higher n-olefins can be polymerized by BF_3 and its coordination compounds to lubricating oils (54). Fontana et al. (31) oligomerized

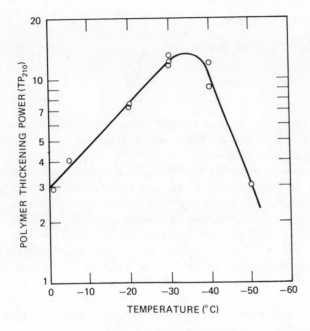

Figure 3. Effect of reaction temperature on Molecular weight of Poly-1-butene (31).

1-pentene, 1-octene, 1-decene, 1-dodecene, and 1-cetene (C_{16}) with $AlBr_3$/HBr at ~$0°C$. Turner polymerized 1-octene with anhydrous (?) $AlCl_3$ to oily liquids in the lubricating oil range (55).

This author knows of no noteworthy cationic polymerization literature on 2-olefins higher than 2-butene (see above) or on nonconjugated n-diolefins containing a vinyl and internal double bond, for example, $CH_2=CH-CH_2-CH=CH-R$, or containing two nonconjugated internal double bonds, for example, $R-CH=CH-(CH_2)_x-CH=CH-R$. This, of course, does not mean that these compounds are not cationic monomers. This is also true with α,ω-diolefins, $CH_2=CH-(CH_2)_n-CH=CH_2$, where $n = 1, 4, 5 \ldots$, except for certain derivatives of 1,5-hexadiene and 1,6-heptadiene which undergo intra-intermolecular polymerization and are discussed separately.

The reason the cationic polymerization behavior of straight chain olefins has not been investigated in scientific laboratories is rather obvious: due to the structure of these compounds numerous side reactions can occur during cationic attack which in turn result in low molecular weight products and questionable ultimate properties.

B. BRANCHED α-OLEFINS

a. Branches α to the Double Bond

1. *3-Methyl-1-butene* $CH_2{=}CH{-}CH\begin{smallmatrix} \diagup CH_3 \\ \diagdown CH_3 \end{smallmatrix}$

This molecule is a reactive cationic monomer. Depending on the reaction conditions it gives low molecular weight oligomers or high polymers having interesting physical properties.

In the presence of strong Brønsted acids 3-methyl-1-butene gives "amylene" oligomers of ill defined structure. According to Norrish and Joubert (56) 70–75% sulfuric acid produces dimers, whereas more concentrated acid gives higher boiling liquid oligomers. Leenderste et al. (57) reacted this monomer with $AlCl_3$ at $-80°C$ and obtained polymers at a slow rate having an average molecular weight of 1440. Thomas and Reynolds described in a patent (58) the cationic polymerization of 3-methyl-1-butene with Lewis acids in alkyl chloride diluent at low temperatures ($-78°C$) that led to the formation of colorless tough semisolids of relatively low molecular weight. Kennedy (59) claimed crystalline polymers of higher molecular weight by a polymerization technique utilizing very low temperatures ($-130°C$).

Recent scientific interest in 3-methyl-1-butene dates back to 1962 when Kennedy and Thomas (60) showed that this monomer can be readily polymerized by cationic techniques to crystalline products of respectable molecular weights ($\sim$50,000 $\bar{M}_v$). In fact this polymer was the first crystalline aliphatic poly-α-olefin obtained by cationic polymerization. Its synthesis generated great interest among polymer scientists since this disclosure came during the time when studies on the preparation of stereoregular crystalline poly-α-olefins with Ziegler-Natta catalysts were at their peak. By 1962 it was known that highly crystalline poly(3-methyl-1-butene) could be synthesized by Ziegler-Natta anionic coordination catalysts and that polymer crystallinity was due to stereoregularity (isotactic order) along the chain. Poly(3-methyl-1-butene) prepared by a typical Ziegler-Natta initiator system, for example, by $Et_3Al \cdot TiCl_3$ in hexane solution at ambient temperatures, is isotactic. It has a conventional head-to-tail or 1,2 structure and the pendant isopropyl groups occupy identical stereochemical positions with respect to the asymmetric carbon atoms in the chain. Such a polymer can be viewed as a succession of *ddd* or *lll* units:

$$-CH_2-\underset{CH_3\ \ \ CH_3}{\underset{\diagup\ \ \diagdown}{\underset{CH}{\overset{H}{\underset{|}{C}}}}}-CH_2-\underset{CH_3\ \ \ CH_3}{\underset{\diagup\ \ \diagdown}{\underset{CH}{\overset{H}{\underset{|}{C}}}}}-CH_2-\underset{CH_3\ \ \ CH_3}{\underset{\diagup\ \ \diagdown}{\underset{CH}{\overset{H}{\underset{|}{C}}}}}-$$

Structure of isotactic poly(3-methyl-1-butene) obtained by
Ziegler-Natta coordination catalysts.

The crystallinity observed in the cationically synthesized polymer, however, is due to a high degree of symmetry in the polymer and not to stereoregularity:

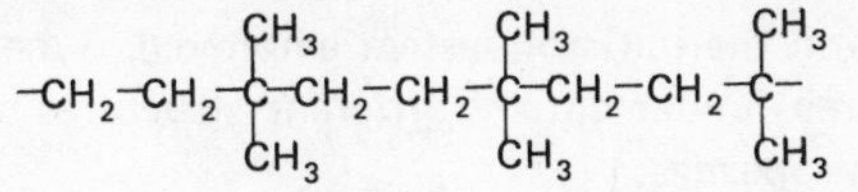

$$-CH_2-CH_2-\underset{\underset{CH_3}{|}}{\overset{\overset{CH_3}{|}}{C}}-CH_2-CH_2-\underset{\underset{CH_3}{|}}{\overset{\overset{CH_3}{|}}{C}}-CH_2-CH_2-\underset{\underset{CH_3}{|}}{\overset{\overset{CH_3}{|}}{C}}-$$

**Structure of poly(3-methyl-1-butene) obtainable by
Lewis acids at low temperature**

This unusual structure is formed by isomerization polymerization. The following set of equations has been proposed (61, 62) to account for the observed structure:

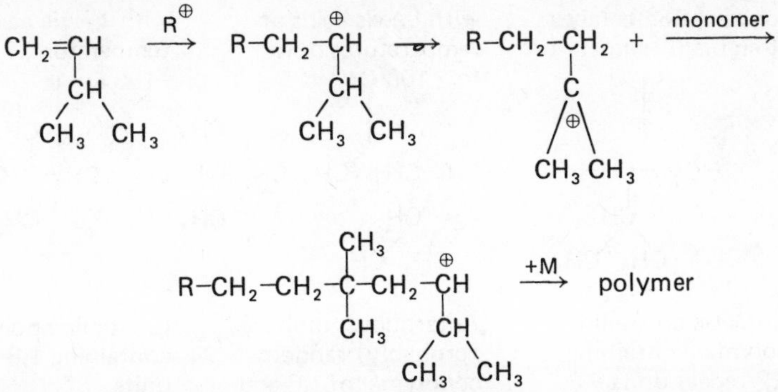

The monomer is attacked by the growing chain end $R^{\oplus}$ (or by $H^{\oplus}$ during initiation) to give the first secondary carbenium ion which immediately rearranges by hydride migration to an energetically more favorable tertiary ion, the true propagating site. The latter attacks a new incoming monomer and the process repeats itself.

At $-130°C$ the rearranged structure is formed exclusively and, due to the high symmetry along the backbone, the product is crystalline. At temperatures higher than $-130°C$ conventional head-to-tail propagation starts to compete with complete rearrangement and polymers containing both structure elements are obtained:

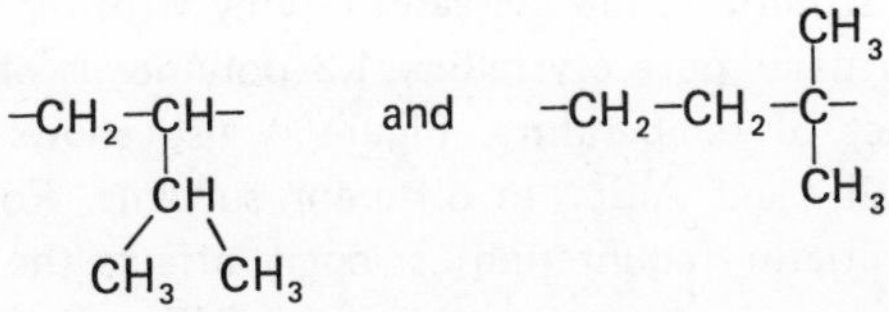

$$-CH_2-\underset{\underset{\underset{CH_3 \quad CH_3}{\diagdown}}{\overset{|}{CH}}}{\overset{|}{CH}}- \quad \text{and} \quad -CH_2-CH_2-\underset{\underset{CH_3}{|}}{\overset{\overset{CH_3}{|}}{C}}-$$

While it has not been established experimentally beyond all doubt, there is strong NMR spectroscopic evidence that the polymer obtained with $AlCl_3$ at say -78 or $-100°C$ is a copolymer of unrearranged and rearranged units and not a blend of polymers of conventional 1,2 structure and of rearranged 1,3 repeat units (63, 62).

Thus, depending on the initiator system employed, 3-methyl-1-butene can be polymerized to three fundamentally different structures having fundamentally different physical properties:

Ziegler-Natta anionic coordination catalysts at ambient temperatures	Cationic techniques with Lewis acids at temperatures down to $-100°C$	Cationic techniques with Lewis acids at temperatures below $-100°C$
Isotactic crystalline polymer containing 1,2 repeat units	Amorphous, rubbery (probably) random copolymer of 1,2 and 1,3 repeat units	Crystalline polymer containing 1,3 repeat units

Polymerization temperature is the most important parameter determining the structure or crystallinity of the cationically obtained polymer. Crystalline polymers cannot be obtained at temperatures higher than about $-115°C$.

The effect of temperature on the structure ratio (the ratio of 1,3 repeat units to 1,2 repeat units) of cationically obtained poly(3-methyl-1-butene) has been investigated (64). A near infrared method that allows the quantitative determination of 1,3 enchainment has been developed to study this effect of temperature (64). It was found that the logarithm of the [1,3]/[1,2] structure ratio increases linearly with reciprocal temperature in the range $+100$ to $-100°C$. Below $-100°C$ the structure ratio increases rapidly with reciprocal temperature and at $-130°C$ virtually pure crystalline 1,3 polymer is obtained (Figure 4).

Besides the effect of temperature, Figure 4 also shows the product ratios obtained with $AlCl_3$ and $AlBr_3$ in different solvents. For some reason the nature of the coinitiator (counterion) strongly affects the slope of the lines: the product ratio rises much more rapidly with $AlBr_3$ (i.e., $AlBr_4{}^{\ominus}$) than with $AlCl_3$. Apparently at temperatures higher than $\sim +10°C$ more rearranged structure

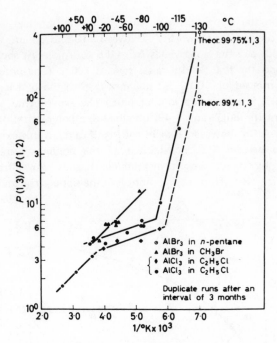

Figure 4. The Effect of temperature on the structure of poly(3-methyl-1-butene) obtained with $AlCl_3$ and $AlBr_3$ in various solvents (64).

is produced with $AlBr_3$ than with $AlCl_3$, whereas below this threshold temperature $AlCl_3$ gives more 1,3 enchainment than $AlBr_3$.

The difference in slopes may be expressed quantitatively. The apparent activation energy ΔE obtainable from Figure 4 has been attributed to the difference between the apparent activation energy of 1,2 enchainment $E_{1,2}$ and that of hydride shift E_H. The ΔE for $AlBr_3$ in methyl bromide is ~1.33, while that for $AlCl_3$ in ethyl chloride is 0.5, giving an estimated 0.8–1.0 kcal/mole for the difference $\Delta E_{AlBr_3} - \Delta E_{AlCl_3}$. This could mean that the larger $AlBr_4^{\ominus}$ counterion interferes more with the conventional 1,2 propagation than the smaller $AlCl_4^{\ominus}$ anion, perhaps by blocking the path of the incoming monomer molecule. Because of this steric effect, the overall rate of polymerization in general and 1,2 propagation in particular may be retarded in the presence of the more bulky $AlBr_4^{\ominus}$ counterion so that the rearrangement from the secondary to the tertiary carbenium ion would become faster.

Incidentally, $AlBr_3$ is a poor initiator as compared with $AlCl_3$ for the polymerization of 3-methyl-1-butene at temperatures lower than ~$-100°C$. Indeed, even at $-100°C$ uncommonly large amounts of initiator are required to initiate the polymerization (64) and therefore the $AlBr_3$ data in Figure 4 are shown only down to $-80°C$.

The change in the slope of the product ratio line obtained with $AlCl_3$ in ethyl chloride diluent is noteworthy. Between +10 and about $-90°C$ a linear relationship holds and the slope yields $\Delta E = 0.5$ kcal/mole. At about $-90°C$ the slope breaks and the line rapidly raises toward 100%. (The product ratio of completely rearranged or 100% 1,3 polymer is $+ \infty$; the plotting of product ratios above 99% is meaningless in a log plot.) The experimental value of 98% 1,3 enchainment (product ratio = 49) obtained in a polymerization carried out at $-115°C$ with $AlCl_3$, however, is valid and significant. In this experiment the conversion and percent 1,3 enchainment of the polymer were determined simultaneously by direct weight measurements and near infrared studies, respectively (Figure 5). Thus the sharp break in the slope is real and may be due

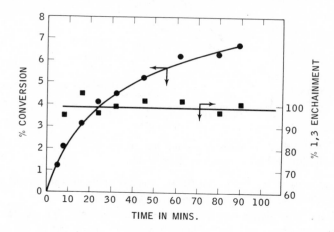

TIME IN MINS.

Figure 5. The effect of monomer conversion on the structure of poly(3-methyl-1-butene) obtained with $AlCl_3$ at $-115°C$. (Experimental conditions: $[AlCl_3] = 7.23 \times 10^{-3}$ mole $AlCl_3$ per one C_2H_5Cl, $[M] = 2.92M$

to diffusion control. Diffusion would decrease the effective monomer concentration around the growing site which would retard fresh monomer incorporation but would not affect the intramolecular hydride shift. Thus at $-130°C$ essentially pure rearranged product would be formed.

The best initiator from the point of view of isomerization and molecular weights is $AlCl_3$. Other Lewis acids such as BF_3, $AlBr_3$, and $TiCl_4$ in ethyl chloride at $-130°C$ either give poorly crystalline polymer ($AlBr_3$, $TiCl_4$) or largely amorphous product (BF_3) or are unable to initiate polymerization at all ($SnCl_4$, VCl_4). The approximate viscosity average molecular weight of the polymer prepared with $AlCl_3$ is 50,000. Higher molecular weights can be obtained by a semicontinuous technique.

Interestingly, the cationically obtained poly(3-methyl-1-butene) exists in at least two and possibly even three crystalline modifications (65, 66). These polymorphs, designated α, β, and γ forms, were defined by infrared spectroscopy differential thermal analysis (DTA), and their respective "d" spacings obtained from x-ray diffraction photographs. Table 4 gives a comparison of some of the data. Only one sample obtained in an experiment using $AlCl_3$ initiator in vinyl chloride diluent at $-130°C$, has been identified in the γ form. The parameter(s) that determine which crystalline form of the polymer will be obtained during synthesis is unknown. However, it has been established that the α modification is the most stable conformer. Thus when a sample of the β polymorph is dissolved and reprecipitated, or melted and recooled, the x-ray diffraction pattern of the α form appears (67). All these forms exhibit the same infrared and NMR spectra in solution. Cast films may show characteristic bands for the α or β modification and infrared spectroscopy can be used to analyze mixtures of α and β polymorphs (67). There is no published information available on the infrared spectrum of the α modification. DTA is also a very convenient method to detect the identity of the polymorphs (66).

Efforts have been made to prepare highly oriented samples for more detailed structure analysis. Dilatometric studies showed that this polymer crystallizes very slowly (weeks) from the melt. Hot pressed films are pliable and somewhat tacky but become leathery and lose their tack completely within 2–4 weeks. Attempts to grow polymer single crystals failed. Cast films could be oriented by stretching and rolling but the best x-ray diagrams were obtained by a combination of these two processes. Freshly made stretched films showed birefringence that rapidly disappeared as the films relaxed. The best method found to investigate oriented films of both α and β conformers was to slightly press and subsequently stretch leathery films.

TABLE 4   CRYSTALLINE MODIFICATIONS OF POLY(3-METHYL-1-BUTENE) OBTAINED BY AlCl$_3$ IN ETHYL CHLORIDE AT $-130°C$

X-ray "d" Spacings in Å Units

α Modification	7.8 (w)	5.4 (s)	4.6 (m)	2.8 (vw)	2.4 (w)	2.1 (vw)
β Modification	9.0 (s)	5.4 (s)	4.8 (s)	3.66 (vw)	2.8 (w)	2.1 (w)
γ Modification	8.2 (vw)	6.7 (vw)	5.8 (s)	5.4 (s)	2.4 (w)	

Melting Points and Transition Temperatures by DTA (°C)

	Melting Point	Other Transitions
α Modification	66 ± 2	$-14.5, +23.5$
β Modification	48.5 ± 1	$-18.3, +19$
γ Modification	60.0 ± 2	$-15, +26.1$

Independent studies (68) suggest that the α polymorph is oriented in the form of a twofold screw or planar zigzag with a fiber repeat of 7.65 Å. Figure 6 shows the polymer from three views as it would appear in the planar zigzag form. This is the most stable conformation since all the C–C bonds are *trans* and the nonbonded atoms show maximum separation, as can be seen on examination of molecular models. Deviation from this conformation is due to side group interaction and results in a loss in the conformational stability. In cationically prepared poly(3-methyl-1-butene) the position of the side groups is symmetrical and the nearest neighbors are on opposite sides of the chain. Groups on the same side of the chain are separated by 7.65 Å.

From the d spacings and their relative intensities a basic unit cell for the α conformer has been proposed (68). The unit cell is tetragonal with the a and b axes being 10.7 Å and the c axis 7.65 Å. From this unit cell all spacings of the orientation diagram may be explained except the 7.65 Å meridianal.

The molecular architecture of cationically polymerized poly(3-methyl-1-butene) and its relation to its physical properties have been investigated by Martin and Gillham (69) who found $-7°C$ for the glass transition temperature by torsion braid analysis (0.2 cps). This relatively high transition temperature (the T_g of polyisobutylene, the structure of which is quite similar, is $-55°C$ by this method) was explained by an ingenious theory assuming intermolecular interlocking between

$$
\begin{array}{ccc}
\text{C} & & \text{C} \\
| & & | \\
-\text{C}-\text{C}-\text{C}-\text{C}- \\
| & & | \\
\text{C} & & \text{C}
\end{array}
$$

units resulting in maximum restriction to segmental motion.

In a fairly recent British patent (70) it was claimed that 3-methyl-1-butene can be polymerized with $TiCl_4/GeR_4$ (1:3) mixtures in hydrocarbon diluent in the range +20 to $-20°C$ and that the repeat unit contains more than 98% 1,3 enchainment $-CH_2-CH_2-C(CH_3)_2-$. It is interesting that this material is 40% crystalline, has a melting point of $108°C$ and molded thick plaques are translucent. To date this product exhibits the highest melting point ever reported for this structure. Unfortunately no molecular weight or other data were disclosed. Theoretically it is of great interest that a 1:3 $TiCl_4/GeR_4$ system gives rise to a rearranged structure since rearrangements of this sort are certainly diagnostic for the presence of carbenium ions and one would not expect cationic character from the catalyst composition (1:3 Ti/Ge) employed. This lead should be followed up.

Some kinetic details of 3-methyl-1-butene polymerizations have been elucidated. Kennedy et al. (71) investigated the polymerization of this olefin using $AlCl_3$ and $AlBr_3$ in various solvents in the temperature range -42 to $-102°C$, that is, in the range where amorphous polymer is formed. The polymerization

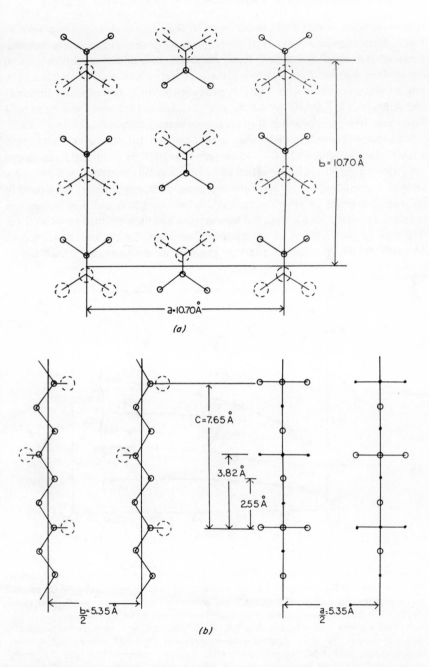

Figure 6. Possible arrangement of poly(3-methyl-1-butene) chains in planar zigzag form. Tetragonal unit cell; axes *a* and *b* 10.70 Å, *c* 7.65 Å. ⊙ = carbon atom in plane, • = carbon atom below plane, ⊙= superimposed atom both above and below plane.

is homogeneous in *n*-pentane and above $-50°C$ in methyl chloride. As shown in Figure 7, the molecular weights of the polymer increase with decreasing temperatures at the same conversion level, however, they increase with increasing conversions and seem to reach a plateau at higher conversions ($>25\%$). The rate of polymerization with $AlBr_3$ in propane solvent is first order in monomer and $AlBr_3$ and $k = 0.610$ l/mole sec with an apparent activation energy of 6.57 kcal/mole. In methyl bromide the rate is significantly higher.

Subsequent kinetic studies have been extended to polymerizations below $-100°C$, for example, $-130°C$ or lower, where crystalline completely rearranged polymer is formed (72). The effect of temperature on conversion and molecular weights was studied mainly with the monomer/$AlCl_3$/ethyl chloride system. In an experiment the polymerization was started at $-130°C$ and was allowed to progress smoothly for 55 min.; the temperature was then dropped to $-150°C$ for 195 min. (polymerization all but stopped, "dormant" period) then brought back to $-130°$ for 95 min (polymerization resumed at the same rate it had before

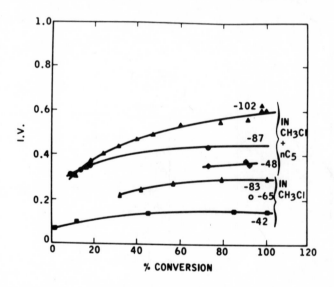

Figure 7. The effect of temperature on the molecular weight of poly(3-methyl-1-butene) obtained with $AlCl_3$ catalyst. At temperatures of -102, -87, and $-48°C$, catalyst concentration was 288×10^{-4} moles/l and monomer concentration was 0. 416 moles, composition of the diluent: 70 vol. % *n*-pentane and 30 vol. % methyl chloride.
At $-83°C$ catalyst concentration was 138×10^{-4} moles/l and monomer concentration was 1.8 moles/l. At $-65°C$ catalyst concentration was 29×10^{-4} moles/l and monomer concentration was 1.94 moles/l. At $-42°C$ catalyst concentration was 31×10^{-4} moles/l and monomer concentration was 0.91 moles/l. The diluent was pure methyl chloride in the runs at -83, -65, and $-42°C$ (71).

the "dormant" period), and finally it was raised to −100°C (at which point the reaction took off and rapidly reached 100% conversion). During the entire experiment the conversions and polymer molecular weights (by inherent viscosities) were followed by periodically removing aliquot samples. The conversion–time curve and the corresponding conversion–intrinsic viscosity profile are shown in Figures 8 and 9. In Figure 10 the first straight line describes the run until the point when the temperature was raised to −100°C. The dormant interval appears as a single point on the line. The eight values obtained at −130°C, four before and four after the −150°C period, all fall on the first straight line. The warming up period from −130 to −100°C is indicated by the dotted portion of the curve and the solid portion indicates the period when sample withdrawing resumed at −100°C. During this latter part of the experiment rubbery amorphous polymer was obtained. Additional experiments established the repeatability of linear molecular weight–conversion plots. All these lines were straight and could be extrapolated to the origin (72). A linear molecular weight–conversion relationship is a strong indication for the existence of "living" polymerization. Indeed, this system seems to be the only carbenium ion polymerization to date leading to respectable molecular weights in which, at least over a limited conversion range, "living" polymerization is indicated.

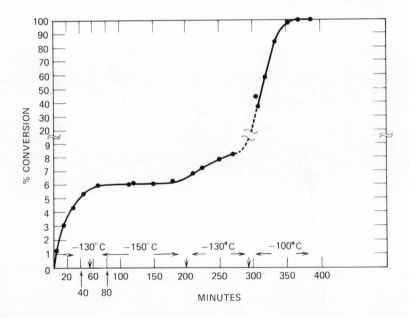

Figure 8. Conversion versus time of 3-methyl-1-butene polymerization at extremely low temperatures (72).

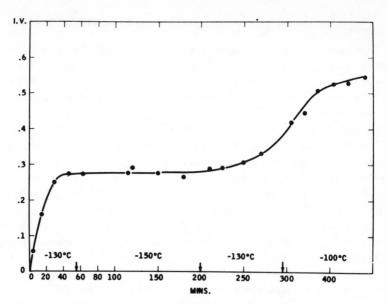

Figure 9. Molecular weights versus time of 3-methyl-1-butene polymerization at extremely low temperatures (72).

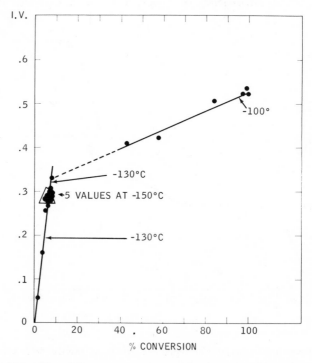

Figure 10. Molecular weight versus conversion (combination of Figures 8 and 9) of 3-methyl-1-butene polymerization at extremely low temperatures (72).

$$CH_3$$
2. *3-Methyl-1-Pentene* $CH_2=CH-CH-CH_2-CH_3$

This olefin is a reactive cationic monomer. It occurs together with other C_6 olefins ("hexylenes") in various refinery streams and as such has been co-oligomerized, for example, with BF_3, to lubricating oils (25). The high polymerization of this olefin has been studied by Bacskai (73) in the course of his investigation of isomerization polymerization of optically active branched monomers. This author has shown that 3-methyl-1-pentene undergoes rearrangement in the presence of $AlCl_3$ in ethyl bromide diluent in the range −75 to −55°C and gives rise to a polymer with 1,3 enchainment:

$$
\begin{array}{ccc}
CH_2\!=\!CH & & CH_3 \\
| & & | \\
HC\!-\!CH_3 & \xrightarrow{\ \sigma\ } & -CH_2\!-\!CH_2\!-\!C\!- \\
| & & | \\
CH_2 & & CH_2 \\
| & & | \\
CH_3 & & CH_3
\end{array}
$$

$$
\begin{array}{c}
CH_3 \\
| \\
CH_2 \\
|
\end{array}
$$
3. *3-Ethyl-1-Pentene* $CH_2=CH-CH-CH_2-CH_3$

This olefin was investigated by Kennedy (74) who found that it can be polymerized to respectably high molecular weights ($\bar{M}_n \sim 10,000$ at −78°C with $AlCl_3$ in ethyl chloride) and that the repeat unit of the polymer is probably a rearranged one:

$$
\begin{array}{ccc}
H_2C\!=\!CH & & Et \\
| & & | \\
CH & \xrightarrow{\ \sigma\ } & -CH_2\!-\!CH_2\!-\!C\!- \\
/\ \ \backslash & & | \\
Et\ \ \ Et & & Et
\end{array}
$$

$$CH_3$$
4. *3,3-Dimethyl-1-Butene* $CH_2=CH-C-CH_3$
$$CH_3$$

Meier reported that this olefin was rapidly trimerized with BF_3 in the gaseous phase at −45°C (51). This author recognized that about 50% of the neopentyl groups have isomerized during reaction. Edwards and Chamberlain also obtained trimers with $AlCl_3$ in ethyl chloride diluent (62). Kennedy, Elliott, and Hudson (75) prepared high polymers of this monomer with $AlCl_3$ in ethyl chloride by lowering the operating temperature below −100°C. A white amorphous powder was obtained that softened between 55 and 66°C and had a DP of −68.

Infrared and NMR spectroscopy showed that the repeat unit was a 1,3 structure formed by an intramolecular methide shift mechanism:

$$\sim CH_2-\overset{\oplus}{\underset{\underset{CH_3}{\overset{|}{C}}-CH_3}{\overset{|}{C}}}H \quad \xrightarrow{\sigma} \quad \sim CH_2-\underset{\underset{CH_3}{\overset{|}{C}}\oplus}{\overset{|}{CH}}-CH_3 \quad \longrightarrow \quad \sim CH_3-\underset{\underset{CH_3}{|}}{CH}-\underset{\overset{CH_3CH_3}{|}}{C}\sim$$

Maltsev et al. (76) confirmed these reports and added· that $TiCl_4$, $SnCl_4$, and $BF_3 \cdot OEt_3$ are less effective in inducing the isomerization polymerization of 3,3-dimethyl-1-butene than $AlCl_3$.

b. Branches β to the Double Bond

    1. 4-Methyl-1-Pentene $CH_2=CH-CH_2-CH\overset{\diagup CH_3}{\diagdown CH_3}$

This olefin is a reactive cationic monomer that readily yields high molecular weight rubbery polymers. Some details of the polymerization of 4-methyl-1-pentene with Lewis acids, in particular with $AlCl_3$ using methyl chloride diluent, have been investigated by Kennedy. Controlled, fast polymerization can be carried out with the above system at $-78°C$ and high molecular weight products are obtained. Interestingly, the highest molecular weight polymer is obtained at $\sim -78°C$ (up to $\sim 2 \times 10^6 \bar{M}_v$), and at higher or lower temperatures low molecular weight products are obtained. The effect of temperature on the molecular weight of poly(4-methyl-1-pentene) obtained with $AlCl_3$ in methyl chloride under a variety of conditions in a dry box (nitrogen atmosphere, $\sim$50 ppm moisture) is shown in Figure 11.

The structure of cationically obtained poly(4-methyl-1-pentene) has been in vestigated by NMR spectroscopy by a number of groups (62, 77–79). The investigators agree that the polymer is in fact a copolymer composed of the repeat units:

$$\sim CH_2-\underset{\underset{\underset{\underset{CH_3 \quad CH_3}{\diagup \quad \diagdown}}{CH}}{\overset{|}{CH_2}}}{\overset{|}{CH}}\sim \qquad \sim CH_2-CH_2-\underset{\underset{CH_3 \quad CH_3}{\diagup \quad \diagdown}}{\overset{|}{CH}}\sim \qquad \sim CH_2-CH_2-CH_2-\underset{\underset{CH_3}{|}}{\overset{\overset{CH_3}{|}}{C}}\sim$$

   1,2 Enchainment   1,3 Enchainment   1,4 Enchainment

of which the most important is doubtlessly the 1,4 enchainment. Recent high resolution NMR (300 MHz) work has shown the presence of significant amounts of 1,2 and 1,3 enchainments (up to ~40%) in polymers prepared under a variety of conditions at low temperatures (80). Table 5 is a compilation of information available on the structure of this polymer.

On the basis of structural information the following reactions have been proposed to explain the polymerization mechanism (78):

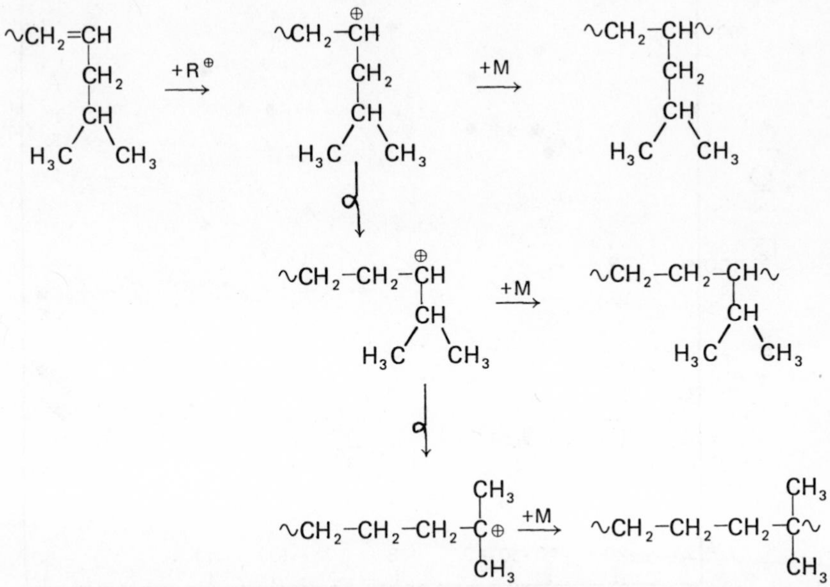

$$\sim CH_2=CH \quad \overset{+R^{\oplus}}{\longrightarrow} \quad \sim CH_2-\overset{\oplus}{CH} \quad \overset{+M}{\longrightarrow} \quad \sim CH_2-CH\sim$$

with pendant groups

$$-CH_2-CH(CH_3)_2$$

$$\sim CH_2-CH_2-\overset{\oplus}{CH} \quad \overset{+M}{\longrightarrow} \quad \sim CH_2-CH_2-CH\sim$$

with pendant $-CH(CH_3)_2$ groups

$$\sim CH_2-CH_2-CH_2-\overset{CH_3}{\underset{CH_3}{\overset{|}{C}}}{}^{\oplus} \quad \overset{+M}{\longrightarrow} \quad \sim CH_2-CH_2-CH_2-\overset{CH_3}{\underset{CH_3}{\overset{|}{C}}}\sim$$

Experiments with deuterated monomer $CH_2=CHCH_2CD(CH_3)_2$ indicated that hydride "jump,"

$$\sim CH_2-\overset{\oplus}{CH}-CH_2-CH(CH_3)_2 \quad \longrightarrow \quad \sim CH_2-CH_2-CH_2-\overset{\oplus}{C}(CH_3)_2 ,$$

does not occur and that the 1,4 enchainment is produced by two consecutive hydride shifts (78).

It is of interest that the major repeat unit, that is, the 1,4 enchainment, of cationically prepared poly(4-methyl-1-pentene) is formally equivalent to a perfectly alternating copolymer of isobutylene and ethylene, a monomer pair that cannot be copolymerized to high polymer by any known techniques.

While the structure of this polymer has been fairly well characterized, very little published information is available on the physical properties of this material. It has been mentioned that the low molecular weight polymers ($\overline{M}_n = 2000$–30,000) are amorphous, viscous, tacky materials soluble in common solvents (77), whereas the higher molecular weight products are rubbery solids with a T_g of $-17°C$ by torsion braid analysis (69) and in their physical appearance resemble natural rubber.

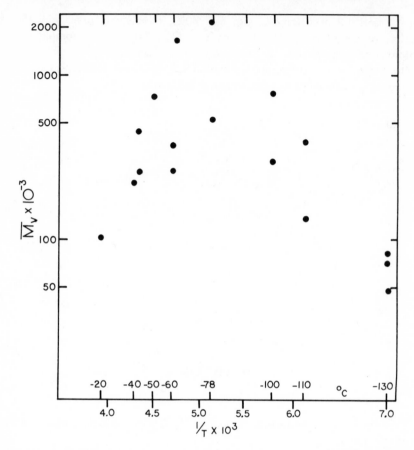

Figure 11. The effect of temperature on the molecular weight of poly(4-methyl-1-pentene) obtained under various conditions.

$$\overset{\displaystyle CH_3}{\underset{\displaystyle |}{}}$$

2. 4-Methyl-1-Hexene $CH_2=CH-CH_2-CH-CH_2-CH_3$

According to the patent literature this olefin can be readily polymerized to high molecular weight products in the presence of Friedel-Crafts acids in alkyl halide solvents (81). The polymerization of 4-methyl-1-hexene with $AlCl_3$ in ethyl chloride diluent in the range +10 to −130°C was investigated by Kennedy, Naegele, and Elliott (82). The highest molecular weight polymer was obtained at ∼−78°C ($\bar{M}_v$ = 273,000, $\bar{M}_n$ > 40,000). This material was an amorphous rubber, T_g ∼−23°C by DTA.

TABLE 5 THE STRUCTURE OF CATIONICALLY SYNTHESIZED POLY(4-METHYL-1-PENTENE)

| Contributing Repeat Units (%) | | | | | |
$\sim CH_2-CH\sim$ | CH | $H_3C-\underset{H}{\overset{}{C}}-CH_3$	$\sim CH_2-CH_2-CH\sim$ | $H_3C-CH-CH_3$	$\sim CH_2-CH_2-CH_2-\underset{CH_3}{\overset{CH_3}{C}}\sim$	Analytical Method Used	Polymerization Conditions	Investigators (Reference)
33	7	60	NMR and IR	$AlCl_3/CH_3Cl$, −78 and −130°C	Wanless and Kennedy (78)
20–22	32–35	43–48	Mass spectroscopy	$AlCl_3/CH_3Cl$, −78 and −130°C	Wanless and Kennedy (78)
Present	May be present	62 (max.)	Composition of pyrolytic gases	$AlCl_3/CH_3Cl$, −78 and −130°C	Wanless and Kennedy (78)
28–54	—	46–72	NMR	Various cationic initiators in n-heptane, 20–60°C	Goodrich and Porter (77)
?	Minor	Major (III/II = 1.3–1.8)	IR	$AlCl_3$, $AlBr_3$, in C_2H_5Cl, n-pentane −78 to −120°	Ketley (79)
?	?	Major	NMR	$AlCl_3/CH_3Cl$, −78 and 0°C	Edwards and Chamberlain (62)
~20	~20	~60	NMR (300 MHz)	Various	Kennedy and Johnson (80)

83

The structure of the rubbery polymer was investigated by NMR and infrared spectroscopy which indicated an essentially completely rearranged structure:

$$CH_2=CH \atop CH_2 \atop H_3C-CH-CH_2-CH_3 \quad \xrightarrow{\sim H^\oplus, \sim H^\oplus} \quad \sim\!\!\sim\!CH_2-CH_2-CH_2-\overset{CH_2-CH_3}{\underset{CH_3}{C}}\!\!\sim\!\!\sim$$

Bacskai came to similar conclusions by investigating the loss of optical activity of the asymmetric carbon in the pendant group (73). Recent work using 300 MH_z NMR indicates significant amounts of unrearranged units in the chain (80).

3. 4,4-Dimethyl-1-Pentene $CH_2=CH-CH_2-C(CH_3)_3$

Edwards and Chamberlain (62) mention that when neat 4,4-dimethyl-1-pentene is polymerized with solid $AlCl_3$ at $0°C$, the NMR spectrum of the polymer indicates conventional 1,2 enchainment. Sartori et al. (83) used $AlCl_3$ in ethyl chloride and decreased the polymerization temperature to -78 and $-130°C$. They determined by NMR and infrared spectroscopy that their product, a sticky solid with $\bar{M}_n \sim 1400$ by VPO, consisted mainly of a repeat structure that must have been formed by successive intramolecular hydride and methide migrations:

$$-CH_2-\overset{\oplus}{C}H \atop CH_2 \atop H_3C-\overset{}{C}-CH_3 \atop CH_3 \quad \xrightarrow{\sim H^\ominus} \quad -CH_2-CH_2 \atop {\oplus}CH \atop H_3C-\overset{}{C}-CH_3 \atop CH_3 \quad \xrightarrow{\sim CH_3^\ominus} \quad -CH_2-CH_2 \atop H_3C-CH \atop H_3C-\underset{\oplus}{C}-CH_3 \quad \longrightarrow \quad \sim\!\!\sim\!CH_2-CH \atop H_3C-CH \atop H_3C-\overset{}{C}-CH_3$$

The polymer is formally equivalent to a copolymer of ethylene and 2-methyl-2-butene.

c. Branches γ, δ, etc. To the Double Bond

1. 5-Methyl-1-Hexene $CH_2=CH_2CH_2-\overset{CH_3}{\underset{}{C}H}-CH_3$

Edwards and Chamberlain polymerized this olefin with $AlCl_3$ in methyl chloride diluent at $-73°C$ and with solid $AlCl_3$ in bulk at $0°C$ (62). From NMR analysis they concluded that the low temperature polymer contained about equimolar amounts of rearranged and conventional repeat units, whereas the high temperature product contained practically no completely rearranged repeat structures:

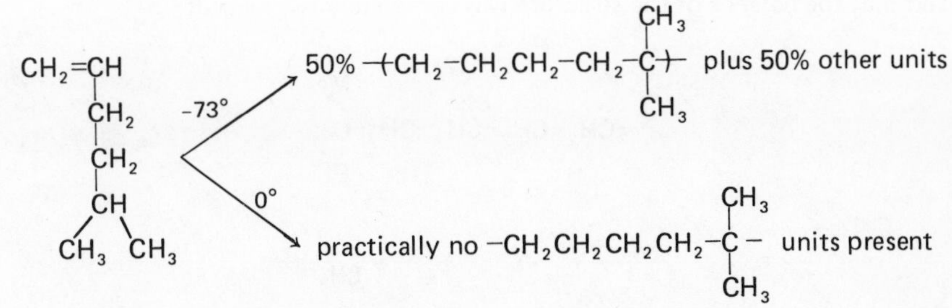

CH$_2$=CH $\xrightarrow{-73°}$ 50% $-\!\left[\!CH_2\!-\!CH_2CH_2\!-\!CH_2\!-\!\underset{\underset{CH_3}{|}}{\overset{\overset{CH_3}{|}}{C}}\!\right]\!-$ plus 50% other units

    |
    CH$_2$
    |
    CH$_2$
    |
    CH
   /   \
CH$_3$ CH$_3$ $\xrightarrow{0°}$ practically no $-CH_2CH_2CH_2CH_2-\underset{\underset{CH_3}{|}}{\overset{\overset{CH_3}{|}}{C}}-$ units present

Ketley (79) used AlBr$_3$ in ethyl chloride at −78 and −85°C to polymerize this olefin and obtained rather low molecular weight polymers ($\bar{M}_n \sim$ 6000). Polymerization did not proceed at −100°C. According to NMR and infrared spectroscopy, the polymer was a mixture of the whole spectrum of possible repeat units:

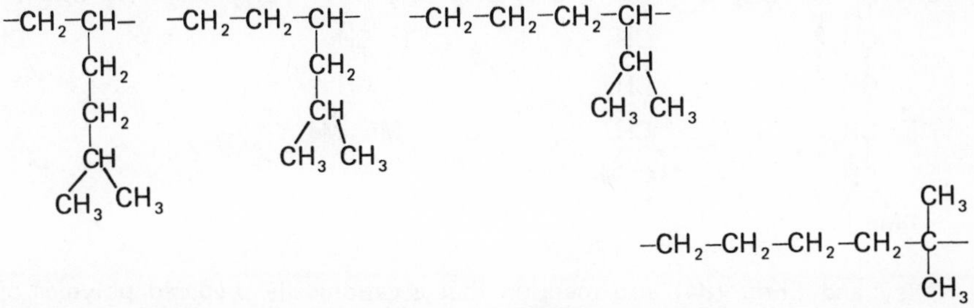

$-CH_2-\underset{\underset{\underset{\underset{CH_3}{}}{\underset{CH}{|}}{\underset{|}{CH_2}}}{\overset{|}{CH}}-$ $-CH_2-CH_2-\underset{\underset{CH_3\;CH_3}{/\;\backslash}}{\underset{CH}{|}}{\overset{|}{CH_2}}-$

-CH$_2$-CH-   -CH$_2$-CH$_2$-CH-   -CH$_2$-CH$_2$-CH$_2$-CH-
   | | |
   CH$_2$ CH$_2$ CH
   | | / \
   CH$_2$ CH   CH$_3$ CH$_3$
   | / \
   CH   CH$_3$ CH$_3$
  / \
CH$_3$ CH$_3$

                                                    CH$_3$
                                                        |
                   −CH$_2$−CH$_2$−CH$_2$−CH$_2$−C−
                                                          |
                                                         CH$_3$

                                CH$_3$
2. 5-Methyl-1-Heptene   CH$_2$=CH−CH$_2$−CH$_2$−CH−CH$_2$−CH$_3$

In the course of his studies on isomerization polymerization, Bacskai investigated the polymerization of this monomer (73). This author used an optically active monomer (+)-5-methyl-1-heptene and AlBr$_3$ in ethyl chloride at −75°C. Differences between the optical rotation of the monomer and the polymer led to the conclusion that 14–20% of the polymer contained unrearranged or partially rearranged structures:

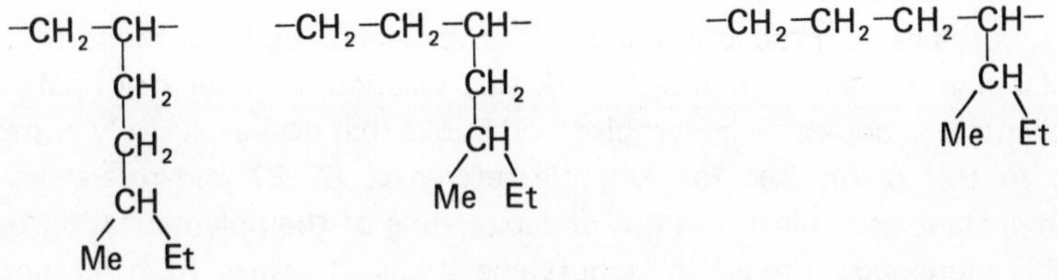

   −CH$_2$−CH−   −CH$_2$−CH$_2$−CH−   −CH$_2$−CH$_2$−CH$_2$−CH−
       | | / \
     CH$_2$ CH$_2$ Me   Et
     | |
     CH$_2$ CH
     | / \
     CH   Me   Et
    / \
  Me   Et

and that the balance of the structure was completely rearranged:

$$-CH_2-CH_2-CH_2-CH_2-\underset{\underset{Et}{|}}{\overset{\overset{Me}{|}}{C}}-$$

3. 6-Methyl-1-Heptene $CH_2{=}CH{-}CH_2{-}CH_2{-}CH_2{-}\underset{}{\overset{\overset{CH_3}{|}}{CH}}{-}CH_3$

Ketley (79) studied the polymerization of this olefin (AlBr$_3$ in ethyl chloride, $-78°C$). The polymer was of low molecular weight ($\overline{M}_n = 2200$) and, according to some infrared data, it contained the whole spectrum of partially rearranged structures:

$$\begin{array}{llll}
-CH_2-CH- & -CH_2-CH_2-CH- & -CH_2-CH_2-CH_2-CH- & -CH_2-CH_2-CH_2-CH_2-CH- \\
\quad\ CH_2 & \qquad\quad CH_2 & \qquad\qquad\quad CH_2 & \qquad\qquad\qquad\quad CH \\
\quad\ CH_2 & \qquad\quad CH_2 & \qquad\qquad\quad CH & \qquad\qquad\qquad\ /\ \backslash \\
\quad\ CH_2 & \qquad\quad CH & \qquad\qquad\ /\ \backslash & \qquad\qquad\qquad Me\ Me \\
\quad\ CH & \qquad\ /\ \backslash & \qquad\qquad Me\ Me & \\
\ /\ \backslash & \quad Me\ Me & & \\
Me\ Me & & &
\end{array}$$

Ketley and Ehrig (84) also mention that a cationically prepared polymer of 6-methyl-1-heptene did not show NMR evidence for a completely rearranged structure:

$$-(CH_2)_5-\underset{\underset{CH_3}{|}}{\overset{\overset{CH_3}{|}}{C}}-$$

C. ISOOLEFINS

1. Isobutylene $CH_2{=}C(CH_3)_2$

Isobutylene is probably the most investigated cationic monomer. Notwithstanding the large amount of work that resulted in hundreds of patents, several chapters in books, review articles, and books that deal exclusively or prominently with this olefin (see for example references 85–87 and references therein), important gaps remain in our understanding of the polymerization behavior of this monomer. Interest in isobutylene stems of course from its modest price, availability in large quantities, and significance in petrochemical operations.

Enormous amounts of isobutylene are used in various alkylation and polymerization processes.

First the monomer itself and the factors that determine its polymerization behavior are examined. Subsequently isobutylene oligomers are discussed, and finally the kinetics and other information available on high molecular weight polyisobutylene are critically surveyed.

If there is such a thing, isobutylene is an "ideal" cationic monomer. It is an extremely reactive species under all kinds of acidic conditions and it is one of the few cationic monomers, indeed, one of the few monomers that can be readily polymerized from the lowest oligomer, through medium molecular weights, up to the highest polymers with molecular weights in the millions.

The reactivity of isobutylene may be attributed to the two methyl substituents on the same carbon atom of the basic ethylene structure. Inductive electron donation polarizes the π electron system and thus electrophilic attack is facilitated:

$$CH_2 = C \underset{CH_3}{\overset{CH_3}{}} \longleftrightarrow \overset{\delta\ominus}{CH_2} - \overset{\delta\oplus}{C} \underset{CH_3}{\overset{CH_3}{}}$$

While the experimental heat of polymerization is lower by ~8 kcal/mole than the calculated value (88), presumably because of steric hindrance, the two methyl substituents do not prevent propagation by steric compression at the tertiary cation:

$$\sim\sim \overset{H}{\underset{H}{C}} - \overset{H}{\underset{H}{C^{\oplus}}} \overset{\overset{H}{C}---H}{\underset{\underset{H}{C}---H}{}}$$

The possibility of disturbing side reactions (e.g., hydride transfer and isomerizations) is minimized by the absence of tertiary hydrogens in the polymer or monomer. Thus from the chemical (but not from the kinetic) point of view the polymerization of isobutylene is a relatively uncomplicated process.

a. Isobutylene Oligomers

The oligomerization of isobutylene is one of the oldest polyreactions known (89). The lowest oligomers of isobutylene, di-, tri-, and tetraisobutylenes, are commercially available products and are produced with sulfuric acid, phosphoric acid, and/or acidic clays. Isobutylene oligomers can also be obtained under suitable conditions with a large variety of other acidic agents. An essentially complete although rather uncritical compilation of technologically significant processes is available in Güterbock's book (85) and a thorough but also uncritical survey of early scientific and patent publications has been published by Eidus and Nefedov (90).

Isobutylene oligomers are used as precursors for the manufacture of high octane gasolines. For example, hydrogenated diisobutylene produces isooctane, an important component of gasoline. Diisobutylene can also be a starting material for isobutylene manufacture. The diisobutylene is obtained by selective absorption of cracked gases in sulfuric acid solutions and further, usually catalytic, cracking gives rise to isobutylene. Other uses of diisobutylenes are in surfactants, lube-oil additives, plasticizers, and pesticides.

The dimerization of isobutylene with Brønsted acids involves protonation, addition, and deprotonation and leads to two isomers:

$$CH_2 = C(CH_3)_2 + H^\oplus \longrightarrow CH_3 - \overset{CH_3}{\underset{CH_3}{\overset{|}{C}}}{}^{\oplus} \quad + \quad \xrightarrow{CH_2 = C(CH_3)_2} \quad CH_3 - \overset{CH_3}{\underset{CH_3}{\overset{|}{\underset{|}{C}}}} - CH_2 - \overset{CH_3}{\underset{CH_3}{\overset{|}{\underset{|}{C}}}}{}^{\oplus} \quad \xrightarrow{-H^\oplus}$$

$$CH_3 - \overset{CH_3}{\underset{CH_3}{\overset{|}{\underset{|}{C}}}} - CH_2 - C\overset{CH_2}{\underset{CH_3}{\diagdown}} \quad \text{or} \quad CH_3 - \overset{CH_3}{\underset{CH_3}{\overset{|}{\underset{|}{C}}}} - CH = C\overset{CH_3}{\underset{CH_3}{\diagdown}}$$

2,4,4-Trimethyl-1-pentene   2,4,4-Trimethyl-2-pentene

The endo/exo olefin ratio in this reaction is usually 4:1. The fascinating chemistry of isobutylene oligomers was first investigated by Butlerov (89) who recognized that the dimer was a mixture of the above isomers. However, it took another 40 years before Whitmore and Church (91) proved the structure by ozonolysis and another 10 years or so until Hughes, Ingold, and their school (92, 93) and Brown (94) became interested in the elucidation of the relative quantities of the isomers formed and tried to explain quantitatively the experimental observations. In general, proton elimination leads to the thermodynamically more stable endo olefin and consequently one would expect to find in the dimer mixture a larger concentration of the internal isomer, 2,4,4-trimethyl-2-pentene, than the exo isomer, 2,4,4-trimethyl-1-pentene. This, however, is obviously not the case and the product ratio is usually about 4:1 in favor of the exo isomer. While our understanding of the parameters responsible for determining the exo/endo ratio in eliminations is far from being complete

(95), it appears that the elimination in this particular system is controlled by steric factors as first pointed out by Brown and Berneis (96). The basis for this argument is that the transition state of the elimination to the endo product is compressed because of unavoidable nonbonding interactions between the methyl hydrogens:

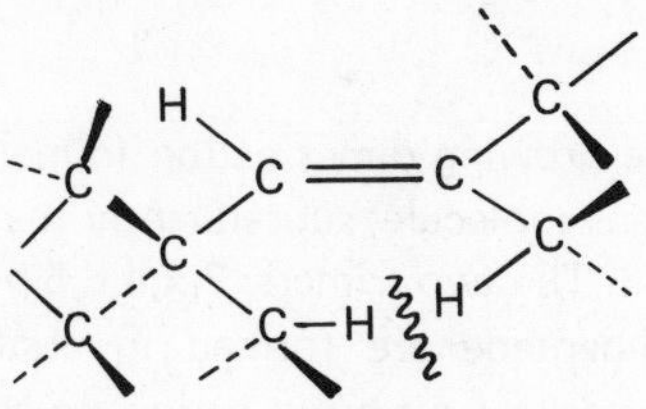

These two hydrogen atoms are forced to approach each other as the double bond is formed. In comparison, the exo transition state is much less compressed, which compensates for the difference in the heat content between the endo and exo olefins. The exo olefin is the major product when internal strain cannot be relieved by rotation of the interfering groups (94).

The situation is similar, though somewhat more complicated, with the trimers of isobutylene. Whitmore and associates (97) studied the isomer distribution of triisobutylene, a product obtained as a by-product of diisobutylene preparation by the dehydration and polymerization of tertiary butyl alcohol. They established the existence of five isomers. The formation of these species can be visualized as follows:

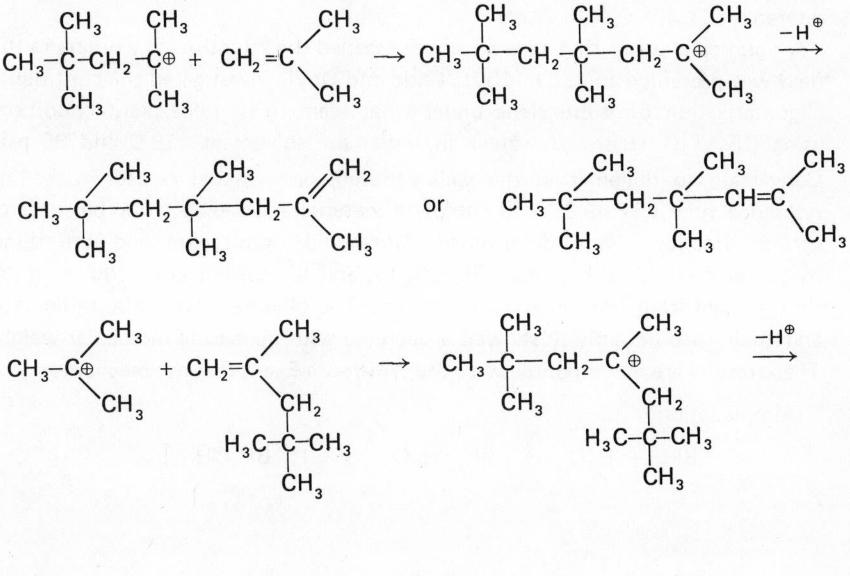

$$
\underset{\underset{CH_3}{|}}{\overset{\overset{CH_3}{|}}{CH_3-C-CH_2-C}}\diagdown\overset{CH_2}{\underset{\underset{\underset{\underset{CH_3}{|}}{H_3C-C-CH_3}}{|}}{CH_2}}
\quad \text{or} \quad
\underset{\underset{CH_3}{|}}{\overset{\overset{CH_3}{|}}{CH_3-C-}}\overset{\overset{CH_3}{|}}{\underset{\underset{H}{|}}{C}}=C\diagdown\overset{CH_3}{\underset{\underset{\underset{\underset{CH_3}{|}}{H_3C-C-CH_3}}{|}}{CH_2}}
\quad \text{or} \quad
\underset{\underset{CH_3}{|}}{\overset{\overset{CH_3\ H}{|}}{H_3C-C-}}C=C\diagdown\overset{CH_3}{\underset{\underset{\underset{\underset{CH_3}{|}}{H_3C-C-CH_3}}{|}}{CH_2}}
$$

In the first reaction the growing dimer cation (dimethyl neopentyl carbenium ion) attacks an isobutylene molecule; subsequently the proton is eliminated from the trimer cation formed. The two trimers, 2,4,4,6,6-pentamethyl-1-heptene and 2,4,4,6,6-pentamethyl-2-heptene are formed in about equal quantities. The second reaction is the attack of a *t*-butyl cation on an exo dimer and results in the formation of three isomers. The isomer distribution favors the exo trimer over those of the endo products 3:2. The *t*-butyl cation cannot attack the endo dimer because of prohibitively large steric compression.

Isobutylene oligomers are also of historical significance. The classical investigations of the Manchester school by Evans, Polanyi, and coworkers were in fact carried out with very low molecular weight polyisobutylenes. These early studies concerned the elucidation of initiation of isobutylene polymerization with BF_3 and protogenic initiators (for earliest references see Chapter 3, and for a good brief summary p. 153 of reference 87).

Kinetic aspects of isobutylene oligomerizations with H_2SO_4/H_2O and BF_3/H_2O initiator systems have been studied by Wichterle and coworkers (98, 99) who found a substantially linear relationship among the logarithm of the rate constant, the molecular weight in the DP = 2–4 range, and the acidity function H_0 of the initiator. This work has been summarized by Plesch (p. 154 reference 87).

A significant paper that seems to have escaped the attention of workers in this field was presented in 1961 (100). Child and Doyle investigated the continuous oligomerization of isobutylene under what seem to be pilot plant conditions using $BF_3 \cdot T_2O$ (tritrated water) in n-pentane solvent at $-18°C$ and 85 psig. Conversion to oligomer in the well agitated, back mixed reactor with 1 hr residence time was 99.3% and complete material and radioactivity balances on certain fractions were determined. Unreacted isobutylene and the dimer by-product exhibited high specific activity, and the solvent also acquired a low though significant level of specific activity. The oligomers were also radioactive and their specific activity showed a decrease with increasing molecular weight. These results are incompatible with the Whitmore-Evans-Polanyi mechanism:

Complexation:

$$BF_3 + H_2O \;\rightarrow\; BF_3 \cdot H_2O \;\rightleftharpoons\; H^{\oplus}\,BF_3OH^{\ominus}$$

Initiation:

$$H^{\oplus} + CH_2{=}C(CH_3)_2 \rightarrow (CH_3)_3C^{\oplus}$$

Propagation:

$$(CH_3)_3C^{\oplus} + CH_2{=}C(CH_3)_2 \rightarrow (CH_3)_3C{-}CH_2{-}C^{\oplus}(CH_3)_2$$

Transfer:

$$\sim CH_2C^{\oplus}(CH_3)_2 \rightarrow \sim CH_2{-}\underset{CH_3}{\overset{CH_2}{\underset{\|}{C}}} + H^{\oplus} \xrightarrow{+M} (CH_3)_3C^{\oplus} \text{ etc.}$$

Termination:

mechanism unknown, probably by impurities

This mechanism which has been generally accepted for the polymerization of isobutylene (or, slightly modified, for other olefins) does not explain the major findings of Child and Doyle, for example, that unreacted isobutylene was found to contain radioactivity, that the molar radioactivity of the dimer was much higher than that of the other oligomers, and that the solvent became radioactive. Analysis showed that the first finding, that concerning radioactivity in the monomer, is of great importance. The other two findings, relative to the dimer and the solvent, may be of lesser importance and may have arisen because of the specific conditions employed.

To explain radioactivity in the unreacted monomer the following exchange reaction was proposed:

$$BF_3{\cdot}OT_2 + C_4H_8 \rightleftharpoons BF_3{\cdot}OHT + C_4H_7T$$

It was theorized that exchange occurs in a concerted fashion. Since fast cationic initiation was not observed, this reaction cannot proceed via a carbenium ion mechanism.

Radioactivity in the *n*-pentane after oligomerization is probably due to, in the opinion of this writer, branched and/or olefinic impurities in the solvent. Tertiary hydrogen containing hydrocarbons readily undergo acid catalyzed exchange.

Certain molecular sieves have been found to contain acidic sites [whether they are Brønsted or Lewis acidic is an unsettled question at the moment (44)] and as such were able to oligomerize isobutylene (43–45).

Isobutylene can readily be polymerized to medium or high molecular weight products. Polyisobutylenes below about 10^4 viscosity average molecular weight are viscous liquids and are used as lube oil additives, tackifiers in rubber adhesives, blends with waxes, and so on. Polyisobutylenes in the molecular weight range

10^3-10^4 are called "polybutenes," whereas the term "polyisobutylenes" is used to designate products in the molecular weight range above $\sim 5 \times 10^4$— a completely unjustified and confusing terminology, nonetheless widely used in petrochemical literature. Low molecular weight polyisobutylenes are manufactured by various companies in the United States (Amoco Chemicals Corp., a subsidiary of the Standard Oil Co., Indiana; Oronite Division of the California Chemical Co.; and Cosden Oil and Chemical Co.) and in West Germany (Badische Anilin and Soda-Fabrik A.G.). The processes involve the contacting of suitably purified C_4 fractions of cracking gases with a Lewis acid such as $AlCl_3$. Since the C_4 fraction always contains along with isobutylene a large number and variety of other olefins that interfere with the polymerization of isobutylene, the final product is in fact a cooligomer of isobutylene and other mainly C_4, but some C_5, components.

Japanese researchers investigated the structure of commercially available polybutenes (products of Amoco, Cosden, and Oronite) by infrared spectroscopy (101) and found trisubstituted ethylene end-groups in these samples. This is contrary to the reports by Dainton and Sutherland (102), Flett and Plesch (103), Tokura et al. (104), and Imanishi (105) who also found by infrared studies predominantly vinylidene end-groups. According to the theory of compressed transition states (see above), vinylidene groups would be expected to form from propagating polyisobutylene cations. This discrepancy between the finding of Iimura et al. (101) and the above group of workers may be due to the fact that while the latter group of investigators prepared their own pure polyisobutylene samples for their study, Iimura et al. (101) used commercial products that are not pure polyisobutylene, but are in fact cooligomers of isobutylene and other olefins, notably *cis*- or *trans*-2-butene. It is conceivable that 2-butene becomes incorporated into the chain to give a trisubstituted ethylene:

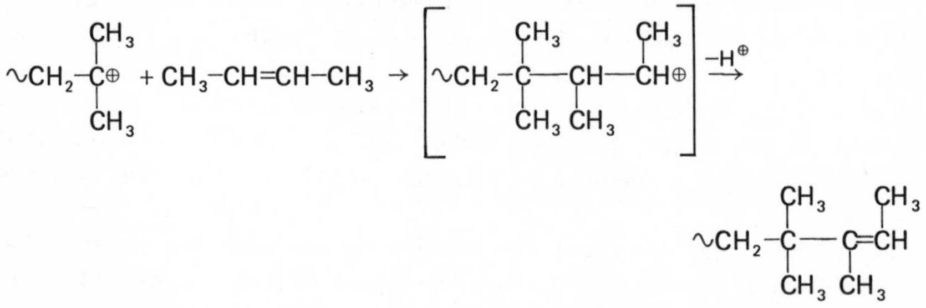

Beyond this study and a little more information in the trade literature concerning overall unsaturation and viscosity levels, the structure of these materials is largely a matter of conjecture.

b. High Molecular Weight Polyisobutylenes

Surveying polymerizations leading to high molecular weight polyisobutylenes, one immediately senses a dualistic approach to the field: on the one side we find "basic" researchers, mostly British workers stemming directly from or inspired by Polanyi's laboratories in Manchester in the forties and recently a few other pockets of investigators, mostly around Professor Sigwalt in Paris, who expended uncommon efforts to gather some basic information mostly on the kinetics of cationic isobutylene polymerizations under the most exacting and rigorous experimental conditions. The results of these studies were a rather meager bounty of still controversial hard core information.

On the other side of the fence are "industrial" researchers, mostly American and Czechoslovakian investigators who have concentrated on the expeditious generation of reproducible data obtained under conventional laboratory conditions.

How does the balance sheet look at the present? On the basic research side, we find a rather confused, ill understood body of fundamental information, bogged down in many details. It would take an immense input of time and effort, certainly unavailable anywhere in the world today, to piece together from this jumble of facts "hard" information, that is, those elements needed to construct a *firm base* for the next clear thrust toward a definitive understanding of this very limited field of polymer chemistry. (For a balance sheet on "industrial" research see p. 111.)

c. "Basic" Research

Since the discovery of coinitiation by Polanyi, Plesch, and coworkers (see Chapter 3), one of the cardinal questions facing basic researchers working with isobutylene polymerizations has been the control and understanding of the role of water in conjunction with Lewis acids. The novice can easily get confused by trying to sift through and unravel all pertinent experiments dealing with this question. Fortunately, over the past two decades experimental techniques have steadily been improved to the extent that most of the older researches and data (i.e., data obtained prior to 1965) have been superseded. This is clearly stated in a key paper by Plesch and coworkers (106) who investigated the isobutylene/ $TiCl_4/H_2O + CH_2Cl_2$ system in the temperature range +18 to $-91°C$. This paper appeared in 1965, some 20 years after that author witnessed the discovery of coinitiation (see Chapter 3), and carried out elegant research continuously in many phases of cationic polymerizations. Nonetheless, this work is still plagued by irreproducibility and other experimental difficulties, for example, the DP data obtained in runs conducted below $-50°C$ are so scattered, they are almost completely useless (cf. Figure 10 in reference 106). The authors write: "The results as a whole, especially those obtained at lower temperatures, show a relatively large scatter, which makes the exact nature of some of the correlations rather uncertain" and "We ascribe the scatter to the effects of small changes in the water concentration and of impurities becoming more important the lower the temperature" (106). This finding invalidated some writings by the same author published prior to 1965.

It is no exaggeration to state that Plesch has taught the scientific community how to carry out cationic polymerizations of isobutylene (and other monomers) with $TiCl_4$ under extreme dryness, a polymerization he found needed water as an initiator. Nonetheless, after about 20 years of work trying to overcome the ubiquitous moisture problem, he himself is still suffering from irreproducible scattered data ascribed to the presence of moisture or other impurities. Moreover, just about the time Plesch's paper (106) appeared, Cheradame and Sigwalt (107) found that under slightly different conditions, $TiCl_4$ alone is able to initiate the polymerization of isobutylene in bulk or in CH_2Cl_2 solution in extremely dry systems. In Plesch's experiments the water concentration controlled the yields and, indeed, the concentration of water could be estimated from yield data. All the polymers analyzed contained some chlorine and the presence of terminal unsaturation and –OH groups was demonstrated by infrared spectroscopy. Initial rates were determined and boldly represented by straight lines (Figure 4 in reference 106) and a variety of other shapes (Figure 5 in reference 106). Below $-60°C$ the order of reactions in which the concentration of water was less than $[H_2O]_c$, that is, the quantity necessary to obtain complete conversion, ranged from 1.2 to 1.65. Above $-35°C$ the order was between 0.0 and 0.5. Due to experimental difficulties the effect of monomer and initiator concentration on the rate was not investigated in depth.

The effect of water concentration on the viscosity average molecular weights was investigated at various temperatures and a series of curves of various shapes was obtained as shown in Figure 12 which is a reproduction of Figure 8 in reference 106. The authors comment that the data at $+18°C$ are "unambiguous" (rectilinear) whereas at the other temperatures "the results are compatible with the same type of curve which has a maximum at very low water concentration, followed by a plateau, or possibly a shallow minimum and then a falling branch at the highest water concentrations..." These authors have also investigated the effect of monomer concentration on DP of polyisobutylene at various temperatures but again obtained largely irreproducible data. Finally, they showed the effect of temperature on the DP at three monomer concentrations and claimed that the slopes of the lines increased with monomer concentration (Figure 13 which is a reproduction of Figure 11 of reference 106) as follows: $[iC_4]$ = 0.4, 0.097, and 0.056 M^{-1} corresponding to E_{DP} = –8.6, –8.2, and –6.6 kcal/mole, respectively.

It is instructive to examine some of the data in detail. For example, the

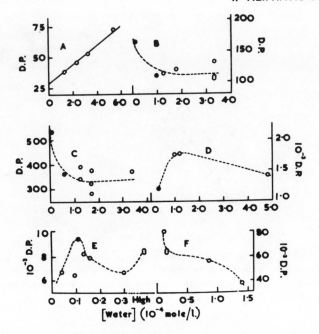

Figure 12. Effect of concentration of water on the degree of polymerization at various temperatures. (●) Reactions were stopped by addition of methanol (106). (Reproduced from reference 106, Figure 8.)

Curve	Temperature	[Isobutene] (mole/l)	Curve	Temperature	[Isobutene] (mole/l)
A	+18	0·085	D	−35	0·092
B	+5	0·087	E	−61	0·096
C	−14	0·089	F	−91	0·100

molecular weight data reproduced in Figure 12 can be reinterpreted by stating that, contrary to the author's contention, the water concentration does not affect the DP of the polymer and, including the +18°C rectilinear plot, there is no discernible trend; the data merely reflect the usual viscosity average molecular weight scatter one might obtain with polyisobutylenes even under less rigorous conditions. This statement is made with the assumption that the individual points in Figure 12 reflect but an experimental variation around a mean. By disregarding the dotted lines in the figure, one may construct a conventional plot of the Arrhenius-type setting at the proper temperature levels the broad ranges of viscosity average molecular weight obtained from Figure 12. (The range of molecular weights obtained at any temperature level shown in Figure 12 is broader than that one can obtain with some practice under conventional laboratory conditions using a dry box.) This operation gives Figure 14 and leads to the not unexpected conclusion that the only parameter that determined the

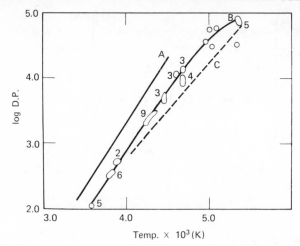

Figure 13. Variation of the DP with temperature ($^{\circ}$K). [Isobutene] = (*A*) 400, (*B*) 97, (*C*) 56 moles/1. Experimental points are shown only for curve B. The numbers indicate the number of results enclosed by the adjacent area. (Reproduced from reference 106, Figure 11.)

The "Activation Energy" of the DP

[Isobutene] (mmole/l)	400	97	56
E_{DP} (kcal/mole)	−8.6	−8.2	−6.6

The Effect of [TiCl₄] on the DP in the Presence of an Excess of Water (Table 6 of ref.106)

$T^a(^{\circ}C)$	Expt. no.	[TiCl₁] (mmole/l)	[Isobutene] (mole/l)	10^{-3}DP	$T^a(^{\circ}C)$	Expt. no.	[TiCl₁] (mmole/l)	[Isobutene] (mole/l)	10^{-3}DP
+19	129	1.05	0.370	0.136	−61	163	1.18	0.0960	8.34
	130	3.16	0.370	0.136		71	2.37	0.0960	8.52
− 5	97	1.07	0.087	0.107		73	2.37	0.0960	6.77
	98	1.61	0.087	0.102		74	2.37	0.0960	7.88
	106	2.15	0.087	0.118	−91	P16	0.35	0.0665	62.2
−14	101	1.10	0.0893	0.282		P37	0.35	0.0665	47.9
	102	1.66	0.0893	0.321		P33	0.89	0.0665	51.0
	107	2.20	0.0893	0.283		P38	0.89	0.0665	55.6
−32	P11	0.32	0.0606	1.06		P59	0.89	0.0665	50.6
	P16	0.81	0.0606	0.995		P8	1.76	0.0665	51.3
	P43	0.88	0.0606	0.975		P53	1.76	0.0665	75.0
	P10	1.60	0.0606	1.09		P12	3.52	0.0665	47.5
	P30	1.60	0.0606	1.10		P35	3.52	0.0665	64.3
	P47	1.60	0.0606	1.11		P52	3.52	0.0665	78.2
−37	92	0.46	0.0924	2.25	−91	110	1.24	0.100	57.0
	54	1.27	0.0924	1.78		10	2.05	0.100	83.4
	50	2.55		2.25		76	2.48	0.100	79.7
						77	2.48	0.100	64.2
						40	2.77	0.100	63.3

[a] Initial temperature.

viscosity average molecular weights in this experiment was the polymerization temperature. The broad molecular weight ranges obtained at individual temperatures as shown in Figure 14 point to an objectionable experimental technique and/or operator (possibilities include insufficient temperature control, too high conversion levels, difficulty in determining very high molecular weights $\bar{M}_v > 3.9 \times 10^6$). The rest of the molecular weight data available in this paper (for example, Figures 9 and 10 and Table 6 in reference 106) can be analyzed in similar terms and one comes to similar conclusions.

The merit of this paper lies in the analysis of the effect of temperature on the rate. The key finding was that the log k_1 versus $1/T$ plots showed a series of curves with minima (Figure 15). It was difficult to explain the observed rate deceleration followed by acceleration with increasing temperature in the range −91 to +18°C. The authors theorized that the rates are determined by (at least) two temperature sensitive processes and that propagation involves two growing species: free ions and ion pairs. The latter necessarily have different kinetic characteristics and their relative concentrations are determined by a temperature controlled mobile equilibrium. Plesch et al. (106) proposed the word "eneidic" to describe polymerizations in which more than one type of active species propagate. Eneidic systems are well known in the field of anionic polymerizations and their study is fairly advanced (108). Interestingly, however, it was first recognized and seriously analyzed in the cationic systems discussed above that log k (rate constant) values may *increase* with *decreasing* temperatures. Later Szwarc and coworkers (109) noted the existence of "negative" activation energy in anionic styrene polymerization as well and came to conclusions fundamentally similar to those of Plesch.

The bend in the log DP versus $1/T$ plot at low temperature was viewed by Plesch as the threshold temperature below which free ions and above which ion pairs are the molecular weight determining species. The bend hinted at around −90°C in Figure 13 is a reproducible strong reality as was shown repeatedly in the early sixties by Kennedy and Thomas (110, 111) in their studies of isobutylene polymerization by $AlCl_3$ using methyl chloride solvent in the range −30 to −50°C.

These findings and thoughts are significant not only for the understanding of isobutylene polymerization, but beyond that, for the whole field of cationic (and for that matter anionic) polymerizations as well. If more than one propagating species is involved in the polymerization, then, by the same token, more than one species may also be involved in the transfer, initiation, and termination. Consequently all the kinetic data gathered in the past must be reexamined as they may be overall or apparent values reflecting the individual contributions of various active species whose concentration depends on the particular experimental conditions employed.

One aspect of Plesch's findings must be questioned. If more than one equilibrating propagating species is involved in the polymerization and if by

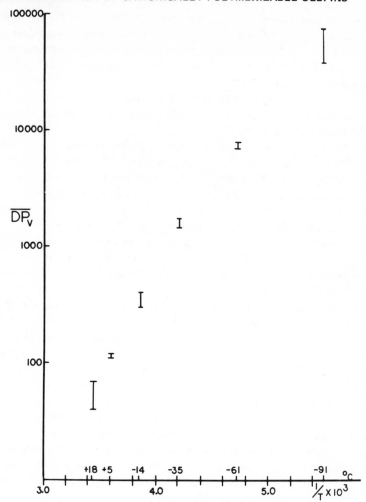

Figure 14. The effect of temperature on $\overline{DP}_V$ of polyisobutylene. Results of Plesch et al. (Figure 8 of reference 106) replotted.

lowering the temperature the relative concentration of the more reactive free ions is enhanced, then there must be an effect of temperature on the molecular weight distribution (MWD). Thus, one would anticipate a narrow MWD at lower temperatures where the rapidly propagating free ions predominate and a broader distribution at higher temperatures because in this range both species are active contributors to the final polymer. A broader MWD is expected at higher temperatures also because the relative concentration of the free ions will be lower and both species will contribute to the overall polymerization, the ion

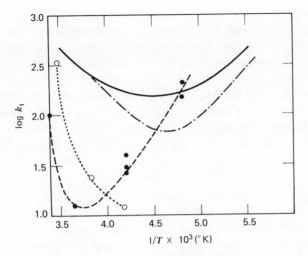

Figure 15. Variation of k_1 with temperature (Figure 6 of reference 106.) (— — — — — —) (R.H.B.) $[TiCl_4]$ = 1.5–1.7 mmole/l, excess of water. (————)(P.P.R.) $[TiCl_4]$ = 2.0–2.5 mmole/l, excess of water. (·····O·····O·····)(P.P.R.), $[TiCl_4]$ = 2.0–2.5 mmole/l, $[H_2O]$ = 0.027–0.030 mmole/l. (— —●— —●— —)(P.P.R.), $[TiCl_4]$ = 2.0–2.5 mmole/l, $[H_2O]$ = 0.010–0.011 mmole/l (106).

pairs because of their higher concentration, the free ions because of their higher reactivity. Experimentally, however, this is not what has been found. Patton and Plesch (112) analyzed the MWD of polyisobutylenes by a gradient elution method. They concentrated on the description of the analytical procedure. On careful analysis of these data, however, one can see that the polyisobutylenes obtained at higher temperatures (e.g., + 17.5, –11.7, and –14.6°C) have much narrower MWDs than those obtained at lower temperatures (e.g., from –60 to –90°C). This writer has examined by gel permeation chromatography the MWD of polyisobutylenes synthesized under various conditions and found that the width of the MWD narrowed with decreasing temperatures. This problem cannot be resolved at the present time and it is hoped that more and precise data will be generated soon by the convenient and fast gel permeation chromatography technique. Researchers must turn their attention to the investigation of kinetic–molecular weight parameters by GPC techniques instead of single point determinations of overall mixtures. The GPC technique is largely used in structure–property characterization studies— it should be adopted for kinetic investigations.

As an epilogue to this interesting development, it might be apropos to describe and offer a possible explanation for a puzzling phenomenon this author observed about a decade ago: one can bubble gaseous BF_3 into a refluxing ($\sim$–10°C) isobutylene—methyl chloride mixture without polymerization taking place; however, when the temperature is dropped to around –78°C immediate polymerization ensues. In line with the above concept, initiation and subsequent polymerization may be due to progressive ionization and/or increasing solvation of the hypothetical initiating entity $H^{\oplus}BF_3OH^{\ominus}$ at decreasing temperatures.

Plesch and his coworkers (113) also investigated the polymerization of isobutylene induced by $AlCl_3$ using methylene chloride diluent in the range 0 to $-60°C$. They found that the $AlCl_3/CH_2Cl_2$ system does not require the presence of water to produce high polymerization and proposed that the solvent is the most likely initiator. This is a reasonable proposition as the cation formed from CH_2Cl_2 in the presence of $AlCl_3$ may have enhanced stability because of contributing structures:

$$\overset{\oplus}{C}H_2Cl \longleftrightarrow CH_2{=}\overset{\oplus}{C}l.$$

Unfortunately, the experiment without the CH_2Cl_2 diluent has not been performed, so we still do not know whether or not $AlCl_3$ alone is able to initiate the polymerization of isobutylene.

In one diagnostic experiment these authors scavenged adventitious moisture by breaking into the isobutylene/CH_2Cl_2 solution, under baked-out high vacuum conditions, a vial containing $TiCl_4$ in CH_2Cl_2 (a limited amount of polymerization ensued). Subsequently a vial with $AlCl_3$ in CH_2Cl_2 was broken into the system which resulted in vigorous and complete polymerization. It was concluded that the presence of water was not required for the polymerization of isobutylene with $AlCl_3$ in CH_2Cl_2. The possibility of self-initiation either by the Hunter-Yohe mechanism

$$Cl_3Al{\leftarrow}CH_2\overset{\oplus}{C}(CH_3)_2$$

or by autoionization

$$AlCl_2^{\oplus}\ AlCl_4^{\ominus}$$

was suggested and discarded on the basis of unfavorable thermochemistry. Barring the unlikely presence of some initiating impurity, Plesch et al. proposed initiation by solvent:

$$iC_4H_8 + AlCl_3{\cdot}CH_2Cl_2 \rightarrow ClCH_2{-}CH_2{-}\overset{\oplus}{C}(CH_3)_2\ AlCl_4^{\ominus}$$

The weakness of this argument is common to all kinetic arguments, where direct chemical information is unavailable. While Plesch's hypothesis is an acceptable one, nonetheless, it is possible to envision alternate explanations for initiation in this system. For example, self-initiation via allylic hydride extraction could occur (114):

$$\underset{CH_3{-}\overset{\|}{C}{-}CH_3}{\overset{CH_2}{}} + AlCl_3 \rightarrow CH_3{-}\overset{CH_2}{\underset{}{C}H}{-}\overset{\oplus}{C}H_2 + AlCl_3H^{\ominus}$$

or one could argue that the moisture requirement of $AlCl_3$ for isobutylene polymerization is much below that of $TiCl_4$, or that the amount of protogenic impurities in the $AlCl_3$ preparation sufficed to aid initiation, or that chemisorbed –OH groups on the glass provided the necessary proton (via |–OH + $AlCl_3$ → |–OAlCl$_2$ + HCl), and so on.

The quantitative data of Plesch et al. requires some comment as well. Figure 5 of reference 113 shows the effect of $AlCl_3$ concentration on the initial rate R_0. The curves are a summary of 15 experiments carried out under "dry" and "wet" conditions, that is, with 6–7 hr of pumping over a wide range of monomer concentrations (from 55 to 186 mmole/l). For some reason, the [iC_4H_8] / [$AlCl_3$] ratios used were generally very low—most of these ratios were 100 or less. The rates seem to be unaffected by [iC_4H_8]; however, the authors see a maximum in the R_0 versus [$AlCl_3$] relationship (Figure 16), although "The reason for the maximum is not clear" (113). It is possible that the lines as drawn in the figure are incorrect and that no discontinuities in fact exist. Assuming that among the seven or eight dry and wet experiments one data point is nonrepresentative, the dry experiments would reduce to a straight horizontal line at about R_0 = 60 mmole/l sec and the wet runs to a straight ascending line.

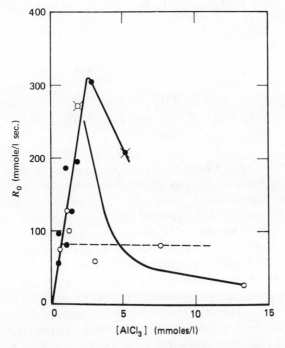

Figure 16. The variation of R_0 with $AlCl_3$ concentration at −60°C. A reinterpretation of Figure 5 of reference 113. (O) "Dry" experiments; (●) "wet" experiments.

This result would be easily understandable from the chemical point of view; in the dry experiments the $AlCl_3$ concentration would be of no consequence as the initial rate would be controlled by the concentration of some other agent (moisture?); in the wet experiments the $AlCl_3$ concentration is rate determining as the other agent (moisture?) is present in abundance.

A series of recent publications by Sigwalt and coworkers, and Marek and his associates concern isobutylene polymerizations with $TiCl_4$ (107, 115, 117), $AlBr_3$ (118, 117), BF_3 (117), $AlEtCl_2$ (119), and mixed Friedel-Crafts halides, for example, $AlBr_3 \cdot TiCl_4$ (120–123), carried out under extreme dryness. According to these reports the polymerization of isobutylene can be effected by the above and certain other Lewis acids in the absence of protogenic impurities, such as water. Indeed the presence of water was claimed to be unnecessary or even deterimental (107, 119).

This and other evidence suggests that the current well entrenched view according to which Friedel/Crafts halides always require protogenic or cationogenic coinitiators must be either modified or abandoned. This author hypothesized (114) that initiation of isobutylene polymerization with certain Friedel-Crafts halides in perfectly dry systems might involve "self-initiation," a process in which the monomer itself provides the first cation by allylic hydride donation to the strong Lewis acid:

$$CH_3\text{-}\overset{\overset{\displaystyle CH_2}{\|}}{C}\text{-}CH_3 + AlBr_3 \longrightarrow CH_3\text{-}\overset{\overset{\displaystyle CH_2}{\|}}{C}\text{-}\overset{\oplus}{C}H_2 + AlBr_3H^{\ominus}$$

$$CH_3\text{-}\overset{\overset{\displaystyle CH_2}{\|}}{C}\text{-}\overset{\oplus}{C}H_2 + CH_2{=}C\overset{\displaystyle CH_3}{\underset{\displaystyle CH_3}{\diagdown}} \longrightarrow CH_3\text{-}\overset{\overset{\displaystyle CH_2}{\|}}{C}\text{-}CH_2\text{-}CH_2\text{-}\overset{\overset{\displaystyle CH_3}{|}}{\underset{\displaystyle CH_3}{C\oplus}} \quad etc.$$

The allylic self-initiation hypothesis (114) accounts for a large body of hitherto unexplained phenomena and observations, particularly those accumulated over the past decade. Briefly, there exists in the scientific literature a large number of carefully executed (high vacuum, extreme dryness, etc.) experiments in which Lewis acids *alone* appear to initiate the polymerization of certain olefins. This writer noted that olefins possessing an allylic hydrogen atom, for example, isobutylene, isoprene, α-methylstyrene, indene, and cyclopentadiene, cannot be "stopped" from polymerizing by exhaustive drying and continue to polymerize, albeit slowly, even after extreme drying. It seems that the presence of an allylic hydrogen is necessary for self-initiation because olefins that do not have such an H atom, for example, styrene and butadiene, can be prevented from polymerizing by drying. Isobutylene and other olefins have been discussed in detail from this point of view (114). Recent experimental facts on the polymerization of isobutylene that have been published since reference 114 appeared are interpreted below from the vantage point of the self-initiation hypothesis.

In a recent paper Lopour, Marek, and coworkers (124) reported that the rate of the self-initiated slow polymerization of isobutylene/AlI_3 in an n-heptane system at $-14°C$ is considerably increased by the addition of n-butyl fluoride (n-BuF), or benzyl chloride, or benzyl bromide. The addition of nBuCl or nBuBr to the slowly propagating isobutylene/AlI_3 systems did not result in a rate increase. According to the authors, the rate increase is due to the catalytic action of *in situ* formed AlF_3 (by $AlI_3 + 3n$BuF $\rightarrow AlF_3 + 3n$BuI) and they theorized that individual AlF_3 *molecules*, before they are able to aggregate and precipitate to form the insoluble, inert AlF_3 crystal network, somehow induce the polymerization of isobutylene. The rate increase on nBuF addition was about 10^3-fold (from $1-5 \times 10^{-5}$ mole/l sec with AlI_3 alone, to $1.6-2.1 \times 10^{-2}$ mole/l sec after nBuF addition). When the addition sequence was changed to $iC_4^=$, nBuF/AlI_3, the faster rate appeared immediately. The addition of aged AlF_3, that is, premixed nBuF and AlI_3, failed to induce the polymerization of isobutylene. It is also well known that AlF_3 is totally inactive as a polymerization initiator for isobutylene.

All these observations are readily compatible with the self-initiation hypothesis. According to this writer's view the $(AlI_3)_2$ in n-heptane in the presence of isobutylene is partially dissociated to AlI_3:

$$(AlI_3)_2 \rightleftharpoons 2AlI_3$$

be it because of the inherent Brownian motion of the dimer or because of the bridge opening ability of the isobutylene acting as a weak nucleophile. The slow polymerization of isobutylene may be due to self-initiation by hydride abstraction by AlI_3, an inherently strong acid. The addition of nBuF increases the rate of opening of the $(AlI_3)_2$, produces the strong acid AlI_3, and thus leads to a faster rate of self-initiation:

$$AlI_3(I_3Al)\cdot n\text{BuF} \rightleftharpoons AlI_3\cdot n\text{BuF} + AlI_3$$

$$AlI_3 + CH_2{=}C(CH_3)_2 \rightarrow \text{self-initiation by } H^{\ominus} \text{ abstraction}$$

Premixing of nBuF + AlI_3 solutions and aging, of course, rapidly gives insoluble AlF_3 that is completely inactive as an initiator.

The inactivity of nBuCl and nBuBr may be due to two reasons: either the basicities of Cl– and Br– are insufficient to open the $(AlI_3)_2$ dimer bridge or steric hindrance caused by the larger size of the –Cl or –Br atoms prevents their sufficiently close approach to the aluminum atom in the $(AlI_3)_2$. In a very similar vein, recent work in the writer's laboratory showed that, compared to CH_3Cl, the molecules CH_3Br and CH_3I are poor "bridge openers" of the $(AlMe_3)_2$ dimer (125).

The self-initiation hypothesis maintains the essentially carbocationic nature of these polymerizations and avoids the *ad hoc* hypothesis that monomolecular AlF_3 is for some reason catalytically active whereas aged AlF_3 is not.

Marek and coworkers also found that with increasing amounts of *n*BuF the initial rate of the polymerization increased until a maximum was reached and then decreased. Thus, in the presence of 0.15 mmole/l AlI_3, the initial rate increased monotonically about 2000-fold until 1.1 mmole/l *n*BuF was introduced, at which point further amounts of *n*BuF resulted in a rate reduction. The observations were explained by assuming that mixed aluminum halides AlF_nI_{3-n} and monomolecular AlF_3 are active catalysts and that these species form rapidly at the start of *n*BuF addition. Later on, with excess *n*BuF present, inactive polymolecular AlF_3 is formed and the polymerization slows down. Again, instead of this *ad hoc* explanation, and keeping in mind the self-initiation hypothesis, one can simply assume that suitable concentrations of *n*BuF increase the rate of self-initiation by increasing the rate of AlI_3 formation from $(AlI_3)_2$ (see the above reactions). In the presence of high concentrations of *n*BuF (nBuF/AlI_3 ~10) the rate of AlI_3 consumption by $AlI_3 + 3n$BuF $\rightarrow AlF_3 + 3n$BuI increases and, because the active AlI_3 initiator is removed from the liquid phase, the rate of polymerization decreases. In other words the ascending rate is due to a breaking up of the $(AlI_3)_2$ by *n*BuF leading to active AlI_3 initiator, whereas the descending rate is explained by the increasingly rapid formation of inactive AlF_3 in the presence of stoichiometrically large *n*BuF concentrations.

Both explanations of the observed facts, that proposed by the original authors and the one advocated by this writer, discount the possibility of an effect of *n*BuI on the rate of isobutylene polymerization. It is conceivable, however, that the *n*BuI produced in the system is not "inert." Indeed, iodine containing organic molecules may inhibit carbenium-ion propagations by dialky iodonium ion formation:

$$\sim\sim C^{\oplus} + IR \longrightarrow \sim\sim C-\overset{\oplus}{I}-R .$$

Evidence for the existence of such ions has been found by Olah and deMember (126). In this writer's laboratory CH_3I was found to be a strong inhibitor of isobutylene polymerization (127). In the system studied by the Czechoslovakian authors, the rate reduction in the presence of large amounts of *n*BuF could also be due to the *in situ* formation of *n*BuI.

The authors also investigated the effect of benzyl halides on the polymerization of isobutylene initiated by AlI_3. According to the data, the addition of large amounts of benzyl fluoride ($C_6H_5CH_2F/AlI_3 = 7$) stopped the polymerization, whereas, benzyl chloride and bromide yielded rapid polymerization to 100% conversion. In addition to the *ad hoc* explanation offered by the authors, the fact that benzyl fluoride inhibits the polymerization can also be explained by assuming a rapid formation of inactive AlF_3.

The addition of benzyl chloride and/or bromide, however, gives rise to $AlCl_3$ and/or $AlBr_3$, active self-initiators on their own, which explains the data obtained with these benzyl halides. Another possibility is the rapid generation of benzyl cation ($C_2H_5CH_2^\oplus$) of finite lifetime capable of initiating the polymerization of isobutylene in the presence of AlI_3. Self-polyalkylation of benzyl halides by AlI_3 to polybenzyl

is also a distinct possibility. In the case of benzyl fluoride, AlF_3 precipitation is instantaneous and the lifetime of benzyl cations is too short (if they form at all) to initiate the polymerization of isobutylene.

At the last International Symposium on Macromolecules, Marek and Toman (128) reported some interesting observations on the polymerization of isobutylene induced by a variety of Lewis acids, for example, VCl_4, $TiCl_4$, $TiBr_4$, and $AlBr_3$. According to these authors the rate of polymerization increases significantly in the presence of light (40 W tungsten lamp). A detailed experiment with the isobutylene/VCl_4/n-heptane system at $-20°C$ showed polymerization under illumination alternating with cessation of polymerization in darkness. Other data indicating the effect of light on conversion are assembled in Table 6.

TABLE 6 EFFECT OF LIGHT ON THE POLYMERIZATION OF ISOBUTYLENE AT $-78°C$ (128)[a]

Initiator	Conversions in Bulk Experiments		Conversions in n-heptane[b]	
	Darkness	Illuminated	Darkness	Illuminated
VCl_4	0	92.5	.1.9	99.0(1.78)
$TiCl_4$	0.3	81.3	0	44.0(3.5)
$TiBr_4$	0	10.3(4.5)	0	0
TiI_4	0	0.1	0	0
$AlBr_3$	—	—	90.5	86.4(1.4)
$SnCl_4$	0	0	0	0

[a] Numbers in parenthesis indicate intrinsic viscosities.
[b] [M] = 1.2 mole/l, [Lewis acids] = 2–16 $\times$ 10^{-3} mole/l.

Evidently in some cases illumination somehow increases the slow rate of polymerization obtained with the transition metal halides. The authors propose the following scheme to account for the effect of light:

$$2MtX_4 \rightleftharpoons MtX_3^{\oplus} \, MtX_5 \xrightarrow{+M} MtX_3 + M^{\cdot\oplus} + MtX_5^{\ominus}$$

where Mt is the metal, M is the monomer, and X is halogen. The radical cation then initiates the polymerization. Experimentation with ESR failed to uncover evidence for the presence of radicals in the system.

Another possibility to account for these interesting observations in the framework of the self-initiation hypothesis would be to assume the photoexitation of transition metal halides to stronger acids that are able to initiate the polymerization directly via hydride abstraction or via a charge transfer complex:

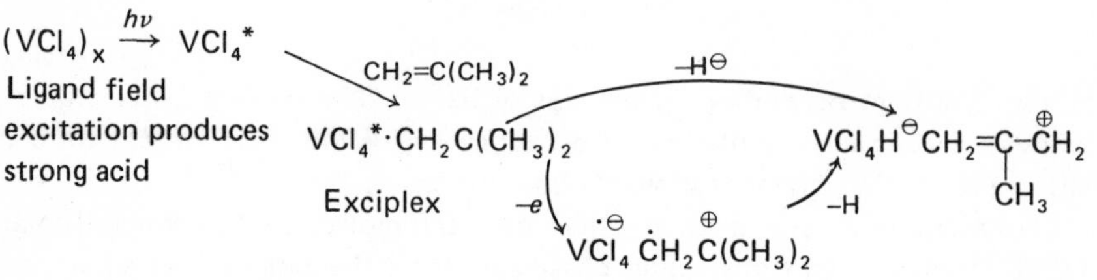

A peculiar publication dealing with aspects of initiation of isobutylene polymerization with AlBr$_3$, BF$_3$, and TiCl$_4$ deserves some comment. Ghanem and Marek (117) polymerized isobutylene under high vacuum ultra purity (dryness) conditions with these Lewis acids in the presence and absence of H$_2$O and/or tritiated water T$_2$O and determined tritium incorporation into the polymer. According to the accepted mechanism of cationic polymerization, initiation starts by protonation of the olefin; thus under the conditions used by these authors, tritium incorporation into the head-group of the polymer can be expected.

Although radioactivity was detected in the polymer, its level was low, for example, the authors calculate ~1.0 g atom H per mole of polymer obtained with AlBr$_3$, BF$_3$, and TiCl$_4$ under a variety of conditions. These calculated quantities are, however, questionable, if for no other reason because the number average molecular weights were computed from viscosity data and arbitrarily using $\bar{M}_v/4 = \bar{M}_n$ and because of the generally low degree of reproducibility of the data. The remarkable finding that the level of T incorporation is strongly dependent on the nature of the Lewis acid used has not been commented upon by the authors.

Interestingly, also, tritium incorporation into "dead" polyisobutylene has occurred by shaking the polymer in the presence of $AlBr_3$ in n-heptane previously saturated with T_2O (the level of T incorporation was ~0.1 g atom H/ per mole of polymer in this experiment). This result led the authors to conclude that all the tritium found in the synthesized polymers must have been incorporated by reactions that have nothing to do with initiation, but by reactions that involve "dead" polymer plus T_2O.

Disturbingly, the authors claim to have induced polymerization with BF_3 alone, in the absence of H_2O or T_2O. For some reason the classical results of the Evans–Polanyi–Plesch group, according to which BF_3 cannot initiate isobutylene polymerization in the absence of H_2O (see Chapter 3), have been completely ignored. It is also difficult to reconcile this finding of isobutylene initiation with dry BF_3 alone, with earlier results of Marek and coworkers (118) who showed by stopping experiments that dry BF_3 alone does not initiate isobutylene polymerization.

Most peculiarly the authors propose that "The role of controlled amounts of water in the cationic polymerization of isobutylene in nonpolar n-heptane is to activate a Lewis acid complex by solvating its ion pair"; "The polymerization is thus proton-less initiated." The nature of the Lewis acid complex that is "activated" by H_2O or how this activation occurs remains open. The possibility of a hydrolytic reaction between BF_3 or $TiCl_4$ and H_2O has not been considered.

Finally some suggestions as to how the T ends up in the polymer are made. The oddest one implies direct initiation by a "cationic catalyst moiety" C giving rise to a head-group in the polymer C∿∿∿. The nature of the species is not specified. Hydrolysis of this cationic catalyst moiety during or after polymerization would then explain T incorporation by the following process:

$$\underline{C}∿∿∿ + H_2O \longrightarrow H∿∿∿ + \underline{C}\text{--OH}$$

It would be interesting to know what head-group(s) the authors would propose for the case of BF_3 initiation. All these results and conclusions require independent confirmation.

Imanishi and coworkers (129) investigated the polymerization of isobutylene with $TiCl_4$, $TiCl_4 \cdot CCl_3COOH$, and $SnCl_4 \cdot CCl_3COOH$ using n-hexane, chloroform, and methylene chloride solvents at −20, −50, and −78°C. They found that (1.) the rate was higher with $TiCl_4$ than with $SnCl_4$, and CCl_3COOH was an accelerator, (2.) the rate was faster in methylene chloride, slower in chloroform, and still slower in n-hexane; (3.) $TiCl_4$ gave the highest molecular weight ($\bar{M}_n$ ~220,000 at −50°C) product among the three initiator systems studied; (4.) increasing solvent polarity decreased molecular weights, and (5.) the polymer end-groups were asymmetric disubstituted double bonds, $R_1R_2C{=}CH_2$, as indicated by the presence of weak absorption at 680, 985, and 1660 cm^{-1}.

The molecular weight data were interpreted by the Mayo equation in the form:

$$\frac{1}{DP} = \frac{k_m}{k_p} + \frac{k_t}{k_p} \; [M]^{-1}$$

The plots were linear and from the intercepts and slopes the authors obtained and analyzed a set of k_m/k_p and k_t/k_p ratios. Plesch severely criticized the Japanese investigators (pages 183 and 185–187 in reference 87) for equating the slope with k_t/k_p by pointing out that, except at lowest temperatures in nonpolar solvent (e.g., at $-78°C$ in n-hexane) where probably only unimolecular termination operates, the slope most probably includes all kinds of chain breaking processes. This criticism was heeded and in the final thesis (105) Imanishi included the necessary correction by considering the k_t as an overall term.

There is a report that diboron tetrachloride Cl_2B–BCl_2 can polymerize isobutylene but no details are given (130). Since B_2Cl_4 decomposes thermally to give among other products BCl_3, it is possible, in the opinion of this author, that the true initiator was BCl_3 (plus protogenic impurities).

Isobutylene polymerizations with $SnCl_4$ have been extensively studied by Norrish and Russell (131) and this work was surveyed and commented upon by Plesch and Biddulph (132, 133). Recently Russell and coworkers (134, 135) continued these investigations with isobutylene under anhydrous conditions by using $SnCl_4$/phenol derivatives as initiating systems and ethyl chloride solvent at $-78.5°C$. Hindered phenols gave π complexes with $SnCl_4$ and showed no initiating activity; unhindered phenols and $SnCl_4$ formed σ complexes via the oxygen atom and were found to be active initiating systems. Interestingly, deuterated phenols resulted in a 2.5 ± 0.5-fold reduction in the polymerization rate, which was viewed as strong evidence for initiation by protonation of isobutylene by the $SnCl_4/C_6H_5OH$ complexes. The overall rate of polymerization was reduced by a factor of 1.7 ± 0.3 with heavy water initiator as compared to a reaction initiated by H_2O. Monomer transfer was negligible at $-78.5°C$ and the main termination mechanism involved the attack of a growing polyisobutylene cation on a free phenol molecule. The presence of phenol end-groups was directly demonstrated in low molecular weight polymers by UV and NMR spectroscopy. NMR also indicated the presence of vinylidene end-groups (CH_2=C), which is in accord with the expectation based on Brown's work (see above).

In sum, the amount of hard scientific information after 20 years of continuous basic research in the field of isobutylene polymerization research is, in the opinion of this writer, not commensurate with the effort expended. The situation is perhaps better with the other highly popular cationic monomer, styrene. Our fundamental understanding of the cationic polymerization of isobutylene has advanced but little since the classical work of Polanyi and coworkers in the

1940s. For example, the question of initiation with Friedel-Crafts halides has not been completely resolved in spite of Plesch's statement in 1963 that "Isobutylene is not polymerized by metal halides unless an ionogenic substance, the cocatalyst, is present" (p. 149, reference 87). Our understanding of the kinetics of iso-butylene polymerization remains fragmentary. Thus we do not fully understand the effects of monomer concentration, initiator concentration, or even the effect of initiator/coinitiator ratio on the rates or molecular weights of most poly-isobutylene systems. We are still at the stage where every system must be examined and scrutinized individually and simple generalized statements should be made only with the utmost care.

This picture does not change much by including the results of another "basic" area, that of the high energy radiation induced isobutylene polymerizations. High polymerization of isobutylene was recently investigated by Kennedy, Williams, and Shinkawa (136) and significant facts emerged. Polymeriza-tions were initiated by cobalt −60 γ irradiation in the range −40 to −78°C. The rate of polymerization was found to be proportional to the 0.55 power of dose rate. The kinetic chain lengths were very large with $G(-m)$ values exceeding 10^8 molecules/100 eV, at a dose rate of 5.3×10^{11} eV/g sec. Importantly, these high rates were nearly proportional to the 0.5 of dose rate predicted by free ion propagation with termination by impurities. The temperature dependence of the polymerization rate was found to be negligible in the range of + 27 to −78°C (or even down to −138°C), indicating that there is practically no activation energy difference. The very high rates obtained under carefully anhydrous conditions were similar to those obtained by Dalton and others (138, 139) who carried out isobutylene polymerizations in baked-out high vacuum systems in the presence of solid ZnO. Thus Williams proposes that the role of ZnO and other solid additives is to scavenge water in the system (137). By making some reasonable assumptions, Williams was able to calculate for the first time the probable rate constant of isobutylene polymerization proceeding via free ions as $k_p = 1.5 \times 10^8$ mole/liter sec. This very large value might be due to the unencumbered nature of the cation. In the presence of charge stabilizing polar solvents and/or counterions a much smaller k_p is anticipated.

This work was continued by careful molecular weight studies (136). Thus $\bar{M}_n$, $\bar{M}_v$, and $\bar{M}_w$ of polyisobutylene have been determined in the range 29 to −78°C and plotted in an Arrhenius fashion; the $\bar{M}_w/\bar{M}_n$ ratio was $\approx$ 3.0 ±0.5. Further, the viscosity average molecular weights $\bar{M}_v$ of polyisobutylenes obtained by chemical initiation (BF$_3$, AlCl$_3$, and AlEtCl$_2$) and those synthesized by γ-irradiation were compared. As shown in Figure 17 (136) the $\bar{M}_v$ of the polymer obtained by radiation initiation and consequently by free ion propagation is almost an order of magnitude higher than by those of products produced by chemical initiation and therefore by ion pair propagation. To date, the highest molecular weight polyisobutylenes are obtained by irradiation induced polymerization. Since this process involves free ions, it is conceivable that these are the highest molecular weight polyisobutylenes that can be obtained.

Sparapany (137a) initiated the free ion polymerization of isobutylene with photon impact (UV radiation) in the vapor phase and directed the positive

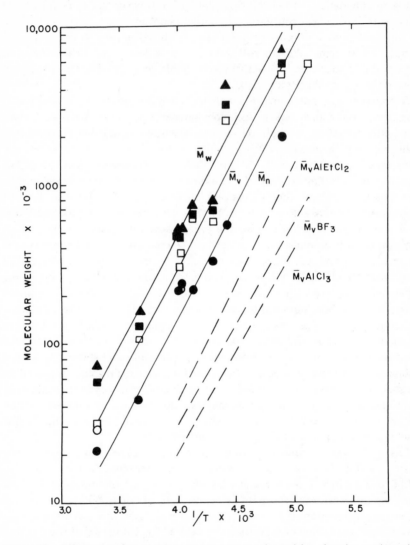

Figure 17. Effect of polymerization temperature on the weight, viscosity, and number average molecular weights of polyisobutylenes obtained by various initiators: (————) molecular weight averages of radiation initiated polymerizations; (— — —) viscosity average molecular weights of catalytically produced polymer; (▲) $\overline{M}_w$ by GPC; (■) $\overline{M}_v$ by GPC; (□) $\overline{M}_v$ by viscometry; (●) $\overline{M}_n$ by GPC; (○) $\overline{M}_n$ by osmometry (136).

fragments by an electric field into liquid neat monomer stirred at -120 to $-135°$C. By this technique the author was able to eliminate the ill understood effects of initiator/coinitiator systems and/or solvents on the polymerization. The molecular weights obtained in these experiments were quite high, for example, $\bar{M}_n$ = $2.5-3.5 \times 10^6$.

d. "Industrial" Research

What does the balance sheet show on the "industrial" research side? Industrial research, in this case, does not mean development or application research, but refers to the group of investigators who decided not to fight the ubiquitous moisture problem but to carry out reproducible experiments in the presence of traces of moisture under "closer-to-life" conditions. Indeed this group includes not only industrial researchers, but also a large number of academic investigators, for example, the Kyoto group (129), Czechoslovakian researchers (140), and Russian authors (40). The standardization of working conditions to attain reproducible data demands a high degree of sophistication. This effort, however, is a worthwhile investment because once reproducible working conditions have been worked out, the exact knowledge of the moisture level becomes less important and a large amount of useful information can be generated readily with relatively little effort. Experience has shown that satisfactory reproducibility for scientific research can be attained by working under inert gas atmospheres in the presence of <50 ppm moisture.

Research along these lines started in prewar Germany in the laboratories of the now defunct I.G. Farbenindustrie A.-G. by Otto and Müller-Cunradi (141) and reached maturity around 1940 with the now classical publication by Thomas et al. (142). The latter paper was written several years after the discovery of butyl rubber (an isobutylene–isoprene random copolymer) and gave the first glimpse of the tremendous complexity of this polymerization. Today butyl rubber together with chlorobutyl rubber and various grades of homopolyisobutylenes and derivatives thereof are important commercial products. In addition to these commercial realities, industrial oriented research contributed greatly to the science of isobutylene polymerizations.

Our understanding of the industrially most important isobutylene polymerization system, the isobutylene/$AlCl_3$/methyl chloride system, is rather sketchy. This particular polymerization system is of great technological interest, as it represents the basis of at least one commercially significant polyisobutylene manufacturing process, that is, that of the Vistanex process by the Exxon Chemical Co., and is very similar to other commercial polyisobutylene processes as well. In addition, this system is closely related to butyl rubber manufacture; butyl rubber is a copolymer of isobutylene and small quantities of isoprene.

Since the pioneering paper by Thomas and coworkers (142) in which the isobutylene/$AlCl_3$/methyl chloride system was first described from the scientific–kinetic point of view, two major groups have spent considerable effort in the further definition of the polymerization of isobutylene induced by $AlCl_3$ in methyl chloride diluent. The first group, the so called ESSO group that included Kennedy and Thomas, was active mostly during the years 1960–1967. These authors continued to build upon the original research of Thomas, Flory, Rehner, and others, carried out at the ESSO Research and Engineering Company during the early forties.

The second major group consists of Czechoslovak researchers, among them Ambroz, Marek, Vesely, Wichterle, and Zlamal, who exhibited a high degree of sustained interest over the last 15 years or so in the isobutylene/$AlCl_3$/methyl chloride system in particular and isobutylene polymerization in general. Their investigations started around 1955, many years after the commercialization of polyisobutylene and butyl rubber by an affiliate of the Standard Oil Co. (New Jersey) and many years after it has become common knowledge that these processes employ $AlCl_3$ in methyl chloride diluent. As research with this particular system is extremely cumbersome, it is possible that efforts have been or are being made in Czechoslovakia to develop a new proprietary butyl polymerization process.

A full reviewing and dovetailing of the many, sometimes conflicting, findings of these two major research groups interested in the isobutylene/$AlCl_3$/methyl chloride polymerization system is beyond the scope of the present modest sized survey. Moreover, Plesch in his 1963 chapter on polyisobutylene (87) and almost verbatim in a later paper (143) reviewed and criticized in some detail the results of these two groups. His interpretation, in turn, was criticized by the Czechoslovak authors (144). In view of this literature, this extremely complex field will be surveyed only briefly by concentrating on a few major findings. Beyond this, the reader is referred to the original literature. References 145–162 constitute a complete list of the publications by the ESSO group, and references 140, 144, and 163–170 are these by the Czechoslovakian workers.

The exploration of the polymerization details of the isobutylene/$AlCl_3$/methyl chloride system is certainly among the most difficult assignments in polymer research. First and foremost looms the experimental difficulty of obtaining meaningful and reproducible data. The sensitivity of this system toward trace amounts of impurities is almost proverbial. Indeed, after more than 30 years of research, the question of coinitiation is still open. We do not know whether or not $AlCl_3$ is able to initiate the polymerization of isobutylene in the absence of protogenic materials (H_2O or HCl) or whether methyl chloride is an initiator or merely an inert (?) polar solvent. The perfect purification of solid $AlCl_3$ is a heroic undertaking. Aluminum chloride is only sparingly soluble in methyl chloride and the molecular structure of the initiating entity in this solution is still largely unknown. The polymerization of isobutylene in methyl chloride diluent is a heterophase process: the polymer precipitates as it is formed which complicates temperature control and kinetic experiments. Successful polymerizations of isobutylene with $AlCl_3$ in methyl chloride proceed extremely rapidly with considerable heat evolution so that meaningful rate studies have not yet been carried out. Polymerizations with soluble Lewis acids, such as gaseous BF_3 or liquid $TiCl_4$ or $SnCl_4$, are more convenient to carry out experimentally, and interesting studies have been published on these systems.

Neither the Esso group nor the Czechoslovakian group chose to work with high vacuum equipment in their investigations on the isobutylene/$AlCl_3$/methyl chloride system. The limitations imposed on high vacuum manipulations in regard to lengthy and diminished number of experiments and a low degree of operational mobility is self-evident. While the chances of experimental reproducibility are greater in high vacuum experiments, this can also be achieved by careful working under a substantially dry atmosphere. As the data attest, both groups have obtained reproducible data in the presence of unknown but controlled amounts of impurities.

Neither group obtained rate data because of the above mentioned difficulties (too fast polymerizations, heterogeneous systems) but were able to accumulate significant yield (ultimate conversion) and molecular weight data. Both groups were interested in the effect of various parameters, such as temperature, monomer, and initiator concentration, on the molecular weights.

The most significant data of the ESSO group concerns molecular weight versus temperature. Indeed Thomas et al. (142) were the first to express quantitatively the inverse effect of temperature on the molecular weight of polyisobutylene. Subsequently these molecular weight data were replotted by Flory (171) according to

$$\log M_V = \log A - \frac{\Delta E}{R} \cdot \frac{1}{T}$$

Since that time this Arrhenius-type relationship has frequently been used to study the effect of temperature on the molecular weight of polymers in general and polyisobutylene in particular. These molecular weight data and similar ones from the ESSO group, particularly those in references 148, 151, and 155, are among the best and most reliable in this field. For example, Figure 18 shows the effect of temperature in the −30 to −150°C range of polyisobutylene obtained with three Lewis acid solutions: $AlCl_3$, $AlEtCl_2$, and BF_3 in methyl chloride. The overall activation energy difference calculated from the slopes is ∼5.6 kcal/mole in the −30 to −100°C range.

The ESSO group also studied the effects of monomer and initiator concentration on the molecular weight. In view of recent findings it is possible that these experiments were carried out without sufficient thermal control so that the

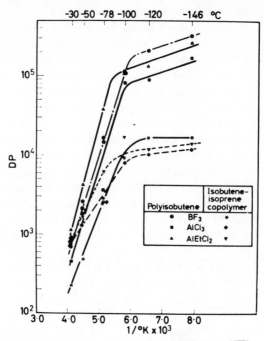

Figure 18. The effect of temperature on DP of polyisobutylene (155).

molecular weights obtained at high catalyst and/or monomer concentrations were artificially depressed. Thus Thomas et al. (142) reported first that the molecular weight of polyisobutylene as a function of dilution with methyl chloride goes through a maximum as shown in Figure 19. This unique lead was later repeated and further studied by the ESSO group (149–152). In view of recent, most carefully controlled experiments in this author's laboratory, it appears that previous molecular weight data obtained with neat isobutylene or high isobutylene concentrations (up to ~30 vol. %) are probably incorrect because of insufficient thermal control. It seems to be experimentally impossible to carry out isobutylene polymerization with fairly concentrated $AlCl_3$ in methyl (or ethyl) chloride solutions (~1.0 to 0.1 wt. % $AlCl_3$ or 7.53×10^{-2} to $10^{-3}M$ $AlCl_3$) in neat monomer without seriously depressing the molecular weights by the rapidly generated heat of polymerization. The extremely rapid polymerization rate coupled with a high degree of exothermicity (~12 kcal/mole) creates a situation such that even a slow rate of initiator addition and high speed stirring are insufficient to dissipate the reaction heat generated in the liquid phase, particularly in the area closest to the propagating site.

Attempts have been made to estimate the "lifetime" of a propagating site and the importance of diffusion in controlling the rate of polymerization and also to estimate the effect of the heat increment on the mechanism (149). These

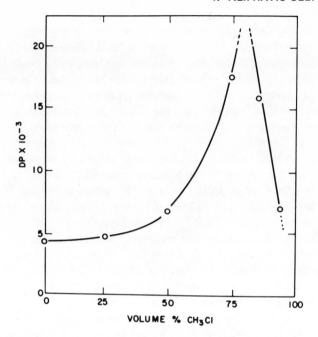

Figure 19. Effect of monomer concentration on DP of polyisobutylene recalculated from Thomas et al. (149).

calculations have shown that diffusion is not rate controlling and the heat buildup during polymerization is insignificant provided one restricts the calculations to a single point charge as the propagating site. Making reasonable assumptions as to the geometry of the sphere of growth r (the sphere within which monomer depletion occurs during propagation), the diffusion constant D, and the concentration gradient dn/dr, one can calculate N, the number of monomer units crossing the surface of the sphere of growth:

$$N = \frac{4\pi r^2 D}{dn/dr}$$

Numerical evaluation of this equation led to the conclusion that the reaction rate is not diffusion controlled (149). Similarly, the amount of heat effect generated by a single propagating site in a real life polymerization system can be estimated. Again, the conclusion was reached that the estimated heat increment ($\approx 0.03°C$) is insignificant and would not explain the data obtained at high monomer concentrations (149). These views are probably erroneous and the calculations based on them incorrect because this treatment focuses only on single events, on the growth of single molecules, and on heat-increments caused by a single point charge, and it ignores the mutual and multiple effects of an array of growing sites with the action radius of the sphere of growth. When a catalyst droplet hits liquid monomer, reaction starts before the initiating entities have sufficient time to diffuse apart. Thus propagation may be limited by sudden monomer depletion and/or affected by a sudden temperature rise in the sphere of growth. The reaction under consideration is so fast that polymerization may proceed long before the two liquids, the monomer and the initiator, are evenly mixed. When a droplet of an initiator solution enters the monomer charge, the reaction is so fast that it seems to be limtied almost exclusively to the area around the entering initiator droplet. High speed cinematography ($\sim$3000 frames/sec) did not resolve the time interval between the arrival of a droplet of an $AlCl_3$ initiator solution on the surface of an isobutylene charge at $-78°C$ and the appearance of solid polymer. These considerations also lead to the conclusion that steady state conditions cannot hold in this system.

In practical terms this means that molecular weights obtained with monomer charges higher than about 30 vol. % may be too low. All the experimental data must be carefully reconsidered by future workers in this field.

Aspects of the mechanism of isobutylene polymerization have been investigated and are described in a series of seven papers by Kennedy and coworkers (156–162). The main goal of these studies was to examine under controlled conditions ($AlCl_3$ dissolved in CH_3Cl initiator solution, n-pentane solvent at $-78°C$) the effect of various chemicals on the polymer yields and molecular weights of polyisobutylene. The analysis of the results led to a new theory of allylic termination in cationic polymerization and to a general classification of a wide variety of chemicals according to their rate (yield) and molecular weight controlling effects. Thus a group of compounds (n-alkenes) was found that reduces only the yields but leaves the molecular weights unaffected; these compounds were denoted as "poisons." Another group of chemicals (e.g., certain alkyl halides) reduced only the molecular weights but did not affect the ultimate yields; these species were termed "transfer agents." A large variety of compounds exhibited both poisoning and transfer activity to a more or less pronounced degree. The magnitudes of the poisoning and transfer activities were cast in quantitative terms by deriving poison coefficients and transfer coefficients. The relative values of these coefficients were compared and explained by a coherent theory.

A summary of the results of a large number of experiments is given in Figure 20. This figure represents a map of poison and transfer coefficients. Pure poisons (e.g., propylene, 1-butene, 1-pentene, etc.) and pure transfer agents (e.g., allyl chloride) fall on the horizontal and vertical axis, respectively; t-butyl chloride and bromide (points 40 and 43) are very effective, almost pure transfer agents and therefore fall very close to the abscissa.

The poisoning activity of n-alkenes was explained by allylic hydrogen transfer

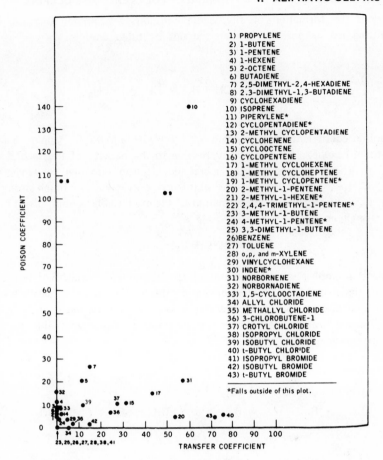

Figure 20. Poison coefficients versus transger coefficients (162).

from 1-olefins to the propagating polyisobutylene cation and by the enhanced stability of substituted allylic cations:

$$\sim C^{\oplus} + \underset{\underset{R}{|}}{CH_2}-CH=CH_2 \longrightarrow \sim \underset{\underset{R}{|}}{CH} + \underset{\underset{R}{|}}{\overset{\oplus}{CH}}-CH=CH_2$$

Stable cation

For example, the poison coefficients of simple alkenes are as follows: propylene, 4.9; 1-butene, 6.2; 1-pentene, 9.7; 1-hexene, 11.8; and 2-octene, 20.8. Propylene is the mildest poison because it gives rise to the least stable unsubstituted allylic cation. With the other 1-alkenes that contain two secondary allylic hydrogens the formation of allyl carbenium ion is facilitated because the substituted ion is more stable than the unsubstituted one:

$$\underset{\substack{|\\R}}{\overset{\delta\oplus}{C}H}\text{---}\overset{\delta\oplus}{CH}\text{---}\overset{}{CH_2} > \overset{\delta\oplus}{CH_2}\text{---}\overset{}{CH}\text{---}\overset{\delta\oplus}{CH_2}$$

The poison coefficients increase from 6.2 for 1-butene to 9.7 for 1-pentene and 11.8 for 1-hexene. This trend may be due to the basicity of the secondary allylic hydrogens caused by the increasing electron donating tendency of CH_3-, C_2H_5-, and C_3H_7- groups in this series. 2-Octene is a much stronger poison possibly because of the formation of the more stable doubly substituted secondary allylic ion:

$$CH_3-\overset{\delta\oplus}{CH}\text{---}CH\text{---}\overset{\delta\oplus}{CH}-C_4H_9 \ .$$

While the simple 1-alkenes do not affect the molecular weight of polyiso-butylene under the selected experimental conditions (i.e., transfer coefficient = 0), 2-octene exhibits appreciable transfer activity, with a transfer coefficient of 12.1. Transfer by internal olefins was proposed to occur according to the following mechanism:

$$\underset{\substack{|\\CH_3}}{\overset{CH_3}{\sim C}}{}^{\oplus} + \underset{\substack{|\\R}}{CH_2}-CH=CH-\underset{\substack{|\\R}}{CH_2} \rightarrow \underset{\substack{|\\CH_3}}{\overset{CH_3}{\sim CH}} + \underset{\substack{|\\R}}{\overset{\delta\oplus}{CH}}\text{---}CH\text{---}\underset{\substack{|\\R}}{\overset{\delta\oplus}{CH}}-CH_2$$

$$\underset{\substack{|\\R}}{\overset{\delta\oplus}{CH}}\text{---}CH\text{---}\underset{\substack{|\\R}}{\overset{\delta\oplus}{CH}}-CH_2 \xrightarrow{-H^{\oplus}} \underset{\substack{|\\R}}{CH}=CH-CH=\underset{\substack{|\\R}}{CH}$$

$$H^{\oplus} + \text{monomer} \rightarrow \text{polymerization}$$

that is, after hydride transfer to the propagating cation, the freshly formed allyl carbenium ion is stabilized by deprotonation to give a conjugated diene, however, the kinetic chain is sustained by the proton which attacks a fresh monomer molecule.

1-Alkenes only affect ultimate yields and consequently are pure poisons; they do not affect the molecular weights. In contrast, 2-alkenes affect both the yields and molecular weights and therefore are at the same time poisons and transfer agents. The poisoning and transfer activity of internal olefins is most likely due to the more favorable energetics in forming the stable bisubstituted allyl carbenium ion (1-alkenes form the less stable monosubstituted cation):

$$\overset{\delta\oplus}{CH}\text{:::}CH\text{:::}\overset{\delta\oplus}{\underset{R}{C}}H \quad > \quad \overset{\delta\oplus}{\underset{R}{C}}H\text{:::}CH\text{:::}\overset{\delta\oplus}{CH_2}$$

and therefore the stable disubstituted conjugated olefin (1-alkenes give rise to less stable monosubstituted conjugated olefin):

$$\underset{R}{C}H{=}CH{-}CH{=}\underset{R}{C}H \quad > \quad \underset{R}{C}H{=}CH{-}CH{=}CH_2$$

It is noteworthy that propylene reduces the rate (yield) of isobutylene polymerization at all. On the basis of solvolysis rates it could be argued that the allyl cation

$$\overset{\delta\oplus}{CH_2}\text{:::}CH\text{:::}\overset{\delta\oplus}{CH_2}$$

is less stable than the t-butyl cation $(CH_3)_3C^{\oplus}$, which is akin to the propagating polyisobutylene cation. However the stability of the dimethyl neopentyl carbenium ion is a much better model for the growing polyisobutylene cation than the stability of trimethyl carbenium ion (t-butyl cation). And there is evidence in the literature that the larger dimethyl neopentyl carbenium ion is

$$CH_3{-}\underset{\underset{CH_3}{|}}{\overset{\overset{CH_3}{|}}{C}}{-}CH_2{-}\overset{\overset{CH_3}{|}}{\underset{\underset{CH_3}{|}}{C}}{}^{\oplus} \quad < \quad \overset{\delta\oplus}{CH_2}\text{:::}CH\text{:::}\overset{\delta\oplus}{CH_2} \quad < \quad CH_3{-}\overset{\overset{CH_3}{|}}{\underset{\underset{CH_3}{|}}{C}}{}^{\oplus}$$

considerably less stable than the trimethyl carbenium ion (172, 173). On this basis it is conceivable that the stability of the unsubstituted allyl carbenium ion is somewhat higher than that of the propagating ion in isobutylene polymerization.

Some branched olefins do not affect the polymerization of isobutylene, while others profoundly alter the course of the polymerization (158). For example, 3-methyl-1-butene and 3,3-dimethyl-1-butene do not interfere with the reaction, the latter because it has no allylic hydrogens to give up, the former because its sole allylic hydrogen is rendered rather inaccessible to attack by the growing polyisobutylene chain by the neighboring two methyl groups and the vinyl group and because the allyl cation that would be formed is sterically quite hindered (unstable):

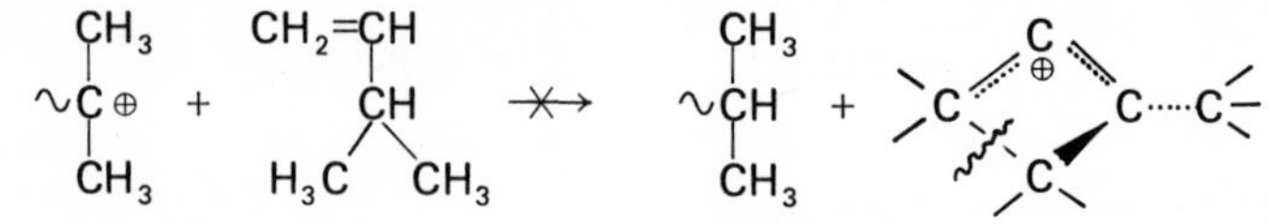

The structurally closely related vinyl cyclohexane has a low but measurable poison coefficient of 3.7, probably because in this molecule the rotation of the two methylene groups that shield the allylic H is restricted:

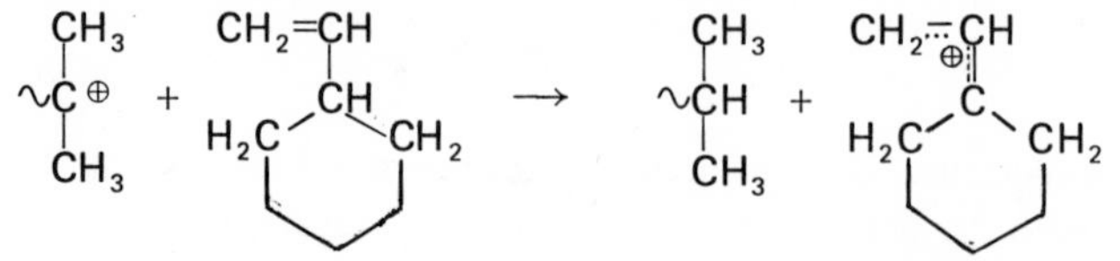

In 4-methyl-1-pentene the shielding of the two allylic hydrogens is somewhat removed and weak poisoning activity appears (poison coefficient = 2.9):

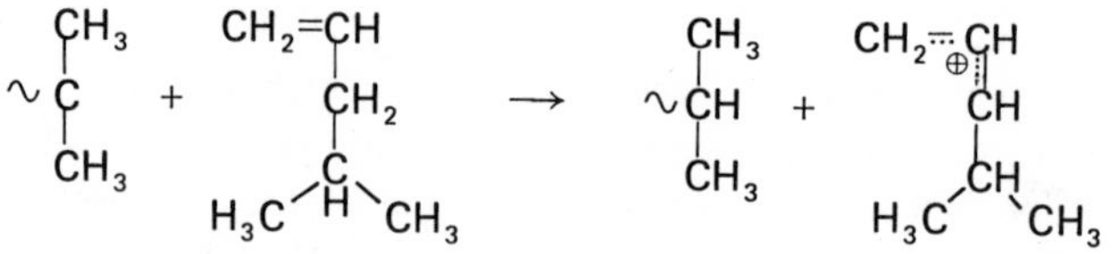

Tertiary hydrogens were found to be inert under the conditions investigated (e.g., isobutane is an inert solvent).

None of the above olefins are expected to incorporate into the polyisobutylene chain (i.e., to copolymerize) since this would mean the formation of an energetically unfavorable secondary carbenium ion. In contrast, 1,1-disubstituted ethylenes are isobutylene homologs and as such would give rise to tertiary carbenium ions and could copolymerize with isobutylene. However, even the next simplest homolog 2-methyl-1-butene gives low molecular weight product in low yield under conventional cationic polymerization conditions. The most likely explanation for this is that the propagating carbenium ion is a "buried" one. Since propagation is retarded by steric compression, the growing ion stabilizes itself by proton loss. Even isobutylene must overcome about 6.7 kcal/mole of steric hindrance to polymerize and the energy requirement must necessarily be larger with higher homologs. Consequently the lifetime of carbenium ions obtained from these homologs is longer and the chances for proton loss are much greater than for isobutylene itself. Thus lowered rates and/or molecular weights are expected from these homologs or from isobutylene polymerizations carried out in the presence of these chemicals.

Diisobutylene is frequently used to control the rate and/or molecular weight of polyisobutylene. Diisobutylene is a mixture of 2,4,4-trimethyl-1-pentene and 2,4,4-trimethyl-2-pentene. The effect of this mixture on the polymerization of isobutylene has been briefly investigated by Thomas et al. (142) and by Horrex and Perkins (174), and the effect of the pure isomers was studied by Kennedy and Squires (158). The latter authors found 2,4,4-trimethyl-1-pentene to be a strong poison and transfer agent, whereas the 2,4,4-trimethyl-2-pentene isomer was found to be a strong poison but only a mild transfer agent:

2,4,4-TM-1-P   P.C. = 66.7   T.C. $\approx$ 700
2,4,4-TM-2-P   P.C. = 66.7   T.C. = 34.6

The allyl carbenium ions expected to form from 2,4,4-trimethyl-1-pentene are

I   II

whereas the one from 2,4,4-trimethyl-2-pentene is:

II

Since cation I is energetically much less favored than cation II, it is likely that 2,4,4-trimethyl-1-pentene yields cation II on allylic termination, which is the same as that formed from the 2,4,4-trimethyl-2-pentene isomer. Thus the allyl carbenium ions arising from the two isomers could conceivably be the same and could explain the fact that the poison coefficients of the two isomers are identical.

Interestingly, however, the 1-pentene isomer is a strong molecular weight depressor, whereas the 2-pentene isomer shows a much milder transfer effect. One explanation may be that 2,4,4-trimethyl-1-pentene can be incorporated into a propagating isobutylene cation:

The resulting strongly hindered carbenium ion is unable to propagate and stabilizes itself by proton elimination, that is, chain transfer occurs. In contrast, 2,4,4-trimethyl-2-pentene contains an unaccessible double bond and cannot enter the growing chain:

$$\sim CH_2 - \overset{\overset{\displaystyle CH_3}{|}}{\underset{\underset{\displaystyle CH_3}{|}}{C}} \oplus \; + \; \overset{\overset{\displaystyle CH_3}{\backslash}}{\underset{\underset{\displaystyle CH_3}{/}}{C}} = CH - \overset{\overset{\displaystyle CH_3}{|}}{\underset{\underset{\displaystyle CH_3}{|}}{C}} - CH_3 \;\; \not\rightarrow$$

Consequently this molecule will be a much less effective chain transfer agent than the 1-pentene isomer.

Conjugated dienes are well known copolymerization partners of isobutylene. However, it is also known that the copolymerization of only a few percent of dienes results in severe molecular weight depression. The theory of allylic termination was applied by Kennedy and Squires (159) to account for these observations. For example 1,3-pentadiene (piperylene) exhibited high poisoning (P.C. = 170) and transfer (T.C. = 327) activity which was ascribed to the stable allyl cations obtained on incorporation:

$$-CH_2 - \overset{\overset{\displaystyle CH_3}{|}}{\underset{\underset{\displaystyle CH_3}{|}}{C}} - CH_2 - \overset{\delta \oplus}{CH} \overset{}{\cdots} CH \overset{\delta \oplus}{\cdots} CH - CH_3$$

and on hydride transfer:

$$\overset{\delta \oplus}{CH_2} \overset{}{\cdots} CH \overset{}{\cdots} CH \overset{}{\cdots} CH \overset{\delta \oplus}{\cdots} CH_2$$

Kennedy et al. (161) continued this work with a study of the effect of hydrogen chloride and simple alkyl and allyl halides on the polymerization of isobutylene. The poison coefficients of the hydrocarbons and halides were obtained from linear plots of $1/W_p$ versus X where W_p is the weight of the polyisobutylene obtained in the presence of poisons and X is the poison or transfer agent concentration. In contrast, the transfer coefficients derived for the hydrocarbons and halides have a somewhat different mathematical basis and therefore are directly not comparable. The transfer coefficients for the hydrocarbons have been derived from plots of $1/MW_p$ versus X, whereas those for the halides have been derived from plots of $1/MW_p$ versus $X^{\frac{1}{2}}$ (where MW_p is the polyisobutylene molecular weight obtained in the presence of transfer agent). That the poison coefficients determined from $1/W_p$ versus X plots were linear for all compounds investigated was viewed as evidence for the similarity of the termination mechanisms for hydrocarbons and halides. This, however, was not true for the transfer coefficients, and it was concluded that the chain transfer mechanisms operating with these two compound classes must be different.

Surprisingly, the poison coefficients of the allylic chlorides are low and are of the same order of magnitude as those obtained with n-alkenes. Furthermore the halides are strong chain transfer agents whereas the alkenes are not. For example, allyl chloride is not a poison but a transfer agent, whereas propylene is a poison but not a transfer agent.

It was proposed that the poisoning activity of allyl halides is due to the chlorine substituted allylic ion:

$$\overset{\delta\oplus}{Cl-CH} \text{::} CH \text{::} \overset{\delta\oplus}{CH_2} \quad \longleftrightarrow \quad \overset{\delta\oplus}{Cl} \text{::} CH \text{::} CH \text{::} \overset{\delta\oplus}{CH_2}$$

whereas transfer is derived from the unsubstituted cation:

$$\overset{\delta\oplus}{CH_2} \text{::} CH \text{::} \overset{\delta\oplus}{CH_2}$$

Simple Hückel molecular orbital calculations indicated that the ion formed by hydride loss from the allyl halide is of comparable or even greater stability than the allyl cation formed by hydride loss from the olefin.

The results of the ESSO group obtained with isobutylene polymerizations carried out in the presence of HCl should be mentioned. Isobutylene polymerization initiated by small aliquots of $AlCl_3$ in methyl chloride solution do not go to completion but stop at low conversions (147). Termination cannot be attributed to simple coinitiator exhaustion because the addition of trace amounts of moisture or HCl to these systems does not reinitiate propagation (175). However, if small amounts of HCl or moisture are added to the system prior to $AlCl_3$ introduction, higher ultimate conversions and lower molecular weight products are obtained than in the absence of added HCl. The results of a representative experiment are shown in Table 7 (175).

TABLE 7   THE EFFECT OF HCl ON THE AMOUNT OF POLYISOBUTYLENE FORMED AND ON ITS MOLECULAR WEIGHT ($\bar{M}_v$) (175)

Amount of HCl (mole/1)	Polyisobutylene (g)	M_v $\times 10^{-3}$
0	0.365	951
0.001	0.436	256
0.01	0.421	157
0.05	0.533	85
0.1	0.573	91

Evidently HCl is an initiator (promoter?) and transfer agent in this system (162, 175).

Cesca et al. (176) claim the synthesis of $\overset{\delta\oplus}{t\text{Bu}}-Cl \rightarrow \overset{\delta\oplus}{AlCl_3}$ complex from tBuCl and $AlCl_3$ in CH_2Cl_2 or CH_3Cl at $-78°C$, and report enhanced conversions and reduced molecular weight polymers (as compared with $AlCl_3$) when the complex is used as an initiator for isobutylene polymerization. These observations are understandable assuming an enhanced rate of initiation (higher carbenium ion concentration) and transfer by tBuCl produced from the partially dissociated complex.

A new class of cationic initiator systems that are very effective in the polymerization and copolymerization of isobutylene (and other cationic monomers) has recently been discovered by Kennedy and Baldwin (177, 178). Thus certain dialkylaluminum halides or trialkylaluminum compounds in conjunction with Brønsted acids (HCl, HBr, etc.) or cation sources (t-butyl chloride, benzyl halides, etc.) initiate a rapid polymerization of isobutylene to high molecular weight products at relatively high temperatures. In a typical polymerization experiment isobutylene can be stirred in methyl chloride solution in the presence of suitable R_2AlX compounds where X is alkyl or halide without reaction occurring. However, as soon as a coinitiator is introduced into the quiescent system, immediate rapid polymerization commences and, depending on the temperature and other variables, relatively high molecular weight product is formed.

In Figure 21 the viscosity average molecular weights of polyisobutylenes obtained with the conventional $AlCl_3$ and $EtAlCl_2$ initiators in methyl chloride are compared with those of polyisobutylenes synthesized with Et_2AlCl/tBuCl, Me_3Al/tBuCl, Et_3Al/tBuCl, and iBu$_3$Al/tBuCl initiating systems (179). Evidently highest molecular weight polymer is obtained with the Et_2AlCl/tBuCl system in the -20 to $-78°C$ range. Below this temperature the conventional initiators give highest $\bar{M}_v$. A possible reason for the large difference in the temperature coefficients between the conventional initiators ($E_{\bar{M}_v} = \sim 6.6$ kcal/mole) and the novel systems ($E_{\bar{M}_v} \sim 1.7$ kcal/mole) is given later in this section.

A very recent development in this field is the discovery (179a) that halogens in conjunction with certain alkylaluminum compounds are most effective initiating systems for the polymerization of isobutylene and other olefins. Kennedy and Sivaram (179b, 179c) report that elemental chlorine in methyl chloride solution in the presence of Me_3Al or Et_2AlCl in the range -35 to $-100°C$ produces high molecular weight polyisobutylene and that the initiator efficiency is very high (grams of polyisobutylene/mole of $Cl_2 \approx 15,000$). Model experiments have shown that the initiating entity is most likely a positive chlorine ion. Baccaredda et al. came to similar conclusions (179d).

Russian authors (40) reported that while Et_2AlCl alone does not initiate the polymerization of isobutylene in toluene at $-78°C$, active polymerization commences with Et_2AlCl in conjunction with hydrated salts, for example, $LiCl\cdot xH_2O$. In ethyl chloride solvent, however $LiCl\cdot xH_2O$ was not required and Et_2AlCl alone induced the polymerization to high molecular weight product. Indeed,

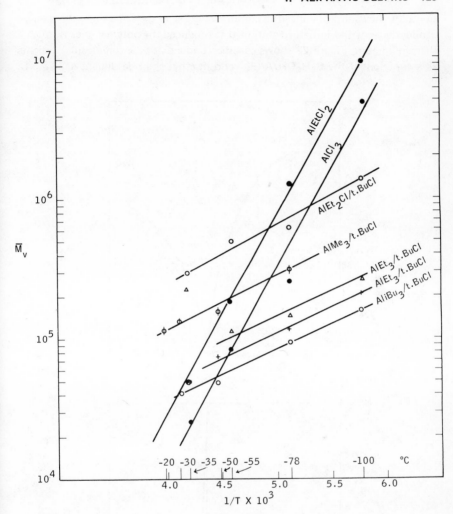

Figure 21. Effect of temperature on the viscosity average molecular weight of poly-isobutylene obtained with various aluminum containing catalyst systems in methyl chloride diluent (179).

isobutylene-butadiene copolymers have been obtained from these systems under suitable conditions.

The major findings of the Czechoslovak group concerned the correlation between electrical conductivity of $AlCl_3$ initiator solutions and the associated molecular weights they produce. In a series of papers (140, 144, 166–170) they demonstrated that the polyisobutylene molecular weight versus $AlCl_3$ concentration plot shows a pronounced maximum whereas, at the same time, the specific conductivity of the initiator solution that produced the polymer goes through a sharp minimum. Figure 22 shows specific conductivities and molecular weights as a function of the $C_2H_5OH/AlCl_3$ ratio in ethyl chloride diluent at $-78.5°C$.

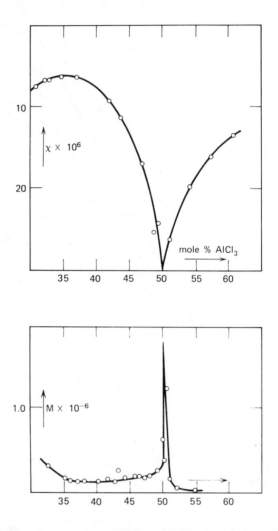

Figure 22. Relation between the specific conductivity χ and the molecular weight of polyisobutylene formed: (*top*) a section of the conductivity graph of the system ethanol/$AlCl_3$; (*bottom*) dependence of the molecular weight of the polymer on the ethanol/$AlCl_3$ ratio. Polymerization in ethyl chloride solutions at $-78.5°C$; the initial concentration of ethanol was 25×10^{-3} mol/l (169).

The highest molecular weight is obtained in the presence of equivalent amounts of $AlCl_3$ and C_2H_5OH, that is, where the conductivity is minimum. These findings have been extensively summarized by one of the original authors (180), criticized by Plesch (143, 181), and rebutted by the original authors (144).

Some aspects of the research by the Czechoslovakian group have been followed up by a group of Russian investigators who, probably while developing their own butyl polymerization process, have also carried out polymerizations with the isobutylene/$AlCl_3$/methyl chloride system (182, 183). The latter group was interested in elucidating the effect of dimethyl ether and ammonia on polymer yield and molecular weight. The experimental technique used was rather crude and heat control was particularly poor with experiments carried out with external cooling at $-78°C$. A better set of experiments was carried out under isothermal conditions with boiling (refluxing?) ethylene used as the internal coolant. The results indicates that dimethyl ether or ammonia severely depress both yields and molecular weights (182).

Plesch analyzed the shapes of the DP versus monomer or catalyst concentration curves exhibiting discontinuities (breaks, minima, or maxima) (143, 181). He discerned four types of curves: (A) the DP goes through a maximum, (B) the DP rises to a maximum constant value, (C) the DP falls to a minimum constant value, and (D) the DP goes through a minimum. According to Plesch, most of these DP effects may be artifacts and many could be due to the presence of one or more chain breaking impurities.

e. Complex Initiators

Another interesting development by Czechoslovakian investigators is the accidental discovery by Wichterle, Marek, and Trekoval of an unusual initiator system for the polymerization of isobutylene, an initiator that produces at high rates high molecular weight products at relatively high temperatures (184–186). The initiator is prepared by first saturating an aluminum alcoholate (preferably a secondary butyl alcoholate) in hexane solution with BF_3 and activating the solid precipitate formed with $TiCl_4$. Aspects of this work have been surveyed and commented upon by Plesch (87) and by Kennedy and Langer (41). The basic findings have been repeated and verified by a team of Japanese and American investigators (187). The latter team found that the level of molecular weights can be further increased by the use of freshly prepared solid initiator. A comparison of viscosity average molecular weights obtained by the original investigators and the latter team, together with some data obtained by this writer with a conventional $AlCl_3$ initiator at $-10°C$ in n-hexane solution, is shown in Table 8. It was shown conclusively that growth occurred on the surface of the catalyst and not in the homogeneous phase (184).

TABLE 8 COMPARISION OF $\bar{M}_v$ OF POLYSOBUTYLENE OBTAINED AT $-20°C$

	$\bar{M}_v$	Reference
AlCl$_3$ (dissolved in ethyl chloride)	~30,000	
Al(OsecBu)$_3$/BF$_3$/TiCl$_4$ (in n-hexane, probably aged)	180,000	184
Al(OsecBu)$_3$/BF$_3$/TiCl$_4$ (fresh)	264,000	187
MgCl$_2$ (ball-milled anhydrous, in n-hexane)	~250,000	197
γ-Ray induced polymerization, in bulk	~290,000	136

To explain high molecular weights it was assumed (187) that chain breaking involves a "solid phase" counterion and that it is difficult for steric reasons to reach the transition state with the counterion in the solid phase. Diminished chain transfer would lead to high molecular weight product.

This interesting lead has been followed up by Japanese investigators in a series of five papers (188–192). The rates of polymerization of isobutylene and molecular weights of polyisobutylenes obtained with the Al(OsecBu)$_3$/BF$_3$/TiCl$_4$ system were studied. The Ti/Al ratio influenced the rates (highest rates were obtained at Ti/Al ~0.3), whereas molecular weights were a function, among other things, of conversion (highest molecular weight produced at 30–40% conversion). Very broad molecular weight distributions have been obtained that were attributed to the long lifetimes of multiple types of propagating sites (190). Attempts have been made to use the same solid catalyst system with styrene and α-methylstyrene (191) and to copolymerize isobutylene with styrene (192); however, only low molecular weight products have been obtained in these studies.

The interesting initiator research of Wichterle et al. has been further developed by Japanese industrial researchers working for the Nippon Oil Company. Thus a series of patents have appeared (193–196) that describe initiator systems very similar to those first investigated by Wichterle, Marek, and Trekoval. For example, a binary system, consisting of iron alkoxides and boron trifluoride, produces unusually high molecular weight polyisobutylenes at relatively modest temperatures. Specifically, by using a mixture of Fe(OnBu)$_3$ and BF$_3$, isobutylene has been polymerized at -10 and $-60°C$ and the molecular weights ($\bar{M}_v$) obtained were $\sim 151,000$ and $1,303,000$, respectively. The initiator system was prepared by absorbing 2 moles of BF$_3$ per mole of Fe(OnBu)$_5$ in benzene solution. In a later patent (194), a variety of alkoxides and alkoxi halides in conjunction with BF$_3$ have been shown to produce very high molecular weight polyisobutylenes and isobutylene copolymers. Some representative systems and conditions are compiled in Table 9. Polymerizations were carried out for ~1 hour and the yields were usually very high (70–90%).

TABLE 9 POLYMERIZATION OF ISOBUTYLENE WITH NIPPON OIL CO. INITIATORS (195)

Initiator System (ratio)	Solvent	Temperature	$\bar{M}_v \times 10^{-3}$
Al(OsecBu)$_2$Cl/BF$_3$ (1:3)	nC_6	-45	1100
Al(OsecBu)$_2$Cl/BF$_3$ (1:4)	nC_6	-75	2400
Al(OiPr)$_2$(OsecBu)BF$_3$	MeCl	-45	400
Ti(OnBu)Cl$_3$/BF$_3$ (3:4)	nC_6	-45	1250
Ti(OiPr)$_2$(OnBu)Cl/BF$_3$ (1:2)	MeCl	-25	510
Ti(OnBu)$_3$Br/BF$_3$	nC_6	-25	200
Al(O C$_6$H$_4$CH$_3$)$_3$/BF$_3$ (1:5)	nC_6	-50	600
Ti(O CH$_2$C$_6$H$_5$)$_4$/BF$_3$	nBuCl	-45	200
Al(OsecBu)$_3$/BF$_3$ (1:3)	nC_6	-45	109
Al(OsecBu)$_3$/BF$_3$ (1:3)	nC_6	-20	258

The Nippon Oil researchers emphasize the importance of the addition sequence of the catalyst ingredients. Best results are obtained when the BF$_3$ component is added last to the quiescent isobutylene/metal alkoxide system. An alternate procedure is to add simultaneously BF$_3$ and metal alkoxide ingredients to the monomer charge. They further point out that the reaction product formed on mixing BF$_3$ and Al(OsecBu)$_3$, a green gel first described by Wichterle et al. (185), is completely inactive as initiator and must be activated by the addition of TiCl$_4$. Evidently it is important to produce *in situ* the active initiator, whatever its nature.

The Nippon Oil Company initiator seems to produce the highest molecular weight polyisobutylenes in the temperature range down to $\sim-45°$C.

Some unusual initiator systems that produce high molecular weight polyisobutylenes at relatively high temperatures have recently been described in two German patents (196a, 196b) by Japanese investigators at the Sumitomo Chemicals Co. In the first case (196a) the initiator was a binary mixture of BF$_3$ and (RO)$_2$=Al—O—Zn—O—Al=(OR)$_2$. The latter compound is readily produced by

$$Zn(OCOCH_3)_2 + 2Al(OisoPr)_3 \rightarrow Zn[OAl(OisoPr)_2]_2 + CH_3COOisoPr$$

and is soluble in aliphatic and/or aromatic hydrocarbons and chlorinated solvents. Besides Zn, the central metal can be Mn, Be, Cd, and so on. The polymerization is carried out by dissolving the alkoxide and isobutylene in n-heptane at $-65°$C and introducing BF$_3$ into the upper part of the stirred reactor. After 1 hr the reaction is quenched with ethanol. Under these conditions (150 ml n-heptane, 33.7 g isobutylene, 0.5 mM alkoxide, 5.3 mM BF$_3$, 1 hr at

$-65°C$) the yield is 29.5 g of polymer with $\bar{M}_v = 1,210,000$. The control experiment without the alkoxide yields 23.1 g product of $\bar{M}_v = 95,000$. Copolymerization of isobutylene and isoprene (2 wt. %), under similar conditions at $-75°C$ gave 28.1 g product of $\bar{M}_v = 720,000$ and 1.15 mole % unsaturation.

The other similar patent (196b) discloses a catalyst comprising a mixture of BF_3 and $R_2N-Zn-NR_2$ or similar compounds in which the central metal atom is Al, Si, Ti or Zr. A polyisobutylene polymerization (33.7 g isobutylene, 150 ml n-heptane, 0.5 mM $Zn(NEt_2)_2$, 4.5 mM BF_3, 1 hr at $-65°C$) yielded 33.0 g with $\bar{M}_v = 1,100,000$.

The two dots in Figure 23 indicate the best molecular weights obtained at $-65°C$ with these uncommon initiators.

On the basis of the available scanty information it is impossible to discuss meaningfully the mechanism of initiation with these complex systems. However, it might be of interest that both systems described in these patents are claimed to give homogeneous initiation.

In this context research by British investigators associated with the ESSO group should be mentioned. Addecott et al. (197) found that finely divided (ball-milled) solid anhydrous $MgCl_2$ produces very high molecular weight polyisobutylene. The principal disadvantage of the $MgCl_2$ initiator is the low rate of polymerization at even moderately low temperatures. For example, after 1–3 hr of polymerization of 100 ml isobutylene in 140 ml n-heptane with 0.1 g $MgCl_2$ at -10 and $-30°C$, the conversions are only ~30 and $\sim12\%$, respectively.

Initiation with solid, substantially anhydrous $MgCl_2$ was suggested to occur only in the presence of traces of moisture:

$$(MgCl_2)_n \text{ (solid)} + H_2O \rightarrow (MgCl_2)_n OH^\ominus \text{(solid)} H^\oplus$$

$$\xrightarrow{+ iC_4^=} (MgCl_2)_n OH^\ominus \text{(solid)} C^\oplus(CH_3)_3$$

and propagation would involve ion pairs.

The molecular weights obtained with ball-milled anhydrous $MgCl_2$ in the temperature range $-20°$ to $-70°C$ are included in Figure 23. In view of these high values, Addecott et al. claimed (197) that "the outstanding feature of $MgCl_2$ catalyst for isobutene polymerization is its ability to produce higher molecular weight polymer under comparable conditions than any other catalyst system described in the literature." In view of the research carried out at the University of Kyoto that demonstrated the synthesis of very high molecular weight polyisobutylene with freshly prepared mixtures of $Al(OsecC_4H_9)_3/BF_3/TiCl_4$ (187), the above statement may not be true; however, these two reports appeared just about the same time so that Addecott and coworkers were unaware of the latter. With the disclosure of the initiators developed by the Nippon Oil Company (see above) Addecott's claim for the highest molecular

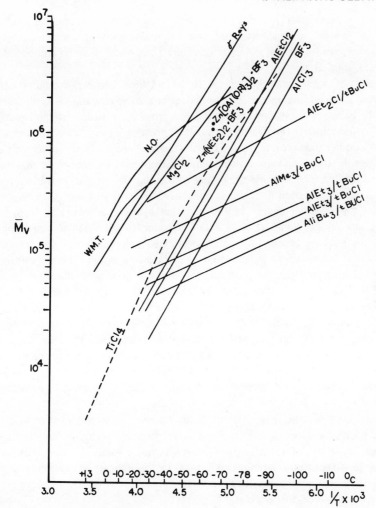

Figure 23. The effect of temperature on the molecular weight ($\overline{M}_v$) of polyisobutylenes obtained by various initiator systems.

weight polyisobutylenes obtained by chemical catalysis became obsolete.

It appears that the common feature of the Wichterle-Marek-Trekoval system (e.g., Al(OsecBu)$_3$/BF$_3$/TiCl$_4$) the Nippon Oil System (e.g., *in situ* prepared Al(OsecBu)$_3$/(BF$_3$) the Sumitomo Chemicals systems (e.g., Zn[OAl(OR)$_2$]$_2$/BF$_3$ and Zn(NR$_2$)$_2$/BF$_3$) and the anhydrous ball-milled MgCl$_2$ system is their complex nature. Of these four complex initiator systems, the first three are definitely solids and only one, the Sumitomo system, appears to be homogeneous. It has been proposed that polymerizations that lead to highest molecular weight polyisobutylenes may involve growing carbenium ions electroneutralized by large, complex counteranions which may be embedded in a separate, solid phase.

In spite of superficial differences between these complex initiators and irradiation induced homogeneous polymerizations, there may exist a fundamental similarity between these systems, that arises from the position of the negative charge (counteranion) relative to the positive growing site in these polymerizations. Assuming that growth occurs by positively charged electrophilic entities, somewhere in these systems negative charges must also be present to maintain electroneutrality. The site and the nature of the negative entities in the complex chemical and irradiation induced systems are far from obvious. Conceivably, however, the negative charge in the complex systems is separated from the growing site and is stabilized by being embedded in colloidal or heterophase species; the negative charge in irradiation induced polymerizations is a solvated electron that is kinetically independent of the growing cation. There is evidence that the propagating entities in irradiation induced polymerizations are "free" cations that exist outside the sphere of influence of the negative species (136, 137). The common denominator between the complex chemical initiators and irradiation induced polymerization is that in either system the counteranion is far removed from the vicinity of the propagating sites which therefore become highly reactive or "free." In addition to enhanced reactivity (faster propagation), free carbenium ions are not assisted by the counteranion in transfer processes. Enhanced reactivity and diminished transfer result in increased molecular weights.

It is conceivable that two kinds of chain transfer to monomer mechanisms exist: a spontaneous or direct one, and a counterion assisted or indirect one. In direct chain transfer the proton is transferred directly to the monomer nucleophile without counteranion assistance. In the indirect or assisted process, the counteranion is the nucleophile that "helps" transfer by lowering the activation energy of the process:

Direct transfer:

$$\sim \underset{\underset{H}{|}}{C}-C^{\oplus} + C{=}C \;\rightarrow\; \sim C{=}C + HC{-}C^{\oplus}$$

Indirect transfer:

$$\sim \underset{\underset{H}{|}}{C}-C^{\oplus} + G^{\ominus} \;\rightarrow\; \sim C{=}C + H^{\oplus}G^{\ominus}$$

$$H^{\oplus}G^{\ominus} + C{=}C \;\rightarrow\; HC{-}C^{\oplus} + G^{\ominus}$$

The counterion assisted mechanism is impeded or completely eliminated with complex initiators because the route to the transition state for transfer is inhibited due to unfavorable steric requirements involving complex counterions, and with irradiation induced polymerization because no suitable counterion is available. Thus the molecular weights of polyisobutylenes obtained by irradiation induced polymerization or by complex chemical initiators may be the highest molecular weights one can obtain by cationic mechanism in the liquid phase from isobutylene.

f. Initiation by Charge Transfer Complexes

In a pioneering report Yamada and coworkers (198, 199) described the polymerization of isobutylene by $VOCl_3$/naphthalene and similar initiating systems. Certain observations led the authors to the conclusion that the polymerization may proceed by a new mechanism involving initiation by charge transfer. The key observation was that $VOCl_3$ did not initiate polymerization of isobutylene, although initiation occurred readily with $VOCl_3$ in conjunction with some aromatic compounds, for example, naphthalene, amines, ethers, acids, and phenols. Representative data are shown in Table 10.

TABLE 10   POLYMERIZATION OF ISOBUTYLENE WITH $VOCl_3$/ACTIVATOR IN n-HEPTANE (198)

Activator	Activator/$VOCl_3$ (moles)	Temperature (°C)	Time (hr)	Yield (%)	$[\eta]$	$\bar{M}_v \times 10^{-3}$
Naphthalene	1.0	0	20	98.7	—	
Naphthalene	0.5	0	20	96.0	0.15	13.5
Naphthalene	0.05	0	20	81.4	0.46	78.0
Naphthalene	0.5	−78	40	93.5	0.42	67.7
Naphthalene	0.5	−78	40	75	0.65	134
Anthracene CCl_4	0.2	0	2	94.5	Rubbery polymer	
Fluorene	0.1	−30	10	77.7	Rubbery polymer	
Phenanthrene	0.1	0	10	85.7	Rubbery polymer	
β-Naphthylamine	0.2	0	12	96.4	Rubbery polymer	
Aniline	0.5	−78	20	74.0	Rubbery polymer	
CS_2	1.0	−78	20	31.1	Rubbery polymer	
Phenol	0.2	0	20	51.0	Rubbery polymer	

According to the available data the rates are slow and the molecular weights are low. While the discovery at the moment does not appear to be ready for industrial oriented exploration, it has considerable fundamental significance and certain aspects could lead to important practical advances. From the fundamental point of view, this report may be the first, claiming isobutylene homopolymerization to respectable molecular weights by a new "nonclassical" mechanism. The authors proposed a mechanism in their brief paper (198) (a somewhat different proposal is made in reference 199):

$$VOCl + C_{10}H_8 \ \rightleftharpoons \ [CT \ complex] \ \rightleftharpoons \ VOCl_3^{\cdot\ominus} + C_{10}H_8^{\cdot\oplus}$$

$$C_{10}H_8^{\cdot\oplus} + CH_2{=}C(CH_3)_2 \ \rightarrow \ C_{10}H_8 + \dot{C}H_2{-}\overset{\oplus}{C}(CH_3)_2$$

$$2\dot{C}H_2{-}\overset{\oplus}{C}(CH_3)_2 \ \rightarrow \ (CH_3)_2\overset{\oplus}{C}{-}CH_2{-}CH_2{-}\overset{\oplus}{C}(CH_3)_2$$

$${+}(CH_2{-}\overset{\oplus}{C}(CH_3)_2)_2 \ \xrightarrow{\ +\ M\ } \ polymer$$

This proposal, which is based on the initiation mechanism worked out for anionic charge transfer systems, for example, $Na^{\oplus}C_{10}H_8^{\cdot\ominus}$, can be used as a starting point for further academic experiments. The key experiment, which apparently has not yet been performed, would be to add styrene to an iso-butylene system when the dark red color disappears and to assay the presence of a block copolymer of polyisobutylene–polystyrene. A corollary to this experiment would be to add fresh isobutylene to the isobutylene system at the fade-out point and to determine the molecular weight of the polyisobutylene formed in the second part of this experiment. If either of these experiments turned out to be positive, this could be construed as evidence for "living" carbenium ion polymerization and a further most thorough exploration of this area would be warranted. Living carbenium ion polymerizations have not yet been described and such an accomplishment would begin a new chapter in carbenium ion polymerizations.

g. A Comparitive Summary of Molecular Weights

Many authors have investigated the effect of temperature on the molecular weight of polyisobutylene, and a large amount of this information has been assembled in Figures 14, 17, 21, and 23 which show the logarithm of molecular weights as a function of the reciprocal temperature (Arrhenius plots). There is always an inverse temperature effect on the molecular weight of polyisobutylene. These plots, particularly those in Figures 21 and 23, reveal important differences in detail and contain important clues as to the mechanism of isobutylene polymerization. Figure 23 is a compilation of available log $\bar{M}_v$ versus $1/T$ data obtained with important representative initiating systems. The most obvious difference among these initiator systems is in the slopes of these linear plots, particularly that which exists between the family of lines associated with the "conventional" Friedel-Crafts halides AlCl$_3$ or BF$_3$ ($E_{\overline{M}_v} \approx 6.6 \pm 1.0$ kcal/mole) and the other family of lines characterizing the Et$_2$AlCl/tBuCl and R$_3$Al/tBuCl systems ($E_{\overline{M}_v} \approx 1.7 \pm 0.5$ kcal/mole). The dotted line in Figure 23 also shows the recalculated data from Plesch's work in reference 106 obtained with TiCl$_4$. (The curve was constructed by averaging Plesch's "viscosity average DP" and multiplying by 56.) While this plot is slightly curved to best fit the data points, with only a minor adjustment the plot could have been a straight line whose slope was identical to the lines obtained with the other conventional Lewis acid systems, for example, AlCl$_3$, BF$_3$, and EtAlCl$_2$. The MgCl$_2$ line and the γ-ray line refer to the data in references 197 and 136, respectively. The lines labeled W.M.T. and N.O. signify the highest molecular weights reported by Wichterle, Marek, and Trekoval (185) and those in the patents assigned to Nippon Oil (194), respectively. The two dots indicate the best values obtained at $-65°C$ by Sumitomo Chemical's scientists (196a, 196b). The composition of the mixed initiators described in the last three sources are too complex for a meaningful discussion of the relation between the nature of the initiator and polymerization mechanism.

The log $\overline{M}_v$ versus $1/T$ plots are diagnostic and seem to define at least two large families of initiators, one, to which most "conventional" Friedel-Crafts halides [AlCl$_3$, BF$_3$, EtAlCl$_2$, TiCl$_4$] (Family A) belong and a second that comprises the recently discovered alkylaluminum systems (Family B).

According to the data in Figures 21 and 23, the temperature coefficient of molecular weights is, for some reason, larger with Family A than with Family B. Within these two families of initiator systems, however, important differences exist as indicated by the number of parallel lines in Figures 21 and 23. For example, while the slopes of the Arrhenius lines are identical (within what is considered to be experimental variation) for the Et$_2$AlCl/tBuCl, Me$_3$Al/tBuCl, and Et$_3$Al/tBuCl systems, the absolute molecular weights produced by these initiators are quite different, the highest values being produced at the same temperature by the Et$_2$AlCl/tBuCl system. Most likely the mechanism of the most important molecular weight determining step, that is, transfer to monomer, is different in these two families. On the other hand, similarity of slopes within one family indicates a very similar chain breaking mechanism within that family.

While the slopes of the Arrhenius lines are similar within one family, their intercepts may be quite different. Similarity in the slopes of the Arrhenius lines but dissimilarity in the intercepts indicates a similarity in the energetic term (the counterion affects $E_{tr,m}$ and E_p by the same amount), but dissimilarity in the steric course of the mechanism (different $A_{tr,m}/A_p$ values).

In summary then, the difference in the slopes of the Arrhenius lines may be due to the different chain breaking mechanisms operating in the two families of initiator systems. Differences in molecular weights within one family of initiators are probably due to different steric factors operating within that family of growing cation–counterion pairs. Very tentatively it is proposed that the difference in slopes of the Arrhenius lines is due to the underlying difference in the "ionicity" of the growing ion pairs. Initiators that belong to Family A, that is, have a higher Arrhenius slope, produce counterions of very low nucleophilicity and probably give rise to polymerizations by relatively free ions. This statement is based on the fact that polymers obtained by γ-ray induced polymerization belong to this family. In contrast, initiators in Family B, which have a lower Arrhenius slope, give rise to counterions with somewhat higher nucleophilicity and probably propagate by relatively tight ion pairs.

h. Epilogue

For completeness sake, a few pathetic publications in the scientific literature dealing with aspects of isobutylene polymerizations should be noted and weeded out. Hamada and Gary (199a) claim to have polymerized isobutylene by a typical Ziegler-Natta system consisting of $Et_3Al/TiCl_3$ (Al/Ti = 8.88) in n-heptane at $0°C$. They fractionated their product into a heptane soluble and heptane insoluble fraction(!), determined yields and molecular weights, and proceeded to study kinetic parameters, that is, individual and overall rate constants, activation energies, and so on. A close examination of the experimental conditions and product characteristics given in this publication, however, convinced this writer that Hamada and Gary used faulty synthesis conditions and could not possibly have polymerized isobutylene, but in fact have produced polyethylene. It is proposed that under their conditions the following displacement took place:

$$Et_3Al + 3CH_2=C(CH_3)_2 \rightarrow iBu_3Al + 3CH_2=CH_2$$

and that the ethylene formed in this exchange was immediately converted to polyethylene by Ziegler-Natta coordinated anionic polymerization, that is, by the $iBu_3Al/TiCl_3$ system.

Other peculiar publications are by Krentsel and coworkers (199b, 199c) who describe the polymerization of isobutylene by some Ziegler-Natta catalysts to a 1,3 structure:

$$\underset{\underset{C}{|}}{\overset{\overset{C}{|}}{C=C}} \rightarrow -C-C-C-C-C-C-$$

Subsequent investigators (199d) failed to reproduce this work and this writer also disagrees with the conclusions of Krentsel et al. (41).

$$\textit{2. 2-Methyl-1-butene}\quad CH_2=\underset{\underset{}{\overset{\overset{CH_3}{|}}{}}}{C}-CH_2-CH_3$$

While the polymerization of isobutylene has been extensively explored and exploited, there is a dearth of information on the polymerization behavior of other 1,1-disubstituted ethylenes.

Early investigators obtained low molecular weight liquids when 2-methyl-1-butene was treated in the absence of solvent with solid $AlCl_3$ at $-80°C$ (57) or by adding BF_3 to a solution of this olefin in propane diluent at $-78°C$ (200). Olah et al. (50) mention that liquid polymers are formed when this olefin is contacted with various complexes such as $HF \cdot BF_3$, $DF \cdot BF_3$, $RF \cdot BF_3$, and $AgBF_3 \cdot RCl$ (where R = isopropyl, t-butyl, and cyclohexyl) at $-80°C$.

High molecular weight polymers (DP = 300–900) of 2-methyl-1-butene were obtained by Dutch researchers (201) who employed $AlCl_3$ or $AlBr_3$ in ethyl chloride diluent at $\sim-140°C$. The highest molecular weight products DP ~ 4600 ($\bar{M}_n \sim 322,000$) were obtained at $-175°C$.

A very high molecular weight polymer ($\bar{M}_v \sim 300,000$ by using the equation developed for isobutylene) obtained at $-45°C$ was described in a recent patent assigned to the Nippon Oil Company (195). The initiator consisted of a mixture of $Al(O i Pr)_2 (O sec Bu)$ and BF_3 and the solvent was methyl chloride.

The physical properties of these high molecular weight products have not been described.

2-Methyl-1-butene has recently been polymerized by $TiCl_4$ in methylene chloride solvent at $-78°C$ (202). Polymerization was accompanied by isomerization to 2-methyl-2-butene and consequently a complex, conjunct polymerization was obtained. The methanol insoluble fraction of the product was of very low molecular weight ($\eta = 0.03$ dl/g in toluene at $30°C$).

$$\overset{\displaystyle CH_3}{\underset{\displaystyle |}{}}$$

3. *2-Methyl-2-butene* $CH_3-C=CH-CH_3$

This internal olefin has been oligomerized with $TiCl_4$ in CH_2Cl_2 solution at $-78°C$ (202). Very high initiator concentrations had to be used and only methanol soluble products were obtained (mixture of dimers plus oligomers with a DP of up to 5).

$$\overset{\displaystyle CH_3}{\underset{\displaystyle |}{}}$$

4. *2-Methyl-1-pentene* $CH_2=C-CH_2-CH_2-CH_3$

Van Lohuizen and De Vries (201) polymerized 2-methyl-1-pentene to high molecular weight solids with $AlCl_3$ or $AlBr_3$ in ethyl chloride diluent at very low temperatures. Thus with $AlCl_3$ in ethyl chloride at $-140°C$ DPs of 300–900 were obtained whereas in mixtures of ethyl chloride and vinyl chloride at $-175°C$ the DP rose to ~4000 ($\bar{M}_n \sim 336,000$). There was some indication that HCl may be a promoter in these low temperature experiments (the author's claim that HCl may be an "essential cocatalyst" cannot be accepted on the basis of the available data). The physical properties of the product have not been described.

5. 2-Ethyl-1-butene

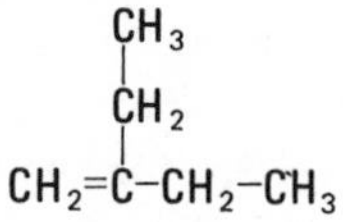

$$CH_2=C-CH_2-CH_3$$
with CH_3 and CH_2 substituents on the carbon

Dutch researchers (201) mention that 2-ethyl-1-butene gives liquid oligomers (mainly dimers as judged from the boiling point and the isomer 2-ethyl-2-butene) in low yield when the monomer is contacted with $AlCl_3$ in ethyl chloride–vinyl chloride mixtures at $-175°C$. The nonpolymerizability of this monomer is considered to be another indication that carbenium ions carrying larger than methyl substituents are unable to sustain effective propagation because of steric compression in the transition state.

6. 2,3-Dimethyl-1-butene

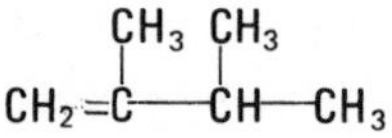

$$CH_2=C-CH-CH_3$$
with CH_3 and CH_3 substituents

Von Lohuizen and De Vries (201) investigated the polymerization of 2,3-dimethyl-1-butene with $AlCl_3$ in ethyl chloride diluent or ethyl chloride/vinyl chloride mixtures in the range -110 to $-175°C$ and obtained respectable molecular weight products, for example, $[\eta] \sim 0.6$–0.68, corresponding to DP ~ 1800–2100 ($\bar{M}_n \sim 150{,}000$–$175{,}000$).

Interestingly, spectroscopic structure characterization indicated a rearranged repeat unit and suggested intramolecular hydride shift isomerization-polymerization:

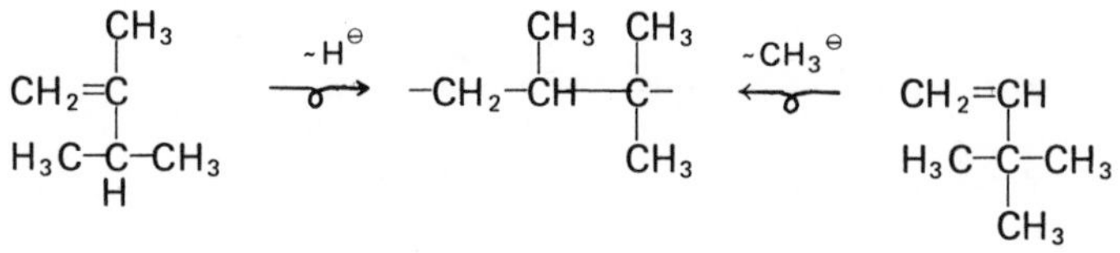

The NMR and IR spectra of this polymer and that obtained from 3,3-dimethyl-1-butene by a similar technique (75) were practically identical. The physical properties of this material obtained from 2,3-dimethyl-1-butene have not been described.

7. 2,3,3-Trimethyl-1-butene

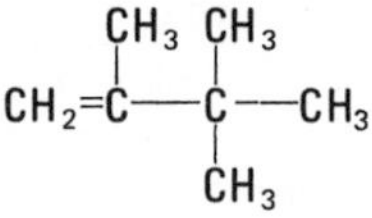

$$CH_2=C-C-CH_3$$
with CH_3, CH_3, and CH_3 substituents

Dutch authors mention (201) that 2,3,3-trimethyl-1-butene yields only a small amount of oil when treated with $AlCl_3$ in ethyl chloride solution at $-140°C$. The nonpolymerizability of this olefin is of interest since methide migration would be possible.

8. 2,4,4-Trimethyl-1-pentene

$$CH_2{=}\overset{\overset{\displaystyle CH_3}{|}}{C}{-}CH_2{-}\overset{\overset{\displaystyle CH_3}{|}}{\underset{\underset{\displaystyle CH_3}{|}}{C}}{-}CH_3$$

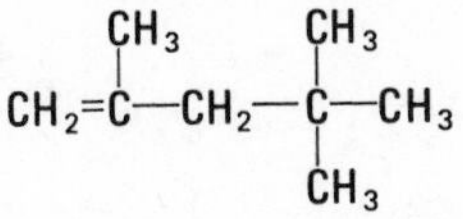

Dutch authors (201) mentioned that 2,4,4-trimethyl-1-pentene yields an oil when contacted with $AlCl_3$ in ethyl chloride/vinyl chloride mixtures at $-175°C$. The polymerization of this olefin is obviously hopeless in view of steric hindrance. The dimerization of this material, however, has been investigated in cationic model polymerization studies by Kennedy and Gillham (202a).

9. 2,5-Dimethyl-1,5-hexadiene

$$CH_2{=}\overset{\overset{\displaystyle CH_3}{|}}{C}{-}CH_2{-}CH_2{-}\overset{\overset{\displaystyle CH_3}{|}}{C}{=}CH_2$$

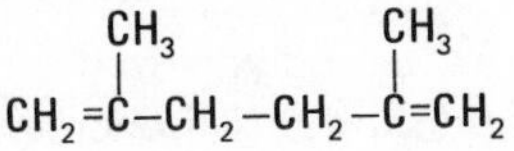

The straight chain dienes 1,5-hexadiene, 1,6-heptadiene, and so on are expected to be cationic monomers, however, no published literature is available on these systems. While the parent compounds have apparently not been studied, their alkyl and aryl derivatives have given interesting, fairly well defined cyclic structures.

The cationic polymerization of 2,5-dimethyl-1,5-hexadiene was studied by Marek and coworkers (203, 204) who used BF_3, $AlBr_3$, and $TiCl_4$ in bulk and various solvents in the range of +20 to $-78°C$. All the polymers were soluble in some common solvents. On the basis of melting point, hydrogenation analysis, and infrared and NMR spectroscopy they concluded that the major structure of the polymer was formed by intra-intermolecular propagation:

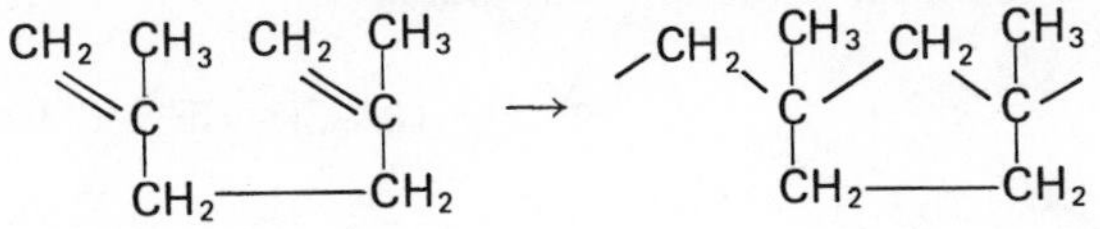

The reason why these branched derivatives of the parent straight chain dienes can be polymerized to more-or-less clean structures and why the parent dienes have not been investigated is that the $-C_6H_5$ or $-CH_3$ substituents prevent isomerization during polymerization and, provided a five- or six-membered cyclic structure can be formed, they direct intramolecular propagation toward clean cyclization:

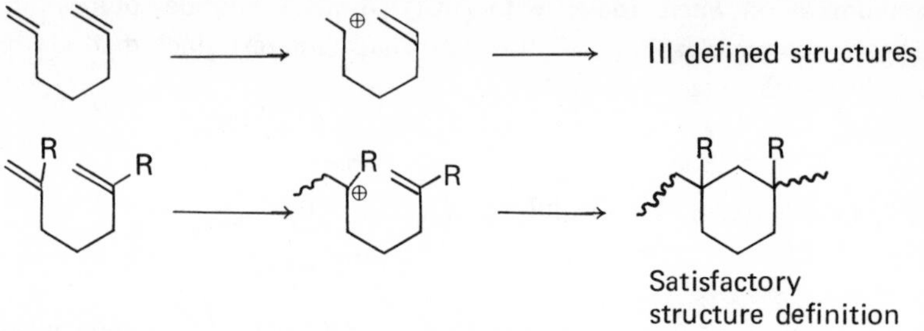

III defined structures

Satisfactory
structure definition

D.   ALICYCLIC OLEFINS, DIOLEFINS, AND DERIVATIVES

1. *Methylenecyclopropane*

Methylenecyclopropane can be polymerized to solid polymer by bubbling gaseous BF_3 into a solution of this monomer in methylene dichloride for 30 min at $-78°C$, yield $\sim$ 58% (205). The structure and physical properties of the polymer have not been determined.

2. *Vinylcyclopropane* $CH_2= CH$

In spite of the scarcity of this monomer, a considerable amount of research has been carried out with it, particularly by Japanese investigators (206–209). Takahashi and Yamashita (206) were the first to polymerize vinylcyclopropane (and its derivatives) in bulk with Lewis acids $AlBr_3$ and $SnCl_4$ at $-50°C$. These authors obtained polymers with molecular weights of 2100–2500. According to NMR and infrared spectroscopy the polymer contained two kinds of repeat units, a 1,2 and 1,5 enchainment. Under cationic conditions the major component was the conventional 1,2 structure, however, with free radicals a 1,5 repeat structure was obtained:

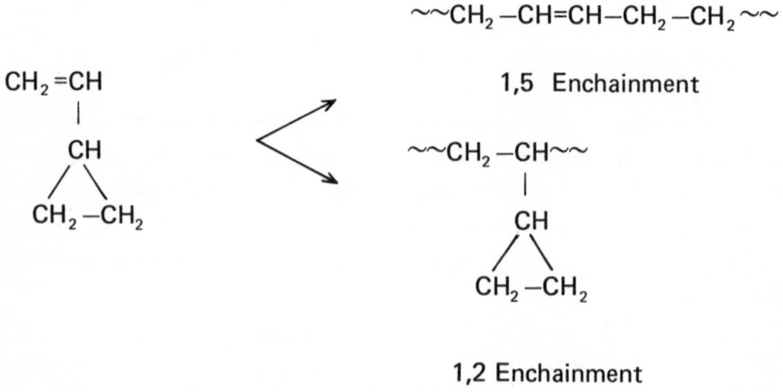

$$\sim\sim CH_2-CH=CH-CH_2-CH_2\sim\sim$$

1,5 Enchainment

$$\sim\sim CH_2-CH\sim\sim$$

1,2 Enchainment

Ketley et al. (210) used AlBr$_3$ in ethyl chloride at $-78°$C to polymerize this monomer. According to these authors the structure of the polymer is a combination of three basic repeat units. On the basis of NMR and infrared data they were able to assign quantitatively the contributions of the following repeat units (210):

$$-CH_2-\underset{\underset{\triangle}{|}}{CH}\raise2pt\hbox{\tiny$\sim\sim$}\qquad \raise2pt\hbox{\tiny$\sim\sim$}CH_2-\underset{\underset{CH_2}{|}}{CH}-\underset{\underset{CH_2}{|}}{CH}\raise2pt\hbox{\tiny$\sim\sim$}\qquad \raise2pt\hbox{\tiny$\sim\sim$}CH_2-CH=CH-CH_2-CH_2\raise2pt\hbox{\tiny$\sim\sim$}$$

60%   30%   10%

$$CH_2=\underset{\triangle}{\overset{\overset{CH_3}{|}}{C}}$$

3. *Isoprenylcyclopropane*

According to a team of Russian scientists (211) who investigated the cationic polymerization of isopropenylcyclopropane with SnCl$_4$ under atmospheric pressure and 14,000 kg/cm^2 pressure at $50°$C, the content of cyclopropane ring containing units was ~25 and ~50%, respectively. Infrared and NMR spectroscopy indicated isomerization and the presence of

$$-CH_2-\overset{\overset{CH_3}{|}}{C}=CH-CH_2-CH_2-$$ units, with a mixture of *cis* and *trans* configurations.

4. *1,1-Dicyclopropylethylene* $$CH_2=\underset{\triangle}{\overset{\triangledown}{C}}$$

The polymerization of 1,1-dicyclopropylethylene has been investigated by Ketley et al. (212) and Kennedy et al. (213) and their major conclusions are identical. Ketley et al. used AlBr$_3$ in dry ethyl chloride at $-78°$C, whereas Kennedy et al. employed AlEtCl$_2$ in dry methyl chloride at -35 and $-100°$C. Both groups found that the cationic polymerization of this olefin yields a polymer whose structure consists of predominantly (70–75%) unrearranged repeat units:

$$-(CH_2-\underset{\triangle}{\overset{\nabla}{C}})-$$

and a lesser amount of rearranged enchainment:

$$-CH_2-\overset{\nabla}{C}=CH-CH_2-CH_2-.$$

It is of interest in this context that the free radical initiated polymerization of this material gives rise to an essentially pure (>90%) rearranged structure (213).

The cationically polymerized, that is, largely unrearranged, poly(1,1-dicyclopropylethylene) is a white powder (213). It is of interest that closely related polymers such as polyisobutylene and poly(2-methyl-1-butene) obtained under similar conditions are rubbers and soft solids, respectively. This difference in the physical properties is presumably due to the presence of relatively bulky geminal cyclopropyl groups in poly(1,1-dicyclopropylethylene) that render the backbone rigid; in contrast, the geminal methyl groups present in polyisobutylene are sufficiently small to allow the interconversion of the three possible conformations (trans and two skewed). This interconvertibility among the possible conformations results in a random coil, which is the basis of the rubbery nature of polyisobutylene.

5. Methylenecyclobutane

In an early publication Pinazzi and Brossas (214) described preliminary polymerization experiments with methylenecyclobutane. These authors employed $AlCl_3$, $TiCl_4$, $SnCl_4$, and BF_3 in various solvents at undisclosed temperatures and obtained some low molecular weight, soluble, colorless, solid oligomers ($\bar{M}_w$ = 1200–1500). Anionic and free radical initiators were found to be inactive. The following structure was proposed for the repeat unit of the products obtained by cationic initiation:

Upon evaporating benzene or chloroform solutions of the oligomer, the authors obtained pliable transparent films. Results of these investigations have recently been verified (215, 216) and structure analysis, mainly by NMR and IR, showed that the polymer is always a mixture of repeat units and that the relative amount of the various repeat units can be controlled to a certain extent by the choice of Lewis acid used. Recently these investigators (216a) used acidic Ziegler-Natta initiators $AlEt_2Cl/TiCl_4$, $\sim$1:1, and obtained oligomers DP=10–15 of mixed structures. Table 11 shows the effect of the Lewis acid on the possible repeat units proposed by the authors.

Dutch researchers (201) have also polymerized impure methylenecyclobutane with $AlCl_3$ in ethyl chloride at $-135°C$ and obtained a semisolid product of specific viscosity 0.05.

TABLE 11   THE EFFECT OF LEWIS ACIDS ON THE STRUCTURE OF THE POLYMER OBTAINED FROM METHYLENECYCLOBUTANE

Repeat Lewis Acid	Structure Obtained (Mole %)			Cyclized Units
$AlCl_3$	80	15	5	
$SnCl_4$	75	Present	—	Present
$TiCl_4$	60	—	—	Present
$ZrCl_4$	Present	Present	Present	Predominating
$Et_2AlCl \cdot H_2O$	60	—	—	Present

6 and 14.   Cyclopentene and Cyclohexene

In a fairly old, short review of his work, Hoffman mentions briefly (217) that cyclopentene and cyclohexene can be readily oligomerized with BF_3 and some HF to dimers, trimers, and tetramers as well as transparent amber solid resins. In contrast, Boor et al. (218) did not obtain polymer of these monomers using $AlCl_3$ in ethyl chloride solution at low temperatures ($-20°C$). An explanation for this contradiction is given below.

The fact that Boor et al. did not obtain polymer at low temperatures may be due to a rapid isomerization of the dimer cation leading to a nonpolymerizable "buried" ion:

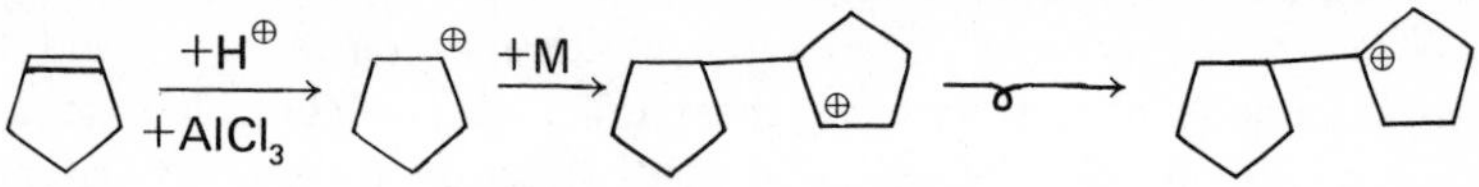

The double bond is reactive to protonation and/or cationation as indicated by the polymerizability of 3-methylcyclopentene (see there); however, in the parent compound propagation is impeded because of steric congestion. It is possible that at a temperature higher than −20°C, as probably used by Hoffman, the reaction continues by elimination and alkylation:

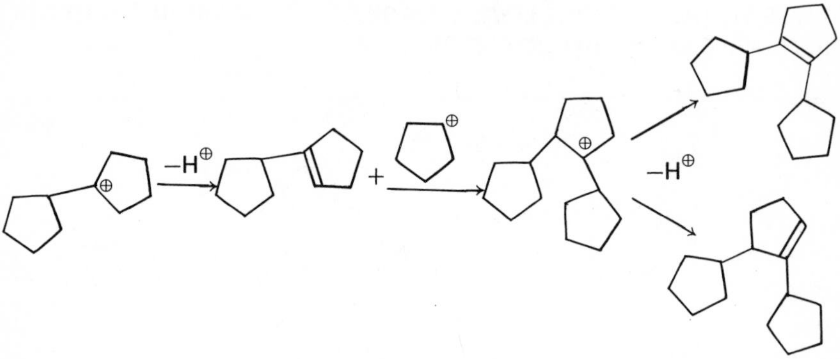

Furthermore, the allylic hydrogens in the trimer are good sources of hydride ions, and hydride transfer coupled with elimination would lead to a reactive diene and ultimately to an ill defined oligomeric polyalkylate.

7. 3-Methylcyclopentene

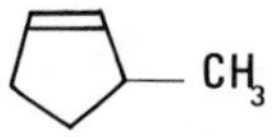

While the parent olefin, cyclopentene, is nonpolymerizable by conventional cationic techniques (218) or gives dimers at best (217), its 3-methyl derivative polymerizes slowly to give a polymer of the unusual repeat:

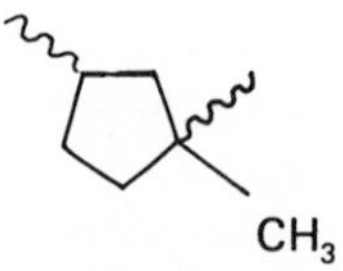

The polymerization is very slow and it is visualized to occur by an intramolecular hydride shift mechanism:

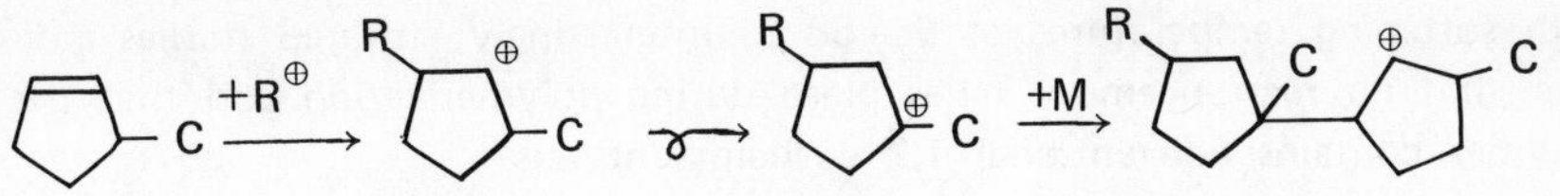

While only 0.2 g of polymer has been prepared, Boor et al. determined that the sample softened below 50°C and was completely soluble (218). The very low yield is probably due to an isomerization yielding a "buried" carbenium ion:

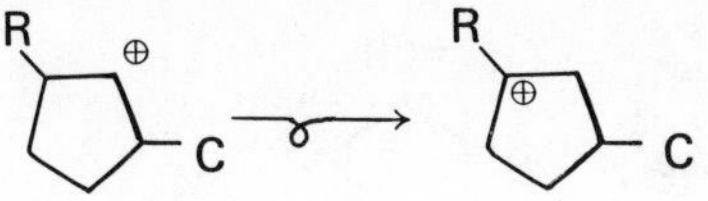

Since this hydride shift can occur at every propagation step, it is not surprising that the ultimate yields are very low.

8. 1-Methylcyclopentene

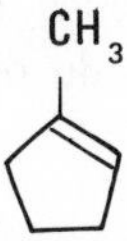

In a research paper (219) whose aim is quite divorced from cationic polymerizations, Schmitt and Schuerich describe high vacuum experiments in the course of which 1-methylcyclopentene was contacted with BF_3 and $BF_3 \cdot$camphor complexes in *n*-hexane and chloroform solvents at 0°C. Polymerization was observed in every instance and a viscous oily product was obtained.

9. Methylenecyclopentane

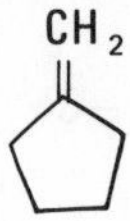

Methylenecyclopentane gave a high boiling liquid, probably trimer, upon treatment with $AlCl_3$ in ethyl chloride solution at −135°C (201).

10. Vinylcyclopentane

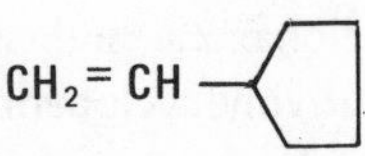

The polymerization of this monomer has been investigated by Ketley and Ehring

(84) in the course of their studies on isomerization polymerization. These authors found that polymerization by $AlBr_3$ in ethyl chloride in the range −78 to −100°C gives low molecular weight products $\bar{M}_n \sim 950$ (DP ~ 10), with softening temperatures at 55–58°C. Interestingly, infrared studies indicate that no rearrangement takes place during polymerization and that the polymer contains conventional 1,2 enchainments:

$$CH_2{=}CH \qquad \longrightarrow \qquad {\sim}{\sim}CH_2{-}CH{\sim}{\sim}$$

11. 3-Vinylcyclopentene

$$CH_2{=}CH$$

Ohara (220) polymerized 3-vinylcyclopentene with various alkylaluminum initiator systems, for example, $EtAlCl_2$, and obtained a partially soluble polymer. Structure studies indicate the presence of large amounts of bicyclic repeat units. Intramolecular hydride shifts do not seem to be involved in the polymerization mechanism.

12. Allylcyclopentane $CH_2{=}CH{-}CH_2{-}$

Allylcyclopentane was polymerized by Ketley and Ehring (84) with $AlBr_3$ in ethyl chloride solvent at −78°C and obtained a product with $M_n \sim 1300$ and a softening point of 78°C. On the basis of infrared data these authors concluded that the major repeat unit formed by a partial rearrangement:

$$CH_2{=}CH \qquad \longrightarrow \qquad {-}CH_2{-}CH_2{-}CH{-}$$
$$\;\;\;\;\;\;\;\;CH_2$$

In view of the facts that vinylcyclohexane and allylcyclohexane give completely rearranged structures and that vinylcyclopentane gives rise to a conventional enchainment, the above result is peculiar and should be verified.

13. *3-Allylcyclopentene*

$$CH_2 = CH$$
$$|$$
$$CH_2 -$$

According to Ohara (220) this monomer can be polymerized with cationic initiator systems $EtAlCl_2$, $EtAlCl_2/PhCH_2Cl$, and so on. The polymer is soluble in common organic solvents and the repeat unit may contain bicyclic structures such as:

Intramolecular hydride transfer does not take place during the polymerization.

15. *1-Methylcyclohexene*

$$CH_3$$

1-Methylcyclohexene when treated with $AlCl_3$ in benzene solution at $40-45°C$ yields a mixture of oligomers, mostly dimers (221).

16. *3-Methylcyclohexene*

While the parent cyclohexene is nonpolymerizable (218) (see comment under Cyclopentene and Cyclohexene) 3-methylcyclohexene has been polymerized very slowly with $AlCl_3$ in ethyl chloride solution at -20 and $-78°C$ (186). At $-20°C$ an oily product (mixture of dimers and trimers) was obtained, whereas at $-78°C$ a small amount (~0.3 g) of solid formed which, after fusion at $280°C$, developed birefringence at room temperature. After annealing and heating, the birefringence disappeared at $\sim250°C$. X-ray analysis showed two diffractions at 5.82 and 5.35 Å. Therefore this material could be described as semicrystalline. The sample was soluble and its density was 0.978.

Structure analysis of the solid polymer by NMR and IR spectroscopy indicated the repeat unit:

which is formed by an intramolecular hydride shift mechanism (see 3-methyl-cyclopentane).

17. Methylenecyclohexane

In the course of their studies on β-pinene (see there), Roberts and Day (221) have treated methylenecyclohexane with $AlCl_3$ in benzene solvent and obtained a small amount of oil.

Methylenecyclohexane (purity 99.99+) is readily polymerized by $AlCl_3$ or BF_3 in ethyl chloride/vinyl chloride mixed solvents in the range -150 to $-180°C$ to give moderate molecular weight products (intrinsic viscosities $\sim$0.1 to 0.4) (170). Efforts were made to increase the molecular weights by a technique discovered by Zlamal and Ambroz (167) that gives highest molecular weight polyisobutylenes, a close relative of methylenecyclohexane. This technique consists of adding equimolal amounts of diethyl ether to the $AlCl_3$ initiator solution (i.e., adding an ether until the specific conductivity of the $AlCl_3$ solution reaches a minimum value) and introducing the monomer to this solution. However, the molecular weights of polymethylenecyclohexane did not increase by the application of this technique.

The structure of the polymer is predominantly of conventional head-to-tail addition character (178, 201). Amorphous polymethylenecyclohexane is insoluble in methanol but is readily dissolved by carbon tetrachloride and other common solvents. Films cast from solution are opaque and brittle. The glass transition temperature is $\sim$100°C by DTA and dilatometry. Interestingly, a moderately crystalline polymer fraction (by x-rays) could be obtained by slow precipitation from benzene. This fraction was insoluble in carbon tetrachloride and its infrared spectrum was different from that of the amorphous modification. The crystalline melting point was $\sim$210°C.

18. 2-Methylmethylenecyclohexane

19. 3-Methylmethylenecyclohexane

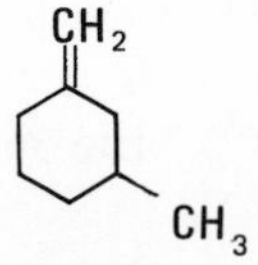

20. 4-Methylmethylenecyclohexane

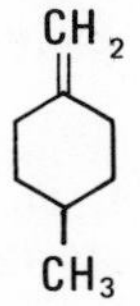

All three monomethylmethylenecyclohexanes can be polymerized by $AlCl_3$ in ethyl chloride/vinyl chloride mixed solvents at $-175°C$ to low molecular weight products (specific viscosities: $0.05 - 0.17$) (201). The poly(4-methylmethylenecyclohexane) crystallized even more readily than the parent compound and exhibited a crystalline melting point of $195°C$. Neither the 2-methyl nor the 3-methyl derivative could be obtained in a crystalline form. Spectroscopy did not suggest any abnormal structures (201, 222).

21. Vinylcyclohexane

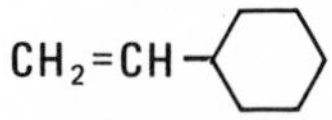

$CH_2 = CH -$

Structurally, this monomer is closely related to 3-methyl-1-butene and, in fact, its cationic polymerization proceeds in an analogous manner, by intramolecular hydride shift:

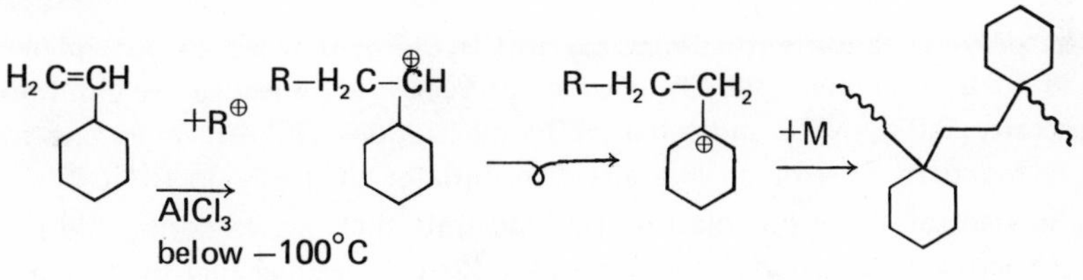

Kennedy et al. (223) showed that vinylcyclohexane can be readily polymerized with $AlCl_3$ in ethyl chloride diluent to amorphous (by x-rays) white powdery materials, the number average molecular weights of which were 2000–6500 depending on the polymerization temperature ($+7$ to $-100°C$). The softening range of the higher molecular weight sample was $112–116°C$, it was soluble in common solvents and on casting gave transparent, brittle films. The structure of the polymer was established by NMR and infrared spectroscopy.

Ketley and Ehring (84) came to similar conclusions on the basis of infrared spectroscopy. These authors used $AlBr_3$ in ethyl chloride and obtained $\bar{M}_n$ = 10,500 at $-50°C$. The material was amorphous by x-rays and its softening point was $110°C$.

22. 4-Vinylcyclohexene $\qquad$ $CH_2 = CH$—

In the course of their studies on the anionic coordinated polymerization of this monomer Marconi et al. (224) also mentioned briefly the cationic polymerizability of 4-vinylcyclohexene . Evidently acid initiators give rise to low molecular weight, amorphous polymers of ill defined structure.

Somewhat later Butler and Miles (225), in the course of their investigation on cyclopolymerization (interintramolecular polymerization, examined the polymerization behavior of this monomer. They used cationic, for example, BF_3, $BF_3 \cdot OEt_2$, and $TiCl_4$, and Ziegler-Natta initiators. The polymerization with BF_3 gas in CH_2Cl_2 diluent at $-70°C$ led to 28% conversion of product, of which 85% was soluble and had an intrinsic viscosity of 0.11. NMR spectroscopy indicated the presence of at least two repeat units in about equal proportions:

$-CH_2-CH-$ $\qquad$ and $\qquad$ $-CH_2-$

The polymerization with $BF_3 \cdot OEt_2$ in CH_2Cl_2 at $0°C$ led to 35% conversion and to a completely soluble product of very low molecular weight (intrinsic viscosity 0.03), NMR indicated ~80% cyclization. $TiCl_4$ in n-heptane solvent was found to be a much less effective initiator than BF_3 in CH_2Cl_2.

In view of the intramolecular hydride shift that occurs during the cationic polymerization of vinylcyclohexane (41, 42) it is peculiar that the formation of structures such as

$-CH_2-CH_2$

has not been considered.

23. *Allylcyclohexane* $CH_2=CH-CH_2-\langle\ \rangle$

Ketley and Ehring investigated the polymerization of this olefin (84) by using $AlBr_3$ in ethyl chloride solvent at $-78°C$. They obtained a solid of $\overline{M}_n \sim$ 13,100 and a softening point of $90°C$. Infrared studies indicated that the repeat structure of the polymer was derived mainly of rearranged units:

$$CH_2=CH \quad \longrightarrow \quad -CH_2-CH_2-CH_2-C-$$

24. *d-Limonene*

Roberts and Day (221) obtained oligomers of *d*-limonene in the course of their studies on α-pinene. These workers concluded that α-pinene, under the influence of Lewis acids for example, $AlCl_3$, rapidly isomerizes to *d*-limonene so that the oligomers they obtained from α-pinene (trimers) were in fact oligomers of limonene. The structure of the oligomers could not be established and the only statement that could be made was that the recurring unit could not have been formed by a simple vinyl-type polymerization of the limonene, that is,

because ozone absorption indicated only 0.4—0.5 double bonds per repeat unit. Marvel and coworkers (356a) who reinvestigated this chemistry somewhat later arrived at the same conclusions.

25. 1,4-Dimethylenecyclohexane

$$CH_2$$

$$CH_2$$

1,4-Dimethylenecyclohexane was polymerized by Ball and Harwood (226) with BF_3 in methylene chloride at $-80°C$ and up to room temperature. The polymer did not melt up to $250°C$, was soluble in aromatic solvents, and did not consume bromine. The authors proposed the following recurring unit:

$$\sim\sim H_2C$$

Transannular bond formation would be quite unexpected in this system in view of the distance between the two methylene groups in the monomer.

26. 1-Methylene-4-vinylcyclohexane

$$CH_2$$

$$HC=CH_2$$

1-Methylene-4-vinylcyclohexane was first prepared and polymerized by Butler and coworkers (227) who used among other initiators gaseous BF_3 in CH_2Cl_2 at $-70°C$. The polymer was a low molecular weight ($[\eta]$ = 0.08 in benzene) solid, soluble in various common solvents, and had a softening temperature of $150°C$. Structure analysis by IR and NMR spectroscopy suggested the presence of an approximately alternating copolymer of bicyclic and monocyclic units with the monocyclic units predominating:

$$CH_2 \qquad and \qquad CH_2$$

$$CH=CH_2$$

27. cis-1,2-Divinylcyclohexane

A German patent (228) by Aso et al. indicates that under a variety of conditions, among them cationic, the polymerization of *cis*-1,2-divinylcyclohexane with $AlCl_3$, $Et_2AlCl/tBuCl$, $Et_2AlCl/PhCH_2Cl$, or Et_2AlCl/CH_3OCH_2Cl yields thermally resistant materials having the repeat structure:

The molecular weight range of the products is 1100–3000 and the softening range is 100–150°C.

The repeat unit of the polymer suggests an intraintermolecular propagation mechanism. In view of the facile intramolecular hydride shift polymerization of vinylcyclohexane under cationic polymerization conditions, this author would expect a large contribution also of this mechanism to the overall structure of this polymer.

28. 2-Allyl-1-methylenecyclohexane

2-Allyl-1-methylenecyclohexane was first synthesized by Butler and associates (229) and polymerized with various initiators. Polymerization with gaseous BF_3 in methylene chloride diluent at −70°C for 46 hr gave 44% yield. The material was essentially soluble (94%). Spectroscopic analysis by IR and NMR led the authors to propose that the structure of the polymer may be represented as a copolymer of the repeat units:

and

in which the cyclized repeat unit predominates.

29. *Methylenecycloheptane*

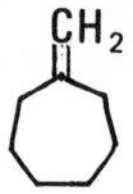

30. *Methylenecyclooctane*

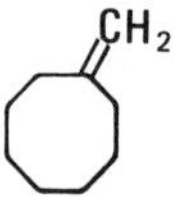

31. *Methylenecyclododecane*

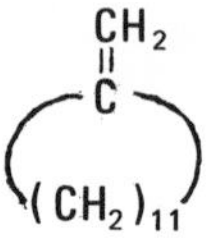

The cationic polymerization of these three large ring methylenecycloalkanes has been investigated by Pinazzi and Brosse (230). While no exact polymerization conditions have been described, it is stated that $AlCl_3$, BF_3, and $TiCl_4$ were tried in *n*-hexane and methylene dichloride for up to 96 hr in the range +20 to $-78°C$. In all cases low molecular weight oligomers were obtained (DP $\sim$ −6). The fastest rates occurred with the C_{12} ring compound.

Structure analysis by infrared and NMR spectroscopy indicated cyclic $-CH_2-$ groups and the presence of $-CH_3$ groups. In addition, NMR showed methyl protons adjacent to C=C and vinylic hydrogens. On the basis of this evidence the authors proposed that these polymerizations involve monomer isomerization that gives rise to the "theoretical" monomers— 1-methylcycloalkenes— which are then consumed by propagation proper:

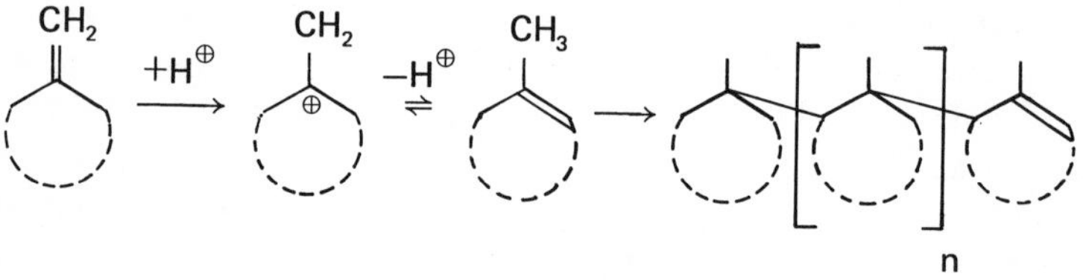

E. CONJUGATED DIENES AND TRIENES

a. Straight and Branched Chain, Conjugated Dienes and Trienes

1. *Butadiene (1,3-Butadiene)* $\qquad CH_2 = CH- CH=CH_2$

Butadiene is a mixture of rotamers with the transoid form by far the major conformer (230a):

plus skewed conformers.

The literature on the cationic polymerization of butadiene, a fairly reactive, inexpensive cationic monomer, is singularly meager. For example, the excellent comprehensive 1971 chapter on "Butadiene" by Bailey (231) has less than a half page out of approximately ~250 and lists only one reference among well over 1000 on the subject of "Cationic Polymerization." Similarly Saltman's chapter (232) on "Butadiene Polymers" spends no more than 14   lines and lists but three references on the subject of "Cationic Processes," and Cooper, in his chapter on the cationic polymerization of conjugated dienes (233), summarized butadiene in about one page and listed less than 10 references. Zlamal who recently published a competent chapter on the "Mechanism of Cationic Polymerization" (180) does not even mention the polymerization of conjugated dienes.

The reason for this low level of interest in such a readily available, inexpensive, cationically reasonably reactive monomer is that the products (homopolymers and copolymers) obtained by conventional cationic techniques are inferior in almost all respects to those synthesized by either free radical, anionic, or anionic-coordination initiators of the Ziegler-Natta type. The only product with some commercial significance is a liquid (Mol. wt. ~ 1500) butadiene oligomer (Budium of the DuPont Company) that is used as a protective interior lining in tin cans in food packaging. This product is prepared with $BF_3/OEt_2/$ $0.4H_2O$ in petroleum ether at 0 to $+5°C$ (234).

Research in the fifties showed that butadiene can be readily polymerized under suitable conditions with strong organic acids (e.g., chlorosulfonic acid)(235) or Friedel-Crafts metal halides (e.g., $AlCl_3$, $AlBr_3$, $SnCl_4$, $BF_3 \cdot OEt_2$, and $TiCl_4$) (235–237) with or without various complexing agents (e.g., nitroethane) (236) to sticky liquids or to soluble, tacky, or tough solids. Insolubility was attributed to crosslinking, particularly at higher conversion (235). Infrared spectroscopy showed predominantly *trans*-1,4 enchainment with some 1,2 linkages (236, 237), however, only 17–35% (Table I of reference 224) of the total theoretical unsaturation could be accounted for, and the balance of the double bonds disappeared. Loss of unsaturation was explained by a cationic cyclization (233):

Low molecular weight ($<$500) telomers of butadiene with benzene were prepared by introducing butadiene into solutions of ethylaluminum sesquichloride in benzene at room temperature, particularly in the presence of tertiary butyl chloride (238):

$$H \underset{}{\text{---}}(CH_2 - CH=CH - CH_2 \overset{}{\underset{n}{\text{---}}} C_6H_5$$

The diene enchainment was again *trans*-1,4. The telomerization of butadiene with

$$Cl- CH_2 - CH=CH- CH_3 \quad \text{and} \quad CH_2=CH-\overset{\overset{\displaystyle Cl}{|}}{C}- CH_3$$

in the presence of $AlCl_3$, $FeCl_3$, $SnCl_4$, and $BF_3 \cdot OEt_2$ was studied by a team of Japanese investigators (238a). They obtained a variety of products whose structures were explained by classical carbenium ion reactions. According to IR 80–85% of the unsaturation was present as *trans*-1,4, whereas the 1,2 and *cis*-1,4 enchainments contributed 9–15% and 2–10%, respectively. The average molecular weight of the telomers obtained with $TiCl_4$ in C_2H_5Br was higher than that produced by $AlCl_3$ or $FeCl_3$ at the same conversion level (239, 240). The addition of various oxygen (e.g., ethyl ether) or nitrogen containing (acetonitrile, nitroethane) solvents on the telomerization was investigated. Among the many complexing agents only the addition of nitroethane to an $AlCl_3$ in ethyl bromide solution increased the rate.

The synthesis of low molecular weight (200–750) polybutadienes has been studied by Runge and Aust (241). These authors found that while high molecular weight but crosslinked and brittle products are obtained with metal halide/hydrogen halide mixtures, low molecular weight mobile oils are produced in the presence of suitable amounts of aliphatic ethers, for example, with the initiating system $AlCl_3/HCl/OEt_2$ or $TiCl_4/HCl/OEt_2$.

Russian workers reported (40) that while $AlEt_2Cl$ alone was inactive, polymerization occurred readily with $AlEt_2Cl$ in conjunction with salt hydrates, for example, $LiCl \cdot nH_2O$, $BaCl_2 \cdot nH_2O$ in benzene solvent at $+20°C$. The disappearance of double bonds was noted and explained by an intramolecular cyclization:

In another set of experiments carried out in ethyl chloride solvent at $-78°C$, polymerization occured with $AlEt_2Cl$ alone, in the absence of salt hydrates. The product was a typical cationically polymerized polybutadiene: a powdery amorphous product, softening point $\sim 80°C$ and degree of unsaturation $\sim 67\%$. The author's conclusion was that in ethyl chloride the solvent itself provides the initiating species:

$$EtCl + Et_2AlCl \longrightarrow Et^{\oplus}[Et_2AlCl_2]^{\ominus}$$

While this proposition appears to be quite acceptable, it is not unassailable since the experiments have not been conducted under vigorously pure conditions and contamination of ethyl chloride by hydrochloric acid from the manufacturing process or moisture from the atmosphere remains a distinct possibility. The problem of initiation with alkylaluminum compounds is discussed in some detail under isobutylene (see above).

Since 1963 a joint group of Czechoslovakian and American workers have been interested in the cationic polymerization of conjugated dienes, among these butadiene. They have published a very large number of papers (242) proposing that cationically synthesized butadiene polymers (or other conjugated dienes) are of cyclized structure, "cyclopolymers." For example, in their first two papers Gaylord et al. (243, 244) described the polymerization of butadiene with a mixed initiator of $C_6H_5MgBr/TiCl_4$ (Mg/Ti = 0.5) in benzene that gives rise to a heat stable (decomposition point $\sim 405°C$) "cyclopolybutadiene." These materials are proposed to be block copolymers containing sequences of cyclic ladder units and linear trans-1,4 units, with perhaps a small contribution of 1,2 enchainment (243). Noting the similarity of infrared spectra of polybutadienes prepared by the conventional cationic initiator system $TiCl_4/H_2O$ and those prepared by the mixed systems, the authors suggest that these polymers have similar structures, that is, both polymers are cyclopolybutadienes containing some linear segments (243). Later it was stated that "practically no differences exist between the infrared spectra of cyclopolymers prepared from monomer and cyclized polyisoprenes prepared by cyclization of linear trans-1,4 and 3,4 polyisoprenes " (245).

In later studies (246) highly acidic $AlEtCl_2/TiCl_4$ (Al/Ti from 0.13 to 8.2) initiator mixtures were used. Using n-heptane as the diluent, the authors obtained insoluble polymer, whereas in benzene diluent soluble products were formed. Insolubility in n-heptane was attributed either to crosslinking or to a regular ladder-type structure with long sequences of fused rings. Solubility in benzene was explained by assuming that the phenyl substitution disrupts the regularity of the ladder structure. The authors attempted to determine quantitatively the amount of phenyl groups in the polymer as a function of solvent composition. With cyclopolyisoprene the maximum value was ~ 3 mole %, obtained in a polymerization in benzene diluent (Figure 1 in reference 246).

While these data are useful indicators for the presence of phenyl rings in the polymer, the low values do not warrant quantitative statements (246) to the effect that the polymers were formed by a nonconventional cationic process. Infrared spectra also indicated 7—9 mole % phenyl groups in cyclo-polybutadiene. The presence of more than one phenyl group in these polymers can be readily explained by alkylations.

There was also infrared evidence for the presence of methyl groups and for internal and some minor vinyl unsaturation in the polymer (246). The effect of the Al/Ti ratio on the CH_3 group content in cyclopolybutadiene was determined. The data indicate that the CH_3 content decreases with increasing Al/Ti (or increases with increasing Ti/Al) ratios. Again, these at best semiquan-titative data (Figures 2 and 3 in reference 247 relating the methyl content to the Al/Ti and Ti/Al ratios fail to show units for the CH_3 groups on the vertical axis) merely indicate a conventional cationic mechanism: with an increasing excess of $TiCl_4$ in the initiator mixture, the cationic character of the system increases, which in turn results in a higher rate of chain transfer and con-sequently higher amounts of CH_3 head-groups in the polymer.

The cyclic structure was recognized to be typical of most cationic polymer-izations in a nonpolar medium (248). It appears that the phenomenon of cyclization of simple conjugated dienes might be construed as a diagnostic tool for the presence of cationic reactions and/or initiators. This might also be useful for the characterization of acidic Ziegler-Natta type catalyst as polyisoprenes obtained at Al/Ti ratios lower than 1.5 were found to be essentially fully cyclized (247, 249).

To explain their major findings, the authors repudiated their earlier hypo-theses (244, 250) and proposed a new mechanism of "monomer activation by the catalyst cation via a one-electron transfer and addition of the activated monomer cation radical to terminal or internal unsaturation, that is, "living" polymerization by the addition of activated monomer to "dead" polymer" (246). This mechanism is unacceptable because in many details it contradicts a number of well established chemical principles; for example, there is little justification to assume that butadiene gives

a cation radical with a primary cation and secondary radical by one-electron removal to the catalyst of the Ziegler-Natta type (this particular cation radical structure is required to satisfy subsequent reactions) and there is little or no evidence to postulate propagation reactions involving primary carbenium ions, for example,

unstabilized alkyl
carbanions, e.g.,

or , etc.

Indeed all the major available facts could just as well be explained by a conventional cationic chain process assuming initiation by protonation with the assistance of ubiquitous protogenic impurities, chain transfer to monomer and/or aromatic solvent, and a conventional cyclization—elimination process. Thus protonation would give rise to methyl groups in the polymer; chain transfer to monomer would result in methyl head-groups plus conjugated double bonds as end-groups that could undergo further secondary reactions:

$$\sim\sim CH_2 - CH{=}CH{-}\overset{\oplus}{CH_2} + CH_2{=}CH{-}CH{=}CH_2 \longrightarrow$$

$$\sim CH{=}CH{-}CH{=}CH_2 + CH_3 {-\!\!-\!\!-}\overset{\oplus}{CH}{=\!\!=\!\!=}CH{=\!\!=\!\!=}\overset{\oplus}{CH_2}$$

Chain transfer to solvent would lead to phenyl end-groups by a well known alkylative transfer plus methyl head-groups:

$$\sim CH_2 - CH{=}CH{-}\overset{\oplus}{CH_2} + \bigcirc \longrightarrow \sim\sim CH_2 - CH{=}CH{-}CH_2 {-}C_6H_5 + H^{\oplus}$$

$$H^{\oplus} + CH_2{=}CH{-}CH{=}CH_2 \longrightarrow CH_3 - CH{=}CH{-}\overset{\oplus}{CH_2} \dots \text{etc.}$$

Most of all, extensive cyclization of *trans*-1,4 enchainments could readily proceed by the conventional cationic cyclization process, particularly in the presence of an acidic Ziegler-Natta type catalyst:

etc.

Last but not least the adventitious presence of 1,2 enchainments (vinyl unsaturation) in the polymer would readily lead to a variety of reactions (crosslink - ing, carbenium ions of the type

$$>CH - \overset{\oplus}{C}H - CH_3$$

that can eliminate, alkylate, isomerize, etc.). Some of the fundamental reactions that would adequately explain the data of these authors in terms of a conventional cationic mechanism were proposed some time ago by Cooper (233). A further, systematic discussion of all the possible pathways that might take place in such an ill controlled polymerization reaction as that of the polymerization of simple dienes like isoprene or butadiene by cationic agents would be but an idle exercise in chemical imagery.

Significant information pertaining to the cationic polymerization of butadiene was published by German investigators in the early sixties. Sinn et al. (251) were interested in details of initiation of olefin polymerization induced by Ziegler-Natta catalysts. In the course of these studies they found that $AlR_3/CoCl_2$ mixtures do not initiate the polymerization of butadiene unless traces of water are added to the system. All these experiments, of course, were carried out under carefully controlled anhydrous conditions (high vacuum). Similarly, $AlEt_3$ did not initiate the polymerization of driest butadiene; however, reaction started upon H_2O or HCl introduction. Similar observations have been described with $AlEtCl_2$. Indeed, observation of sealed tubes containing liquid butadiene and $AlEtCl_2$ for up to 5 months duration showed little if any polymerization. In the presence of moisture this polymer-reaction is complete within a few minutes.

Even Ziegler-Natta initiator systems, for example, mixtures of $AlRCl_2/TiCl_4$, require traces of water for initiation. In these experiments the Ziegler-Natta catalyst was prepared under the exclusion of moisture in high vacuum.

Takahashi et al. (252) reported in the scientific literature some of their findings that are disclosed in a Japanese patent application. These investigators found that when cobalt salts or chelates are added to mixtures of $AlEt_3$, organic halides, and butadiene in toluene solvent at $0°C$, polymerization occurs and high molecular weight poly(cis-1,4-butadiene) is formed. The authors contrast these findings with earlier reports according to which cis-1,4 polymer is obtained with initiator systems such as $AlEt_2Cl/cobalt$ compounds and syndiotactic 1,2 polymer is formed with $AlEt_3/cobalt$ compounds. According to this writer a conceivable interpretation of these observations is the in situ synthesis of $AlEt_2Cl$ by the reaction $AlEt_3 + RX \rightleftharpoons REt$ and polymerization initiation by the cobalt system. The polymerization as such is probably a cationic coordinated process. Some of the active initiator systems and representative products are listed in Table 12.

TABLE 12 POLYMERIZATION OF BUTADIENE WITH AlR₃/RX/CoHQ* SYSTEMS (252)

R_3Al	RX	Conversion %	$[\eta]$	cis 1,4	Microstructure trans 1,4	1,2
AlEt₃	$(C_6H_5)_3C\ Cl$	100	2.05	98.1	1.1	0.8
AlEt₃	$(CH_3)_3C\ Cl$	92	2.28	98.8	0.9	0.3
AlEt₃	$(CH_3)_3C\ Cl$	50	3.51	99	0.6	0.4
Al/Bu₃	$CH_2=CHCH_2Cl$	93	1.26	92.5	3.8	3.7
Al(2Et·Hex)₃	$(CH_3)_3C\ Cl$	99	1.65	98.1	0.9	1.0
Al/Bu₂H	$CH_2=CHCH_2Cl$	60	2.22	98.2	1.6	0.2
Al/Bu₃	—	12	—	65.0	1.5	35.5
Al/Bu₂H	—	14	9.58	65.8	2.2	32.0

*The cobalt compound in these runs was bis (8-hydroxyquinoline)Co(II).

Two significant papers by Gippin (253, 254) concern the polymerization of butadiene with alkylaluminum based initiators in conjunction with cobalt compounds and other chemicals. This research is important because of the highly stereoregular products obtained: all the polymers contain more than 90% cis-1,4 enchainment.

Gippins work started with the discovery that $AlEt_2Cl/CoCl_2$, a completely inactive system for the initiation of butadiene polymerization, can be activated by the addition of small amounts of water. The active system produces up to 98% cis-1,4 polybutadiene (253). Moreover, the polymer is of high molecular weight and is gel-free. The water is introduced last in a saturated benzene solution at $4-10°C$. (Gippin mentions that air oxygen may also function as an initiator, albiet much less efficiently than water; however the drying of injected air has not been specified.) Additional experiments showed that while the heterogeneous $CoCl_2$ system can be brought into benzene solution by using $CoCl_2$·pyridine complexes, the $AlEt_2Cl/CoCl_2$·pyridine system remains inactive in the absence of small amounts of water. However, in conjunction with water, butadiene is rapidly polymerized at $5°C$ to 100% conversion to clear, gel-free high molecular weight rubbers of ~98% cis-1,4 microstructure. Compounds such as C_6H_5OH, C_2H_5OH, C_2H_5SH, and $(C_2H_5)_2NH$ could not replace H_2O.

The effect of water was explained by the assumption of the formation of $EtAl(OH)Cl$ from Et_2AlCl and H_2O, and the proposal that the active initiator is a mixture of Et_2AlCl plus $EtAl(OH)Cl$ or, in general, $EtAlX_2$ where X = electronegative substituent. These ideas led to the next discovery, namely that $AlCl_3$ can be used in lieu of H_2O to start the polymerization and this was interpreted to mean that the true initiator is a mixture of $AlEt_2Cl$ and $AlEtCl_2$, the latter arising by the interaction of $AlEt_2Cl$ and $AlCl_3$.

On the basis of this hypothesis, a series of other chemicals were tried and a series of active initiator systems have been discovered, for example (254, 421), organic hydroperoxides by:

$$Et_2AlCl + ROOH \longrightarrow EtAl(OH)Cl + ROR$$

halogens by:

$$Et_2AlCl + Br_2 \longrightarrow EtAlBrCl + EtBr$$

alcohols by:

$$Et_2AlCl + ROH \longrightarrow EtAl(OH)Cl + REt$$

organic halides by:

$$Et_2AlCl + RX \longrightarrow EtAlClX + REt$$

and even aluminum metal by:

$$2Et_2AlCl + Al \rightleftharpoons Et-Al \underset{Et}{\overset{Cl}{<}} \underset{Cl}{\overset{Et}{>}} Al \underset{Cl}{\overset{Et}{<}} Al-Et$$

In agreement with these views, the most active initiators were found to be alkylaluminum sesquichloride and mixtures of Et_2AlCl and $EtAlCl_2$.

Polymerizations in more polar media, for example, chlorobenzene, o-chlorotoluene, and o-dichlorobenzene, reduced the rate but did not affect the microstructure (254).

The analysis of a large body of experimental findings led to the conclusion that the best initiator in terms of rates, microstructure purity, and solubility is the ion pair $Et_2Al^{\oplus}$ $EtAlCl_3^{\ominus}$ in conjunction with cobalt. This ion pair can be obtained by mixing either Et_2AlCl with $EtAlCl_2$ or Et_3Al with $AlCl_3$. Other combinations of aluminum alkyls result either in completely inactive systems or in systems producing gelled polymers characteristic of conventional cationic systems (i.e., $AlCl_4^{\ominus}$ counterions). These data have been summarized in Table 13 (254).

The stereoregular propagation induced by these initiator systems is visualized as a two step process. In the first step the cobalt is complexed with a butadiene monomer in a cisoid conformation which is subsequently incorpor-

TABLE 13 BUTADIENE POLYMERIZATION WITH ALKYLALUMINUM-COBALT INITIATORS* (254)

	Conversion(%)	cis-1,4(%)	Dilute Solution Viscosity	gel(%)
$Et_3Al + Et_2AlCl \rightleftharpoons Et_2Al^{\oplus} + Et_3AlCl^{\ominus}$	0			
$Et_3Al + EtAlCl_2 \rightleftharpoons Et_2Al^{\oplus} + Et_2AlCl_2^{\ominus}$	64.4	94.6	1.7	0
$Et_3Al + AlCl_3 \rightleftharpoons Et_2Al^{\oplus} + EtAlCl_3^{\ominus}$	93.2	97.5	4.4	0
$Et_2AlCl + EtAlCl_2 \rightleftharpoons Et_2Al^{\oplus} + EtAlCl_3^{\ominus}$	98.4	97.7	4.6	0
$Et_2AlCl + AlCl_3 \rightleftharpoons Et_2Al^{\oplus} + AlCl_4^{\ominus}$	87.8	–	–	90
$EtAlCl_2 + AlCl_3 \rightleftharpoons EtAlCl^{\oplus} + AlCl_4^{\ominus}$	28.4	64.4	–	67.5

*Cobalt octoate used.

ated by a 1,4 addition into the growing allylic carbenium ion with subsequent expulsion of the cobalt:

$$\text{Co} \underset{\text{C}}{\overset{\text{C}}{\longrightarrow}} \quad + \; R_2Al^{\oplus}RAlCl_3^{\ominus} \quad \longrightarrow \quad \underset{RAlCl_3^{\ominus} \; \oplus C}{\overset{R_2Al - C}{\diagdown C \atop \parallel C}}$$

While cationic polymerizations of butadiene with conventional Friedel-Crafts halides or highly acidic Friedel-Crafts halide mixtures usually give rise to difficultly characterizable resinous, partially crosslinked and cyclized products (255), recent research particularly at using alkylaluminum based initiators that probably induce cationic coordinated polymerizations, is encouraging and should be continued.

For completeness sake it should be mentioned that butadiene has been reported to polymerize with certain clays (256). It can also be induced to polymerize by γ-rays and the polymerization proceeds by a cationic mechanism. This field has been adequately reviewed by Pinner (257). Recent work by Stepan et al. (258) indicates that γ-ray polymerization gives rise to cyclopolybutadiene.

2. *Isoprene (2-Methyl-1,3-butadiene)* $CH_2 = \overset{\overset{\displaystyle CH_3}{\displaystyle |}}{C} - CH=CH_2$

There are great similarities between the cationic polymerization behavior of isoprene and butadiene, and many statements made in connection with the latter (see there) are also valid for isoprene.

The cationic polymerization literature on isoprene is limited. Bailey (259) in a 1971 chapter on isoprene barely mentioned its cationic polymerizability and Holden and Mann (260) in their review did not mention its cationic polymerizability at all. The physical properties of cationically synthesized polyisoprenes are unattractive, which explains the low level of interest in this readily available, fairly reactive cationic monomer. Interest in the cationic polymerization of isoprene exists in the main because of butyl rubber, a copolymer of isobutylene and small amounts (<2%) of isoprene.

Pertinent information available up to 1963 in regard to the polymerization of isoprene under cationic conditions is surveyed by Cooper (233). Isoprene is more active *vis-à-vis* cations than butadiene because of the inductive effect of its methyl substituent. Conventional Lewis acids, particularly in chlorinated solvents at low temperature, readily polymerize isoprene at low conversions to low molecular weight products and at high conversions to crosslinked, insoluble resins. The products contain predominantly trans-1,4 enchainment and there is always a substantial loss of unsaturation. For example, Richardson (237) used BF_3, $SnCl_4$, and $AlCl_3$ in various solvents and could account only for 50–80% of the theoretical unsaturation. Of the double bonds ~90% were trans-1,4 and the balance was distributed about equally between 1,2 and 3,4 enchainments. There was no evidence for cis-1,4 bonds. Cooper explained these and some additional data from other literature sources by proposing a set of conventional carbenium ion reactions including initiation by protogenic agents, alkylative grafting onto unsaturation in the polymer, hydride and proton transfers, and cyclizations. Indeed, the propagation and cyclization steps of isoprene cyclopolymerization described in a large number of publications by a joint group of Czechoslovakian and American authors (248–250; see also 239) can be explained by Cooper's concepts. A somewhat more detailed criticism of this work is given by this writer in conjunction with butadiene (see there). Very recently Czechoslovakian authors polymerized isoprene with $TiCl_4$/$tBuCl$ and $TiCl_4$/CCl_3COOH initiator systems in CH_2Cl_2 at $20°C$ (261) and with $EtAlCl_2$/$TiCl_4$ and $AlBr_3$ in benzene at 61 and $20°C$ (262) and analyzed their products by viscometry and light scattering. The major conclusion of this work was the proposition that above a molecular weight limit of ~10^3, these cyclopolymers were essentially intramolecularly very tightly crosslinked spherical microgels and that intramolecular gelation is very difficult to avoid or modify. The presence of cyclized units was also proposed on the basis of NMR (263) and independent viscosimetry, flow birefringence, and light scattering studies (262, 264). Recently Gaylord and Svetska (265, 266) also reported that the use of nitrobenzene as the diluent greatly enhances the yield of isoprene polymerization initiated with conventional Friedel-Crafts halides, for example, $TiCl_4$, $SbCl_5$, and PF_5, but not with $AlEtCl_2$. At the same time, in diluents like nitroethane and acetonitrile very little if any product was obtained. Similarly, carbenium salts, such as $(C_6H_5)_3C^{\oplus}SbCl_6^{\ominus}$, gave good yields in nitrobenzene but poor conversion in methylene chloride. Investigations are currently being carried out to define this interesting lead. In this context it is worth mentioning that Varion and Sigwalt (267) were readily able to initiate the polymerization of another, more reactive, conjugated diene, cyclopentadiene, in methylene chloride with the same carbocation salt $(C_6H_5)_3C^{\oplus}SbCl_6^{\ominus}$.

The polymerization of isoprene by $AlEt_2Cl$/salt hydrates such as $LiCl \cdot n\ H_2O$ in benzene at $20°C$ has been described by Russian authros (40). No polymerization occurred in the absence of the hydrates (see also under Butadiene). The products were typical, cationically synthesized isoprene polymers—low molecular weight amorphous powders.

Sinn et al. (251) carried out polymerization experiments of various olefins under carefully anhydrous conditions (high vacuum) and found that isoprene does not polymerize with $AlEt_3$, however, reaction can be induced by admitting hydrochloric acid. It is also a matter of record that typical Ziegler-Natta catalysts for example, $AlEtCl_2/TiCl_4$, initiate only in the presence of moisture (251).

In the course of their investigations on the polymerization of isoprene with Ziegler-Natta initiators, Yamazaki et al. (52) also examined the reactions initiated with $TiCl_4$, $AlCl_3$, and $AlEtCl_2$ in hexane diluent at $0°C$. Polymerizations were carried out in sealed glass tubes with highly purified chemicals. Under these conditions isoprene did not polymerize with $AlEt_2Cl$, $AlEt_3$, $TiCl_3$, and $TiCl_2$ and gave very small amounts of resinous products with $AlCl_3$ and $TiCl_4$ (traces); however, high polymerization activity was reported with $AlEtCl_2$. The microstructure of the polymer obtained with $TiCl_4$ was (normalized) 38% cis-1,4, 57% trans-1,4, and 5% 3,4.

Interesting studies have been described by Kössler et al. (250) who polymerized isoprene with $AlEtCl_2$ in n-heptane and benzene under high vacuum conditions. Highly purified materials were used and in the opinion of the authors "water is not involved in the formation of the catalytic complex" and "the polymer contains less than 5% of usual structures 1,4 or 3,4" (discussion remarks in reference 250). The polymerization of isoprene in bulk with $AlEtCl_2$ (isoprene/$AlEtCl_2$ = 100:1) was very rapid and nonstationary. When after 20 hr of storage fresh monomer was added, rapid polymerization resumed and a hard, insoluble orange product was obtained. The authors theorized that initiation occurs by the self-dissociation of the alkylaluminum ($AlEt_2^{\oplus}$ $AlCl_4^{\ominus}$). Another possibility is self-initiation by allylic hydride abstraction as proposed by this writer (114).

Matyska et al. (268) carried out polymerization experiments of isoprene with $AlBr_3$ in toluene and benzene under highest purity conditions (high vacuum) and found that polymerization proceeded even in the absence of purposely added protogens like H_2O or HCl. Significantly, upon addition of isoprene to a solution of $AlBr_3$ in toluene or benzene, polymerization starts immediately and simultaneously the electrical conductivity of the system increases by one to two orders. The shape of the conductivity curves indicated ion formation. While product analysis was not the purpose of these investigations, the results may be of considerable significance for our understanding of initiation in cationic polymerization in the absence of an apparent or obvious protogen (114).

3. *2,3-Dimethyl-1,3-butadiene*
$$CH_2=C-C=CH_2$$
with CH_3 groups

One of the earliest reports on the cationic polymerization of 2,3-dimethylbutadiene is by Whitby and Gallay (269) who mention the formation of a solid upon treatment of this diene with $SnCl_4$ in chloroform at room temperature. More recently a joint team of Czechoslovakian and American investigators polymerized 2,3-dimethylbutadiene with $AlEtCl_2$ in n-hexane and benzene (270) with acidic initiators of the Ziegler-Natta type, for example, mixtures of $AlEtCl_2/TiCl_4$ (Ti/Al=1, 2, and 6) and $iBu_3Al/TiCl_4$ (Ti/Al=1 and 5) in n-hexane and benzene (245, 246), and also by γ-ray irradiation (258). The simple Lewis acid $AlEtCl_2$ in n-hexane at 20°C produced at a moderate rate (21% conversion per hour) a white powdery polymer with the highest molecular weight, $\overline{M}_v$ = 8000. All the products contained much less than the theoretical amounts of unsaturation. All the polymers exhibited surprisingly simple infrared spectra indicating essentially identical structures. Interestingly, x-ray diffraction patterns suggested a measure of crystallinity. The melting point of the extracted polymers prepared by the acidic Ziegler-Natta initiators was 198°C. It was observed that the order of mixing of the ingredients strong-ly influenced polymer properties. For example, aged catalyst mixtures gave polymers with sharp melting points and no crosslinked product, whereas when the $TiCl_4$ was added last, crosslinked cyclopolymer with a high melting point (325°C) was obtained. The structure of these materials is largely unknown. Infrared analysis seems to indicate that the acid initiated products contain both cyclized units and $trans$-1,4 enchainment (260). (The chemical equation proposed to explain the formation of the monocyclic ring structure (eq. 3 in reference 270) is untenable.) It is surprising that cyclopolymers with rather ill defined recurring units are able to give even moderately crystalline x-ray diffraction patterns and sharp melting points.

Sinn et al. (251) mention that dimethylbutadiene cannot be polymerized with $AlEt_3$ unless hydrochloric acid is admitted. This observation can be explained by assuming the in $situ$ formation of $AlEt_2Cl$ (or $AlEtCl_2$), which in the presence of HCl is able to induce polymerization of olefins in general.

Evidence that this cationically quite reactive diolefin can be copolymerized is found in the patent literature. For example, Young and Hineline (271) claim the copolymerization of dimethylbutadiene with propylene in the presence of $AlBr_2Cl$ at −78°C; Thomas and Dahlke (272) claim the copolymerization of this diene with isobutylene with $AlCl_3$ in CS_2 at −103°C, Thomas and Sparks report (273) the copolymerization with isobutylene with $AlCl_3$ in CH_3Cl solvent at −103°C, and Sparks and Young with $TiCl_4$ (274).

4. 1,3-Pentadiene (Piperylene) $CH_2 = CH-CH = CH-CH_3$

The cationic polymerization of piperylene has not been systematically investigated. The patent literature discloses (275) that small amounts of piperylene can be copolymerized with isobutylene to elastomeric products by the use of BF_3 in methyl chloride solutions at −78°C.

b. 1,4-Dimethyl-1,3-butadienes:

5. trans,trans-2,4-Hexadiene

$$
\begin{array}{l}
CH_3 \\
| \\
CH=CH \\
\qquad | \\
\qquad CH=CH \\
\qquad\qquad \backslash \\
\qquad\qquad CH_3
\end{array}
$$

6. cis,trans-2,4-Hexadiene

$$
\begin{array}{c}
\qquad\qquad CH_3 \\
\qquad\qquad | \\
CH_3 \quad CH=CH \\
\backslash \quad / \\
CH=CH
\end{array}
$$

7. cis,cis-2,4-Hexadiene

$$
\begin{array}{l}
CH_3 \quad CH=CH \\
\backslash \quad / \quad \backslash \\
CH=CH \quad CH_3
\end{array}
$$

Whitby and Gallay (269) noted the formation of reddish oils when 1,4-dimethyl-butadiene was treated with $SnCl_4$ in chloroform at room temperature. Kamachi et al. (276) studied the homo- and copolymerization of 2,4-hexadiene isomers. First the isomerization of the geometrical isomers under the influence of various Friedel-Crafts acids was investigated. Whereas $BF_3 \cdot OEt_2$ did not exhibit isomerization activity, $SnCl_4$, $TiCl_4$, $SbCl_5$, and WCl_6 caused some isomerization. Polymerizations were carried out with a syringe technique under nitrogen atmosphere in toluene and nitroethane solvents. All the polymers were essentially of the same structure with trans-1,4 enchainment predominating, however, no analytical details (spectra) are given. There is no metnion of cyclization, branching, or insolubility. The similarity of the polymer structures led the authors to conclude that the propagating carbenium ions are similar. The formation of an all trans-1,4-poly(2,4-hexadiene) from the three geometric isomers was explained by making two assumptions: (1.) The trans,trans and the cis,trans isomers would readily give a transoid propagating cation (2.) To explain the formation of the transoid propagating ion from the cis,cis isomer it was assumed that while the cisoid cation is formed initially, somehow rotation must occur during propagation. To be more precise, the statement made by the authors is that since the activation enthalpy of the cisoid → transoid rotations is 4—5 kcal/mole, rotation prior to propagation is unlikely. However, their line of reasoning is that this rotation must occur somewhere along the reaction path:

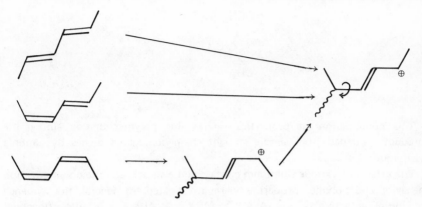

The intrinsic viscosities of the polymers were in the range 0.16–0.26.

Rate studies showed that by changing the solvent from toluene to nitroethane , the polymerization initiated by $BF_3 \cdot OEt_2$ becomes much faster. Interestingly, the overall rate of polymerization initiated by $BF_3 \cdot OEt_2$ in nitroethane and that by WCl_6 in toluene were about the same, which led the authors to the con - clusion that in the presence of the larger counteranion obtained from WCl_6 polymerization is faster because the propagating cation is more independent. This conclusion is controversial and the effect of the counteranion on overall rate is far from clear.

8.   trans-*2-Methyl-1,3-pentadiene (1,3-Dimethyl-1,3-butadiene)*

$$CH_2 = C \underset{\displaystyle CH_3}{|} \text{———} CH \underset{\displaystyle CH}{\diagup} \diagup CH \text{——} CH_3$$

Whitby and Gallay (269) mention the formation of a reddish oil of this diene upon treatment with $SnCl_4$ at room temperature. Cuzin et al. (277) found that 2-methyl-1,3-pentadiene can be quantitatively polymerized by moderately strong Lewis acids to completely soluble, rubbery products of reasonably high molecular weights ($\overline{M}_n$ = 67,000) (278). Spectroscopic analysis indicates that the products contain essentially one *trans*-1,4 double bond per repeat unit, consequently they are not cyclized. The glass transition temperature is $-75°C$ and the products are amorphous and do not crystallize, which was explained by irregularities on the asymmetric carbon in the main chain:

$$\begin{array}{c}
\sim\sim CH_2 \\
\backslash \\
\underset{/ \quad \backslash}{C=CH} \\
CH_3 \qquad \overset{H}{\underset{*}{C}}\sim\sim \\
\qquad CH_3
\end{array}$$

This combination of properties render this polymer unique among the products obtained from simple straight chain conjugated dienes by cationic techniques.

The effects of various aluminum containing Lewis acids on the polymerization behavior and polymer properties were investigated. In general, the sequence of initiating activities was $AlX_3 > AlRX_2 > AlR_2X \gg AlR_3$ (inactive) and $AlEt_2I > AlEt_2Cl > AlEt_2F$ (inactive). $AlBr_3$ and $AlEtCl_2$ did not require initiator, while the weaker acids were found to be active only in the presence of initiators, for example, water or t-butyl chloride. Polymer yield increased with the $H_2O/AlEt_2Cl$ ratio between the limits 0 and 0.5, then it leveled off and did not change even when the ratio reached 2 (279). The molecular weights strongly increased with decreasing temperatures. Figure 24 shows this effect. From the slope of the line in Figure 24, the overall activation energy of molecular weights can be calculated, $\Delta E_{\overline{Mn}} = -1.2$ kcal/mole. This value is close to the -1.7 kcal/mole obtained with isobutylene and the $AlEt_2Cl/tBuCl$ system in methyl chloride solvent by Kennedy and Milliman (179). Some of the other findings of the French authors are shown in Table 14.

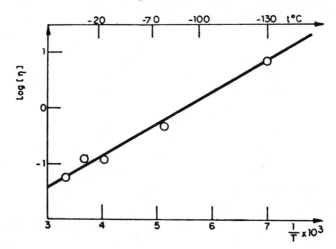

Figure 24. The effect of temperature on the intrinsic viscosity of poly(2-methyl-1, 3-pendadiene).

TABLE 14 CATIONIC POLYMERIZATION OF 2-METHYL-1,3-PENTADIENE(277)

| | | | Yield | $[\eta]$ | Microstructure | | |
Initiator System	Solvent	Temperature(°C)	(%)	(dl/g)	cis-1,4	trans-1,4	1,2
AlEt$_2$Cl/H$_2$O	Benzene	0	70	0.4	—	>90	>5
AlEt$_2$Cl/H$_2$O	Benzene	25	53	0.31			
AlEt$_2$Cl/tBuCl	n-Heptane	−80	100	0.7	—	~93	<7
AlEt$_2$I/H$_2$O	Benzene	25	94	0.23	—	~90	>5
AlBr$_3$	CH$_2$Cl$_2$	0	100	0.23	~5	~90	<6
TiCl$_4$	Benzene	~25	100	0.19	<5	>90	<5

Some discrepancy in the statements made in reference 277 and reference 279 in regard to solubilities and double bond contents should be noted. In the former reference all polymers are described as entirely soluble and it is stated that they contain one double bond per recurring unit. In the latter reference it is claimed that while in general the polymers obtained by cationic technique are soluble, those prepared by a strong Lewis acid do not exhibit theoretical amounts of unsaturation; for example, with AlBr$_3$ only 47% of the unsaturation is accounted for by NMR. In contrast, polymers obtained with weaker acids, for example, AlEt$_2$Cl/tBuCl or AlEt$_2$Cl/H$_2$O, at low temperatures contain more than 90 mole % unsaturation up to the limit of experimental accuracy and exhibit more than 95% trans-1,4 enchainment.

The absence of crosslinking and evidence for cyclization (277) is striking. One explanation may be that propagation is by the transoid secondary allylic carbenium ion

and that the rate of "back-biting" and/or crosslinking by this species is strongly diminished because of steric hindrance.

9 and 10. cis and trans -*3-Methyl-1,3-Pentadiene*

$$CH_2=CH-\underset{\underset{CH_3}{|}}{C}=CH-CH_3$$

3-Methyl-1,3-pentadiene when treated with $SnCl_4$ in chloroform solution at room temperature gives rise to a reddish oil (269). In the course of their investigation of the cationic polymerization of cyclic dienes, Imanishi and coworkers (280) copolymerized *cis*- and *trans*-3-methyl-1,3-pentadiene with 1-vinylcyclohexene by BF_3OEt_2 and $SnCl_4/CCl_3COOH$ in CH_2Cl_2 at $0°C$.

11. *4-Methyl-1,3-Pentadiene (1,1-Dimethyl-1,3-Butadiene)*

$$CH_2=CH-CH=\underset{\underset{CH_3}{|}}{\overset{\overset{CH_3}{|}}{C}}$$

Whitby and Gallay (269) mention the formation of a reddish oil of this diene upon treatment with $SnCl_4$ in chloroform at room temperature. French authors (277, 279), in the course of their work with 2-methyl-1,3-pentadiene, carried out a limited investigation on the cationic polymerization of 4-methyl-1,3-pentadiene. Initially they mentioned (277) that this olefin can be readily polymerized with $AlEt_2Cl/tBuCl$ initiator system in *n*-heptane at $-80°C$ and that the polymer repeat is predominantly 1,2 enchainment (81% 1,2 and 19% *trans*-1,4). In a later publication (279) it was disclosed that the polymer gives much lower molecular weights (e.g., $[\eta] = 0.07$ dl/g for a polymer synthesized at $-80°C$) than the 2-methyl-1,3-pentadiene isomer.

12. *2,4-Dimethyl-1,3-pentadiene (1,1,3-Trimethyl-1,3-butadiene)*

$$\underset{\underset{CH_3}{|}}{\overset{\overset{CH_3}{|}}{C}}=CH-\overset{\overset{CH_3}{|}}{C}=CH_2$$

Whitby and Gallay (269) treated this diene with 80% sulfuric acid at $0°C$ and obtained an oily product.

13. *2,5-Dimethyl-2,4-hexadiene (1,1,4,4-Tetramethyl-1,3butadiene)*

$$\underset{H_3C}{\overset{H_3C}{>}}C=CH-CH=C\underset{CH_3}{\overset{CH_3}{<}}$$

Moody reported (281) that 2,5-dimethyl-2,4-hexadiene can be readily polymerized by BF_3 in petroleum ether solution in the range -78 to $-50°C$. The white powdery polymer of high molecular weight ($\eta_{inh} = 1.44-1.79$) was soluble in high boiling solvents, for example, decahydronaphthalene and xylene above $120°C$, and it was found to be highly crystalline by x-ray diffraction and birefringence. The melting point was $263-265°C$, at which temperature the birefringence caused by polarized light disappeared on a hot stage microscope.

The polymer rapidly depolymerized to monomer above its melting point. Solution casting or compression molding resulted in brittle films.

While conclusive proof of the structure has not been obtained, on the basis of overall physical properties, it was suggested that the product was essentially a stereoregular polymer, most likely an all *trans*-1,4-poly(2,5-dimethyl-2,4-hexadiene). The strongest evidence came from x-ray crystallography: the polymer identity period of 4.8 Å was very close to values obtained with similar conjugated dienes having all *trans*-1,4 polymer structures.

Stereoregularity in this case was attributed to steric control during polymerization, that is, steric compression of propagation by the secondary allylic carbenium ion:

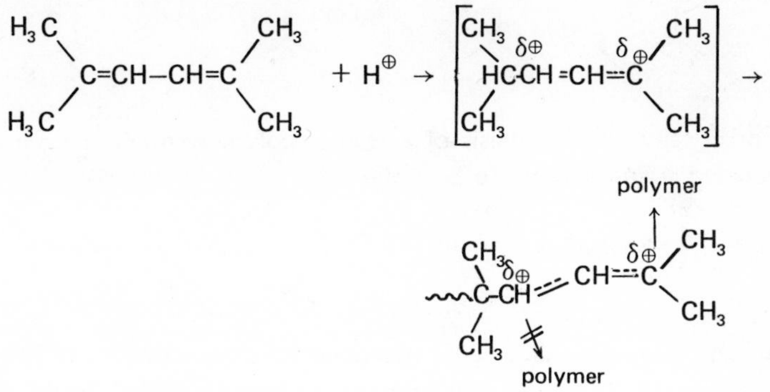

It is amazing that repeat structures containing neighboring geminal methyl groups are able to form at all:

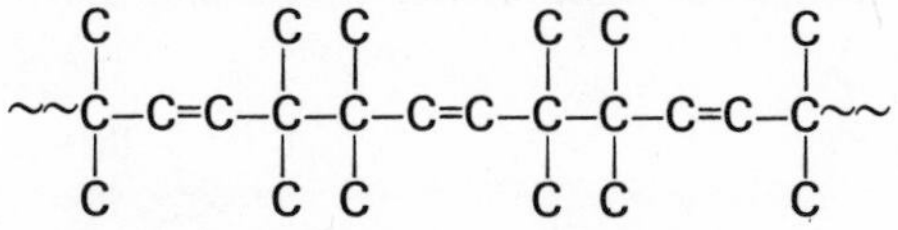

Moody's attempts to hydrogenate this polymer remained unsuccessful: either depolymerization occurred or unchanged polymer was recovered. Evidently the steric strain is too large to allow the existence of a head-to-head, tail-to-tail polyisobutylene.

14. *Myrcene (2-Methyl-6-methylene-2,7-octadiene)*

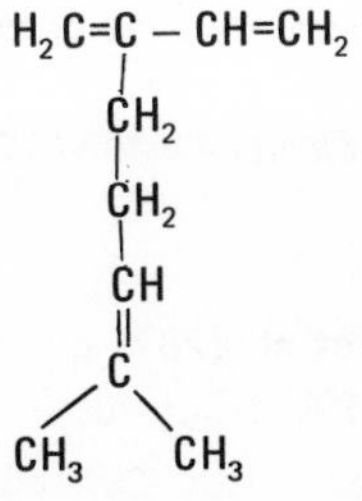

Marvel and Hwa (284) report that myrcene can be polymerized with $BF_3 \cdot OEt_2$ and $TiCl_4$ in *n*-heptane in the range 0 to $-78°C$ to low molecular weight (DP 4—6) white solids, softening range 86—90°C. The products contained an average of 1.09 double bonds per repeat unit. The authors believe that the structure of this polymyrcene is identical to that obtained from β-pinene , however, no evidence for this hypothesis is given. Theoretically this proposition is appealing:

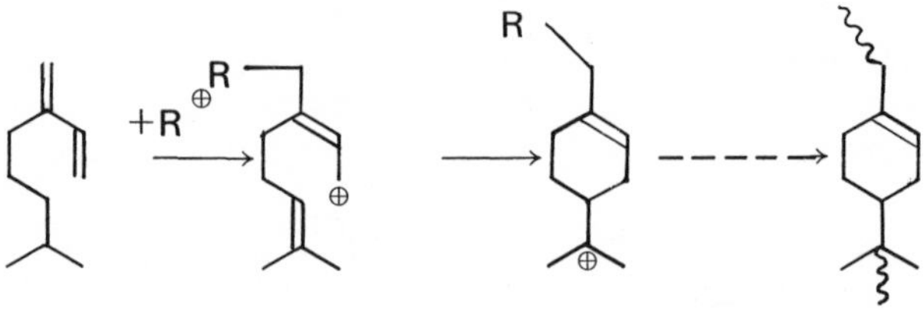

but practically, the synthesis of a regular polymer structure from a partially conjugated triene appears to be unfeasible.

15. 1-Phenyl-1,3-butadiene

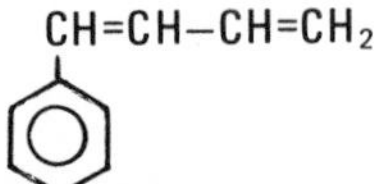

$$CH=CH-CH=CH_2$$

Smets and coworkers (285, 286) described the polymerization of this monomer by the use of $TiCl_4$ in ethyl chloride solvent at $-70°C$ (yield, 57%). The polymer showed less than 1 mole of unsaturation per repeat unit on ICl analysis. Chemical and spectroscopic data suggest a predominantly 1,4 enchainment, however, there is evidence for the presence of 1,2 and 3,4 repeat units as well.

16. 2-Phenyl-1,3-butadiene

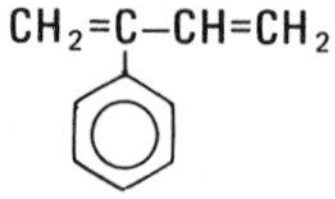

$$CH_2=C-CH=CH_2$$

Smets and coworkers (285, 286) polymerized the 2-phenyl-1,3-butadiene by $SnCl_4$ in $CHCl_3$ at $-70°C$ and in petroleum ether at $-24°C$ and obtained soluble and partially soluble products, respectively. Residual double bond analysis with ICl indicated a loss of unsaturation. Further chemical and spectroscopic analysis showed a predominantly 1,4 enchainment.

17. 1-Phenyl-4-methyl-1,3-butadiene

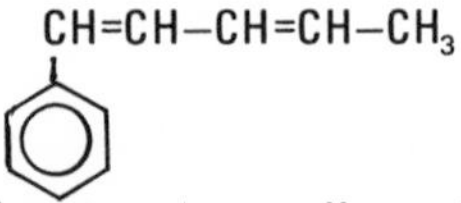

$$CH=CH-CH=CH-CH_3$$

Aliev et al. (287) claim the polymerization of this monomer to linear 85—90% *trans*-1,4 structures by cationic technique. In further research Aliev et al. (288)

have polymerized this monomer by using Lewis acids in conjunction with optically active groups. These workers wanted to synthesize optically active poly(trans-1-phenyl-4-methyl-1,3-butadiene) by the use of certain Lewis acids and Lewis acid complexes in which either an organic substituent in the Lewis acid, or the complexing agent in the Lewis acid complex was optically active: $SnCl_4 \cdot$ 2-(−)menthylethyl ether, $Et_2AlCl \cdot$ 2-(−)menthylethyl ether, $SnCl_4 \cdot$ 1.5-(−)menthol, $BF_3 \cdot$ 2-(−)menthol, $BF_3 \cdot$ 2-(−)menthylacetate, and so on. Among the various initiator combinations they found that $SnCl_4$ complexed with 1.5 moles of (−)menthol gave optically active polymer; however, optical activity was not obtained with $SnCl_4 \cdot$ 2-(−)menthylethyl ethers. The optical activity increased somewhat by lowering the temperature from 25 to $0°C$ ($[\alpha]_{390}^{20}$ = −0.6 to −1.25°C). The intrinsic viscosities of the products were very low: 0.033 and 0.05.

Slightly optically active polymers have also been obtained with $BF_3 \cdot$ 2-(−) menthol at $0°C$: $[\alpha]_{390}^{20}$ = −1.25°C, $[\eta]$ = 0.161. The authors postulate that optical activity arises because of the sterically restrictive nature of the propagating carbenium ion and not because of the orienting effect of the counteranion. Evidently the growing cation cannot, for some restrictive circumstance, assume a completely planar sp^2 structure and due to the remaining sp^3 character, the attack of the cation will be sterically directed toward one preferential side. This then may result in optical activity.

The polymerization of 1-phenyl-4-methyl-1,3-butadiene was also investigated by Mamedyarov et al. (289) who used $SnCl_4$/menthol and Et_2AlCl/menthol complexes. Optically active polymers were obtained. Optical activity was due to the asymmetrical counterion and not to the incorporation of menthoxy end-groups as higher molecular weight polymers had higher specific rotation than those of lower molecular weights. Ultraviolet and infrared spectroscopy indicated the presence of 6−15% units by 3,4 enchainments.

18. *1,2,3,4-Tetraphenyl-1,3-butadiene*

$$CH = CH - CH = CH$$

Whitby and Gallay (269) mention that treatment of this diene with $SbCl_5$, gives a reddish-yellow amorphous powder of a trimer, the melting point of which is higher than $350°C$.

19. *Chloroprene (2-Chloro-1,3-butadiene)*

$$CH_2=\overset{\overset{\displaystyle Cl}{\displaystyle |}}{C}-CH=CH_2$$

Only the cationic copolymerization of chloroprene has been investigated and no meaningful data exist in the scientific literature on the homopolymerization of this monomer. Foster (290) copolymerized chloroprene with styrene using

$BF_3 \cdot OEt_2$ in cyclohexane at $-18°C$ and obtained a very low molecular weight product ($\overline{Mn}$ = 1930-1960). The reactivity ratios were r_{ClP} = 0.24 and r_{St} = 15.6. Overberger and Kamath (291), who used n-hexane and carried out the copolymerization at $0°C$, obtained a similar set of data: r_{ClP} = 0.06 ± 0.01 and r_{St} = 12.8 ± 0.3. Somewhat different values for the reactivity ratios have been obtained with $AlBr_3$ in benzene and nitrobenzene and with $BF_3 \cdot OEt_2$ in nitrobenzene by these authors. The very low reactivity of chloroprene is due to the electron withdrawing influence of the chlorine.

20.   *trans-1,3,5-Hexatriene*

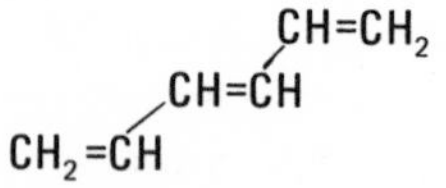

$$CH_2=CH-CH=CH-CH=CH_2$$

Bell (291a) obtained extensively crosslinked polymers upon treating 1,3,5-hexatriene with BF_3 or $AlCl_3$, and low molecular weight products using $BF_3 \cdot OEt_2$ or $SnCl_4$.

21.   *Alloöcimene (2,6-Dimethyl-2,4,6-octatriene)*

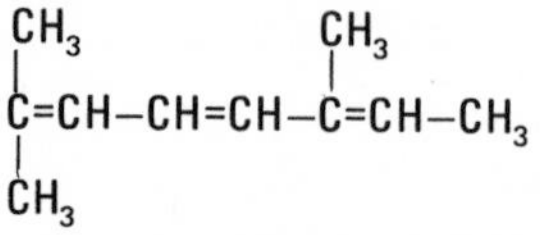

$$
\begin{array}{ccc}
CH_3 & & CH_3 \\
| & & | \\
C=CH-CH=CH-C=CH-CH_3 \\
| \\
CH_3
\end{array}
$$

Jones (282) reacted alloocimene with $BF_3 \cdot OEt_2$ at $0°C$ and obtained a soluble [benzene, carbon tetrachloride, ethyl chloride, and dioxane (!)] and fusible products. He reported a softening range at $68-70°C$ and melting points from 85 to $87°C$. Brittle fibers could be extruded through a fine orifice. On the basis of iodine numbers and by analogy with other cyclizations, the author surmised that the soluble product formed by a cyclopolymerization mechanism, such as:

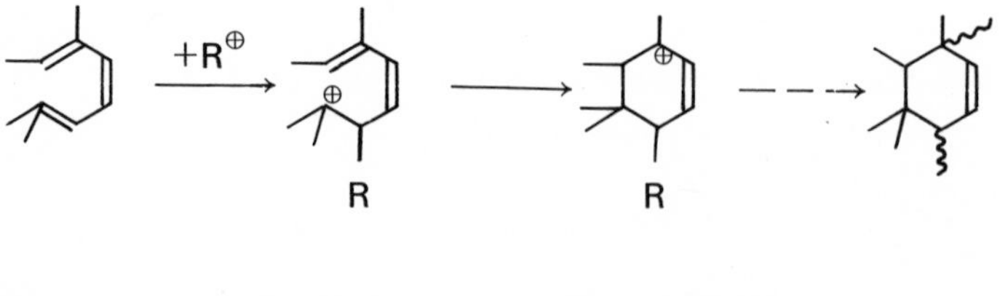

Regular cyclopolymerization of a complicated conjugated diene by a conventional cationic mechanism is highly unlikely. Indeed, Marvel and Kiener (283) reinvestigated the polymerization of this compound by a variety of initiators such as $BF_3 \cdot OEt_2$ in methylene chloride in the range -10 to $-70°C$

and $TiCl_4$ in n-heptane at -15 and $-70°C$. A soluble polymer, η_{inh} = 0.795, melting point $136\text{-}140°C$, was obtained with $TiCl_4$ in n-heptane at $-70°C$. Thorough chemical—spectroscopic analysis indicated that little if any cyclic products were formed. This statement also includes polymers obtained with certain initiators of the Ziegler-Natta type, for example, iBu_3Al/VCl_3 and $iBu_3Al/TiCl_4$.

b. Cyclic Conjugated Dienes

1. Methylenecyclobutene

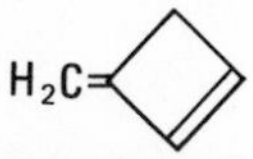

Methylenecyclobutene was first synthesized by Applequist and Roberts (292) who also described its extremely rapid polymerization in the presence of BF_3. The polymer, a brittle material, was insoluble in common solvents, and turned brown soon after its preparation. Infrared spectra showed the absence of exocyclic methylene groups and it was concluded that the predominant repeat structure formed by 1,5 addition

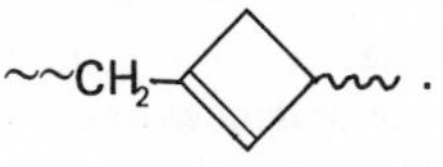

These studies have recently been reopened by Wu and Lenz (293) who polymerized methylenecyclobutene (and 1-methyl-3-methylenecyclobutene) by cationic and other initiators. Successful polymerization was obtained with BF_3, $BF_3 \cdot OEt_2$, and $AlEt_2Cl$, however, the low molecular weight $(\overline{M}_n$ ~3000—3500$)$ products rapidly oxidized after synthesis. In agreement with the earlier workers, Wu and Lenz also concluded that the cationic polymerization of 3-methylenecyclobutene gave predominantly the 1,5 structure.

In this context it is of interest that the anionic polymerization of this conjugated small ring diene gives rise to a 1,2-addition product:

2. 1-Methyl-3-methylenecyclobutene

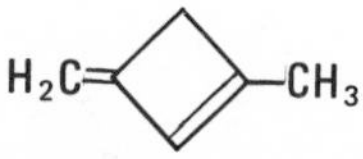

Wu and Lenz investigated the cationic (and Ziegler-Natta) polymerization of this monomer using BF_3 in bulk, $BF_3 \cdot OEt_2$ in n-hexane, and Et_2AlCl/HBr in toluene at $-78°C$ and obtained modest molecular weight polymer $(\overline{M}n$ = 3800 —19,200$)$ in high yields (~70%) (293). These molecular weights and yields are significantly higher than those obtained under similar conditions with the unsubstituted derivative, methylenecyclobutene (see there). Evidently the methyl group at carbon 4 stabilizes the propagating allylic cation. On the basis of IR and NMR studies of their polymers these authors propose a mechanism akin to that of a conjugated diene:

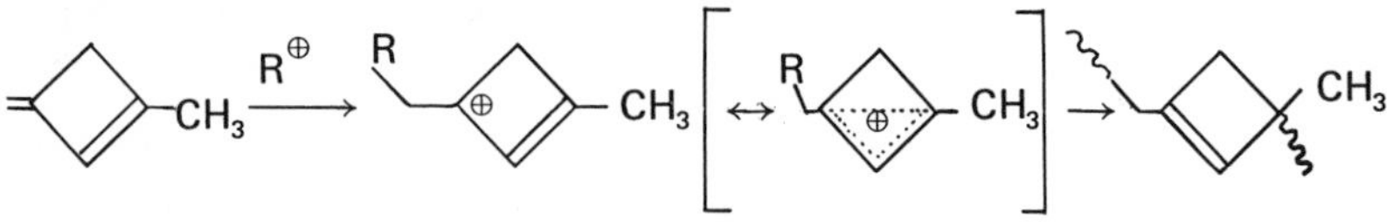

3. Cyclopentadiene

Cyclopentadiene is one of the most reactive cationic monomers known. Although cationically synthesized polycyclopentadiene homopolymers exhibit rather unattractive physical properties and can not be developed into commercially useful products, a large amount of research has been carried out with this inexpensive monomer over the past three decades.

The first systematic investigation of the polymerization of cyclopentadiene was carried out by Staudinger and coworkers (294, 295) who recognized and outlined many of its important polymerization characteristics. These workers obtained polymers by treating cyclopentadiene with a very large number of metal halides, for example, $SnCl_4$, $AsCl_3$, BCl_3, $TiCl_4$ and $FeCl_3$, at low temperatures and noted the rapid exothermic polymerization that ensued and the intense color development during the reaction. The latter observation was much later further investigated independently by Wasserman and coworkers (see below). Bruson and Staudinger also noted the rubbery appearance and vulcanizability with sulfur chloride of freshly prepared polycyclopentadiene and its rapid oxygen absorption with concurrent gelation (insoluble product formation) on prolonged standing on air. These early workers correctly recognized the high polymer nature of this product (294), measured its molecular weight (1 to 7 x 10^3 by cryoscopy), and from the fact that one mole of polymer consumed one mole-equivalent of bromine, proposed a formula for the polymer that has been proven essentially correct by later investigators:

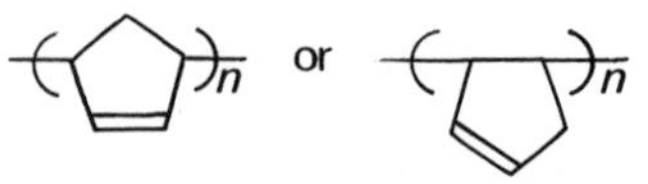

During the early fifties Wasserman and coworkers (296, 297) became interested in the intense color phenomena associated with the polymerization of cyclopentadiene and the color that appears when polycyclopentadiene is treated with strong acid [also noted by Burson and Staudinger (295)], in particular trichloroacetic acid. The strong brown—red color was attributed to conjugated double bonds of the polycyclopentadiene—trichloroacetic acid esters (296, 297) with the following structure:

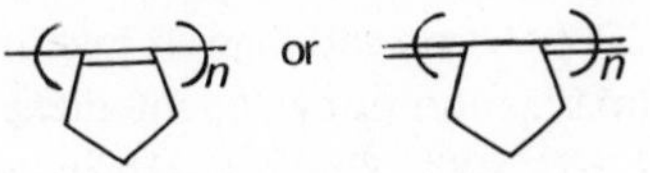

where n = 18 ± 2. A hydrogen atom migration during polymerization (297) was proposed to explain the presence of conjugated double bonds. Aspects of Wasserman's work have been summarized by Cooper (233) and Cheradame and Vairon (298).

These early investigations laid the foundation for further work by one French and two Japanese groups who, in more recent times, made important contributions to the understanding of the cationic polymerization of cyclopentadiene and other cyclic conjugated dienes.

Sigwalt and Vairon (299) observed that all previous workers obtained rather low molecular weight, largely crosslinked products on acid polymerization of cyclopentadiene and therefore embarked on a research project to prepare high molecular weight and soluble polycyclopentadienes. By examining their own preliminary results and those available in the literature, in particular Staudinger's early reports, these authors theorized that research should be carried out with "weak initiators that do not attack the second double bond of the monomer and allow linear propagation" (298, 299). They reduced the electrophilic character of TiCl$_4$ by replacing one of the electron withdrawing chlorine atoms in the TiCl$_4$ molecule by an electron releasing substituent —OnBu, that is, they synthesized TiCl$_3$OnBu. By using this initiator they obtained rather high molecular weight, soluble polycyclopentadienes ([η] = 1.4). Trichloroacetic acid also gave soluble high molecular weight product, although at much lower rates. The presence of small quantities of dicyclopentadiene in the charge was found to be a strong molecular weight depressor (299).

While this reasoning in selecting the less active Lewis acid for the synthesis of high molecular weight, soluble polycyclopentadiene appears to be rather simplistic (contrary to the contention of the French workers (299) there is no "traditional hypothesis by cocatalysis with water [according to which weak electrophiles] ...allow the opening of only one double bond of cyclopentadiene and linear propagation"), the technique was successful and Sigwalt et al. set out to characterize in depth the newly synthesized initiator and the high polymers.

The polymers were readily soluble in aromatic and chlorinated solvents, but they rapidly oxidized in air and became insoluble. To avoid gelation, the products were handled and stored in vacuum. Molecular weights were characterized by intrinsic viscosities and by light scattering, however, the relationship between $[\eta]$ and $\overline{M}_w$ could not be established because $\overline{M}_w$ varied with the polymerization temperature. This was attributed to diminished branching at increasingly low temperatures.

Following their initial report (299) Vairon and Sigwalt investigated the kinetics of the $TiCl_3OnBu$ initiated polymerization of cyclopentadiene under conditions of highest purity and dryness (267, 300). Greatest efforts were made to dry the system and to determine the effect of water on the rate. While polymerization could not be completely stopped even after most rigorous drying, the accelerating effect of water and hydrochloric acid was clearly demonstrated.

Careful experiments with the $TiCl_3OnBu$ initiator in methylene chloride diluent at $-70°C$ showed that the number average degrees of polymerization decreased with increasing initiator concentrations and decreasing monomer concentration (267). The rates increased with increasing initiator and monomer concentrations (267). Indeed, the initial rate was found to be first order in $TiCl_3OnBu$, H_2O, and monomer. Calculations indicated a quasistationary concentration for the active centers. Initiation was slower than propagation and was independent of initial monomer concentration. The rate constant for initiation was $k_i = 11.1$ 1/mole sec at $-70°C$. The enthalpy of polymerization was determined independently, $-\Delta H = 14.1 \pm 0.0$ kcal/mole (301).

The microstructure of the high polymer was investigated by NMR which showed that 1,2 and 1,4 enchainment contribute equally to the overall polymer structure.

Sigwalt and coworkers (302) also investigated the initiation of cyclopentadiene polymerization by a stable carbocation salt, $(C_6H_5)_3C^{\oplus}SbCl_6^{\ominus}$, and postulated for the initiation mechanism a direct addition of the trityl ion to the monomer in methylene chloride diluent:

$$\phi_3C^{\oplus} + \bigwedge \rightarrow \phi_3C\text{---}\bigwedge{}^{\oplus}$$

These authors (303) also found that this initiator yields entirely soluble, quite high molecular weight polymer in CH_2Cl_2 solvent. The range of $\overline{M}_n$ between $+20$ and $-70°C$ was 10,000–120,000.

In the sixties two independent Japanese teams of investigators, one headed by Aso and Kunitake, the other by Imanishi and Higashimura, became interested in the cationic polymerization behavior of cyclopentadiene. While Aso and associates were more interested in the microstructure of the high polymers, Imanishi et al. concentrated on the elucidation of kinetic details. Both groups are still active and have made important contributions to this field.

Aso et al. started their work by a study of oxidation (304) and bromination (305) behavior of polycyclopentadiene. Interestingly, bromination was quantitative in 1,1,2,2-tetrachloroethane but only two-thirds of the double bonds reacted in the THF solvent. The decreased reactivity of bromine in THF was attributed to the formation of $Br_2 \cdot THF$ charge transfer complexes that are sterically hindered to attack isolated double bonds flanked by two brominated repeat units:

Statistical considerations supported the chemical evidence (305).

After a preliminary communication in 1966 (306), Aso et al, published two key papers on the structure of polycyclopentadiene (307, 308). Their basic approach was to synthesize the model compounds

for the recurring units in polycyclopentadiene and to investigate their NMR spectra and compare them with those of high polymers. Methods were developed to determine the relative quantities of methine, methylene, and olefinic protons. In the first paper it was demonstrated that cationically synthesized ($TiCl_4$/ Cl_3CCOOH, $AlBr_3$, $SnCl_4$, etc. in toluene and methylene dichloride solvent at -78 and $0°C$) polycyclopentadienes contain both 1,4 and 1,2 enchainment (307). In their companion paper Aso et al. (308) proceeded to investigate the effect of reaction conditions on the structure of the high polymers.

In general these authors found intrinsic viscosities in the range 0.1–1.7, and higher molecular weights with weaker Lewis acids at lowest temperatures. These results are in agreement with Sigwalt's thinking (see above). While the addition of CCl_3COOH or H_2O to $TiCl_4$ or $SnCl_4$ accelerated the rates, the structure of the polymers remained largely unaffected. Gel formation was more frequent in methylene chloride than in toluene solvent; also there was evidence for some isomerization in the polar diluent as revealed by the ratio of olefinic protons to the sum of methine and methylene (aliphatic) protons (this ratio should be 0.5 if the polymer contains only 1,4 and 1,2 enchainments and is less than 0.5 in the case of double bond migration, e.g.,

).

While this ratio was in general ~0.5, a closer examination of the data suggest some isomerization (<0.5 ratios), particularly with $AlBr_3$ and $TiCl_4$, that is, "stronger" initiators in polar solvents at higher temperatures. Further analysis indicated that the most likely structure formed by isomerization is

and that isomerization occurs during propagation and not as a postpolymerization step under the influence of excess or unreacted Lewis acid. Compelling evidence for this proposal came from independent experiments that showed only little change in the structure of a preformed polycyclopentadiene upon treatment with $TiCl_4$ under simulated polymerization conditions (in methylchloride or in toluene diluents, $0°C$).

It was also impressively demonstrated that the nature of the Lewis acid affects the relative contribution of the two repeat units in the polymer and that the effect of temperature on the structure is negligible in the range 0 to $-100°C$. Some of Aso's data are reproduced in Figure 25.

Evidently the highest amount of 1,2 enchainment is obtained with $TiCl_4$ and progressively lesser amounts with $AlBr_3$, $SnCl_4$, and $BF_3 \cdot OEt_2$. To explain these results the authors postulate that propagation is controlled by the distance between the counterion and the carbenium ion, and that tight ion pairs preferentially direct the incoming monomer toward 1,4 enchainment, whereas loose ion pairs are less selective and consequently allow both 1,2 and 1,4 entries. They argue that tightly associated counteranions from strong Lewis acids localize the positive charge in the 4 position, whereas the charge is delocalized with looser ions obtained from weaker acids:

Tight ion pair

Loose ion pair

The authors do not justify their assumption that the counterion in a tight ion pair occupies the 4 position in preference to the 2 position.

The authors also elucidated the effect of the nature of the initiator and temperature on the intrinsic viscosities. Their data are shown in Figure 26. The viscosity values are rather high, which indicates that high molecular weight polycyclopentadienes ($[\eta] > 1.0$) can readily be obtained under suitable reaction conditions with conventional Lewis acids. It is interesting that under comparable conditions the highest polymer was obtained with the "weak"

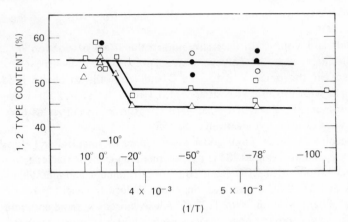

Figure 25. Structure of polycyclopendadiene obtained in methylene chloride. (●) TiCl$_4$; (○) AlBr$_3$; (□) SnCl$_4$; (▲) BF$_3$OEt$_2$ (308).

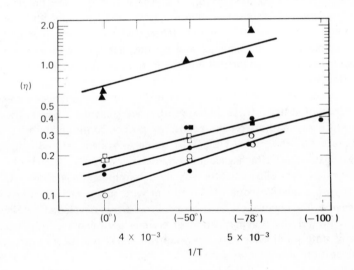

Figure 26. Variation of intrinsic viscosity with polymerization conditions for polycyclopendadiene obtained in toluene. (●) TiCl$_4$; (○) AlBr$_3$; (□) SnCl$_4$; (△) BF$_3$OEt$_2$.

initiator BF$_3$·OEt$_2$ and the lowest with the stronger AlBr$_3$. In light of the well established finding that highest molecular weights are usually obtained with the freest ions [i.e., radiation induced free ion propagation usually produces the highest molecular weights (136) which is also true for polycyclopentadiene (309)], it would follow that BF$_3$·OEt$_2$ would give rise to the freest propagating species. This conclusion, however, is in disagreement with Aso's proposition (see above) that the loosest ion pair is obtained from $AlBr_3$ and the tightest from $BF_3 \cdot OEt_2$.

Schmidt and Kolb (310) have also studied the effect of reaction variables on the molecular weight of polycyclopentadiene produced by $BF_3 \cdot OEt_2$ (H_2O) in toluene in the range +18 to $-78°C$ and were able to obtain products in good yield with intrinsic viscosities > 1.0 ($[\eta] \sim 1.4$ at $-78°C$). Indeed, with $BF_3 \cdot$ anisol or $BF_3 \cdot 2CH_3COOH$ at $-78°C$ intrinsic viscosities as high as 2.0 and 1.6 were obtained, respectively.

The second Japanese group under the guidance of Imanishi and Higashimura started their kinetic studies (311) by polymerizing cyclopentadiene with $TiCl_4/CCl_3COOH$, $SnCl_4/CCl_3COOH$, and $BF_3 \cdot OEt_2$ in toluene and methylene chloride solvents at $-78°C$, but obtained only low molecular weight product ($[\eta]$ = 0.1–0.5). Polymerization with $TiCl_4$/TCA was extremely rapid and nonstationary, and conversions stopped at low values; however, when fresh $TiCl_4$ was added to a "dormant" system, another burst of rapid polymerization occurred (312). Further, more detailed experiments have been reported recently and a kinetic theory was proposed (313). Imanishi found that the polymerization induced by $SnCl_4$/TCA involves two stages: at the beginning the reaction is nonstationary and extremely rapid (characterized by an inverse $[\eta]$ vs. initiator concentration dependence and by the fact that water has practically no effect on $[\eta]$) and then the rate slows down and becomes stationary. With $BF_3 \cdot OEt_2$ the rate is much slower and, except for the very beginning, it is stationary ($[\eta]$ is unaffected by the initiator concentration and it is much depressed in the presence of water). These and other data are treated by a complex kinetic theory which, the authors readily admit, needs to be confirmed by further data.

Very recently the $CPD-TiCl_4$/TCA—toluene system at -69 to $-77°C$ was investigated under high vacuum anhydrous conditions. The exothermicity of the reaction was followed accurately by determining the change in resistance by means of an oscilloscope of a platinum wire submerged directly into the liquid charge. The theoretical treatment followed that developed by Hayes and Pepper (314) for the polymerization of styrene assuming fast initiation and a nonsteady state (diminishing concentration) of active species. The rate constant for propagation was derived to be k_2 = 3500 ± 71 l/mole sec and that for termination k_5 = 0.11 ± 0.06 sec^{-1}. This propagation constant is very high considering various vinyl polymerizations, however, it is at least five to six orders of magnitude less than that obtained by the γ-radiation induced polymerization of cyclopentadiene (309).

Williams and associates showed that under conditions of highest purity, γ-radiation effectively induces the polymerization of cyclopentadiene by a reaction involving free carbenium ions and produces highest molecular weight products (309). For example, irradiation at $-78°C$ gave rise to a polymer with $[\eta]$ = 2.76 (309). Unfortunately the structure of the polymers has not been examined in detail and therefore we do not know the predominant recurring unit. In a similar vein Aso et al. (315), who studied the polymerization of cyclopentadiene induced by γ-rays in thiourea channel complexes, found that under such restrictive complexing conditions propagation was mainly by 1,4 enchainment.

In a recent publication Japanese authors (673) claim the batch copolymerization of cyclopentadiene with 2-chloroethyl vinyl ether. The synthesis involved the addition of $BF_3 \cdot OEt_2$ to the monomer charge in various solvents at $-78°C$. The overall composition of the methanol insoluble polymer fraction was analyzed by NMR spectroscopy. Reactivity ratios and some physical properties of (sulfur) vulcanizates are reported. While the authors imply the synthesis of random copolymer, the structure of the product and/or its homogeneity has not been investigated. Keeping in mind the difficulty of copolymerizing monomers belonging to different compound classes, further investigation should be carried out to determine the randomness (e.g., by sequence distribution studies using high resolution NMR spectroscopy) and homogeneity (by GPC) of the product. At the present time the data may be used to indicate the overall characteristics of a complex system consisting of a wide spectrum of components: homopolymers and copolymers of various compositions.

4 and 5. Methylcyclopentadienes

The separation of 1- and 2-methylcyclopentadiene isomers is very difficult. Thus investigations of the polymerization of this very reactive monomer have been carried out with isomer mixtures.

Aso and Ohara (316) investigated the polymerization of methylcyclopentadienes using a series of various isomer ratios (from 27/73 to 89/11 1-methylcyclopentadiene isomer/2-methylcyclopentadiene isomer) with $BF_3 \cdot OEt_2$, $SnCl_4$, and $TiCl_4$ under a variety of conditions (various solvents, concentrations, and temperatures). Preliminary experiments established that under polymerization conditions the rate of isomerization to an equilibrium composition was insignificant; no 5-methylcyclopentadiene was detected. Medium to high conversions were readily obtained under most conditions, however, product intrinsic viscosities remained relatively low (>0.8). The polymers were soluble white powders, with softening points of $120-160°C$. The lack of gelation is of interest and may be due to the presence of trisubstituted double bonds in the polymer (see below). The products readily oxidized on standing in air.

Copolymerization studies with the two isomers indicated $r_1 = 0.01 \pm 0.05$, and $r_2 = 1.1 \pm 0.2$, that is, the 2-methylcyclopentadiene isomer was far more reactive than the 1-isomer.

The microstructure of the products was analyzed by comparing the infrared and NMR spectra of the polymers with the model compounds 1- and 3-methyl-cyclopentane and 3,5- and 3,4-dimethylcyclopentene. Analysis indicated that the most important contributing repeat unit was

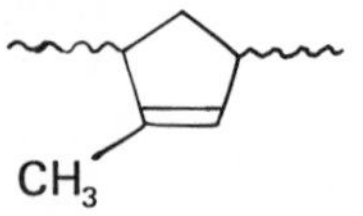

with less important contributions by

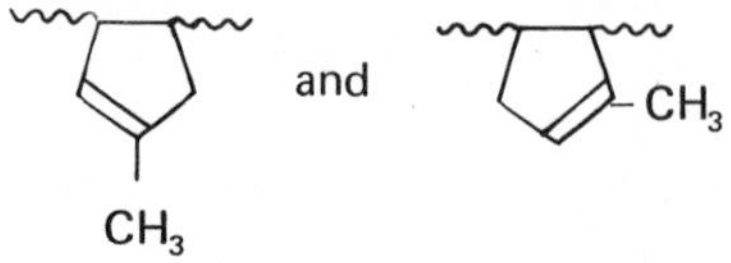

and

and practically no evidence for

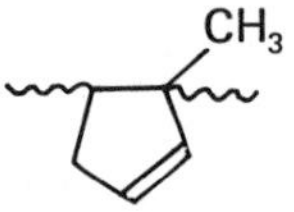

These structures dovetail with the results of the rate and copolymerization studies according to which 2-methylcyclopentadiene is much more reactive than the 1-isomer. Evidently, the copolymer obtained in these polymerization studies has been derived mainly from the 2-isomer, giving rise preferentially to the sterically least hindered 1,4 repeat unit:

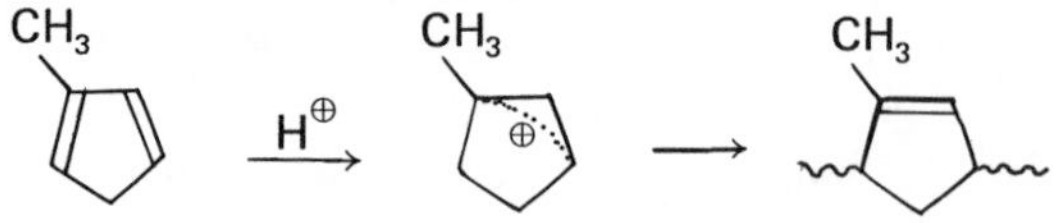

The other two recurring units for which spectroscopic evidence is available would appear to form under sterically much less favorable conditions. It is noteworthy that 1,4 enchainment is preferentially obtained by the use of $SnCl_4$ or $TiCl_4$ (~90% 1,4 structure) and much less with $BF_3 \cdot OEt_2$ (~76%). This finding is contrary to that with cyclopentadiene, where 1,4 enchainment was preferred with $BF_3 \cdot OEt_2$ (see there). An explanation for this finding was given in a later paper by Aso and Ohara (317), who believe that in the presence of a propagating "loose" ion pair ($SnCl_4$ or $TiCl_4$), the sterically less hindered and more reactive secondary allylic position will propagate preferentially by attacking the more negative 1 position of the incoming monomer:

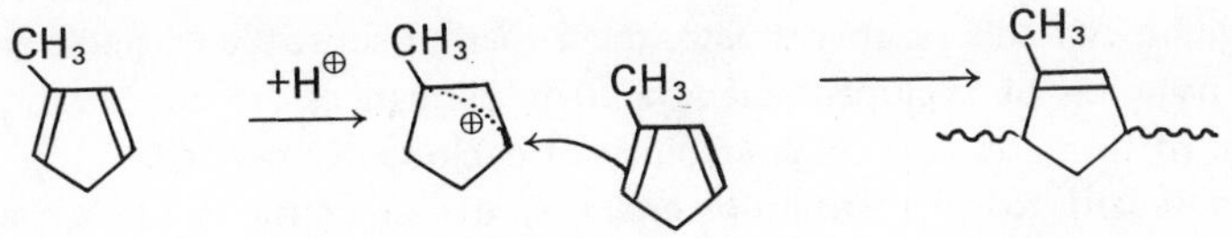

1,4 Structure

Thus when steric restriction is minimum, attack will occur preferentially on the 1 position leading to 1,4 enchainment. With a "tight" propagating ion (BF$_3 \cdot$OEt$_2$) steric compression directs the approach of the monomer so as to have the methyl substituent away from the growing site, that is, the monomer approaches the active site with its 4 position:

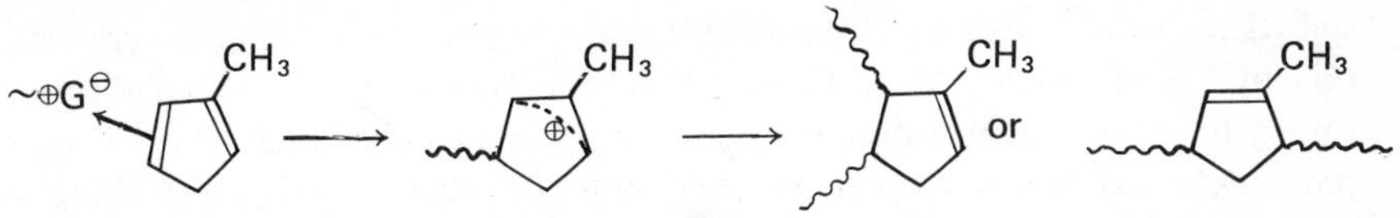

Imanishi and coworkers (318) have also been studying the polymerization of methylcyclopentadiene with SnCl$_4$/CCl$_3$COOH, TiCl$_4$/CCl$_4$COOH, and BF$_3$ $\cdot$OEt$_2$ at $-78°$C. Their monomer was a mixture of 45% 1-methyl-, 52% 2-methyl-, and 3% 5-methylcyclopentadiene. The polymers were white powders, [η] = 0.1–0.5. The polymerizations with TiCl$_4$/TCA and SnCl$_4$/TCA were typically nonstationary: upon initiator addition a burst of polymerization ensued that was followed by rapid termination. Fresh initiator addition was needed to produce higher conversions. In contrast, with BF$_3 \cdot$OEt$_2$, after the rapid polymerization, a stationary phase obtained that slowly progressed to 100% conversion. The level of conversion did not influence intrinsic viscosities however, the nature of the catalyst did affect these values, that is, [η] $\sim$ 0.4, $\sim$ 0.3, and $\sim$ 0.15 with BF$_3 \cdot$OEt$_2$, SnCl$_4$/TCA, and TiCl$_4$/TCA, respectively.

Copolymerization experiments of methylcyclopentadiene with cyclopentadiene indicated a higher reactivity for the former. The reactivity ratios were as follows:

	MeCPD	CPD
In toluene	8.5±3.5	0.36±0.26
In methylene chloride	14.9±5.6	0.42±0.23

The nature of the initiator had little effect on these values.

The enhanced reactivity of methylcyclopentadiene over that of cyclopentadiene is explained by the inductive effect of the methyl group and, particularly in light of Aso's reactivity data, the 2-methyl substituent. Rapid initiation and propagation of cyclopentadiene and its derivatives are conceivably due to the ease of formation and high stability of cyclopentyl cations.

In regard to microstructure analysis, the conclusions of Aso and Imanishi are essentially in agreement, although Aso's analysis is more comprehensive. Because of their judicious use of the model compounds, Aso and Ohara (316) were able to analyze the contributing structures convincingly and completely.

6. *1,3-Dimethylcyclopentadiene*

In the course of their extended studies on the microstructure of cyclopentadiene and its derivatives, Aso and Ohara (317) also examined the cationic polymerization of 1,3-dimethylcyclopentadiene. This monomer is the most stable isomer among the seven possible disubstituted isomers and can be obtained in the pure state. Polymerizations were undertaken with $BF_3 \cdot OEt_2$, $SnCl_4$, and $TiCl_4$ in toluene and methylene chloride at 0 and $-78°C$. High conversions and reasonable intrinsic viscosities ($[\eta] = 0.1-1.0$) were obtained. All the polymers were soluble. As in the case of the polymerization of cyclopentadiene and its monosubstituted derivatives, the nature of the initiator influences intrinsic viscosities, highest values being obtained with $BF_3 \cdot OEt_2$ and, in descending order, with $SnCl_4$ and $TiCl_4$. Lowering the temperature and the use of the less polar solvent also enhances viscosities.

Copolymerization of 1,3-dimethylcyclopentadiene with cyclopentadiene in toluene with $BF_3 OEt_2$ at $-78°C$ gave the reactivity of ratios of $r_{1,3} = 6.85 \pm 1.10$ and $r = 0.30 \pm 0.10$ for these two components, respectively. Thus the 1,3 isomer is much more reactive than the parent diolefin. Evidently methyl substitution enhances the stability of the cyclopentyl cation.

The polymer may be dissolved in benzene and the solution could be cast to yield transparent films. The softening points, $150-155°C$, were somewhat higher than those of poly(methylcyclopentadiene), $120-150°C$, and poly (cyclopentadiene) $90-120°C$. The rate of oxidation or oxygen absorption upon exposure to air was found to be slower with this polymer than with the poly (methylcyclopentadiene) which in turn was slower than that of the poly (cyclopentadiene). The authors attributed this enhanced oxidation resistance to the decreasing number of tertiary allylic hydrogens in this series of polymers.

The microstructure of the polymers was analyzed spectroscopically. Infrared studies indicated trisubstituted double bonds and two kinds of methyl groups. Sharp NMR absorptions revealed the same number of methyl protons on saturated and unsaturated carbons (CH_3-C at 0.96 ppm and $CH_3-C=C$ at 1.66 ppm) and a nearly theoretical ratio of 10 for the total number of protons to the number protons at an unsaturated position. These data indicate the absence of isomerization during or after polymerization.

On the basis of these hard facts and some theoretical arguments concerning the charge distribution and steric course of the polymerization, the authors concluded that the polymerization proceeds predominantly by the following route:

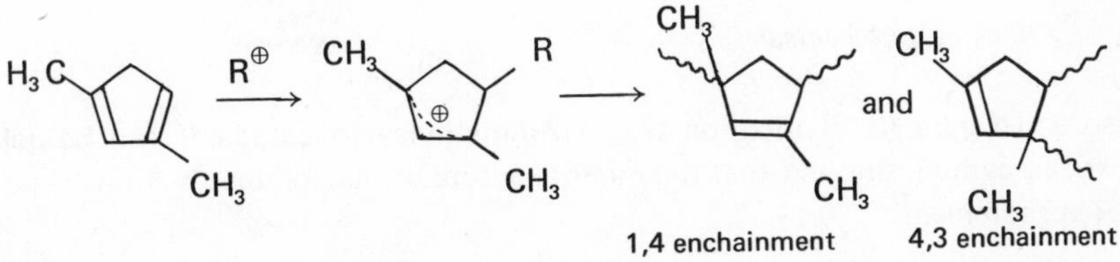

1,4 enchainment   4,3 enchainment

Evidently the steric hindrance exerted by the tertiary methyl substituents does not prohibit propagation. The mode of enchainment was affected to a certain extent by the particular initiator employed.

It is of interest to compare the propagation step of this monomer with that of 3-methylcyclopentene investigated by Boor et al. (218). The latter polymerization proceeds by intramolecular hydride migration:

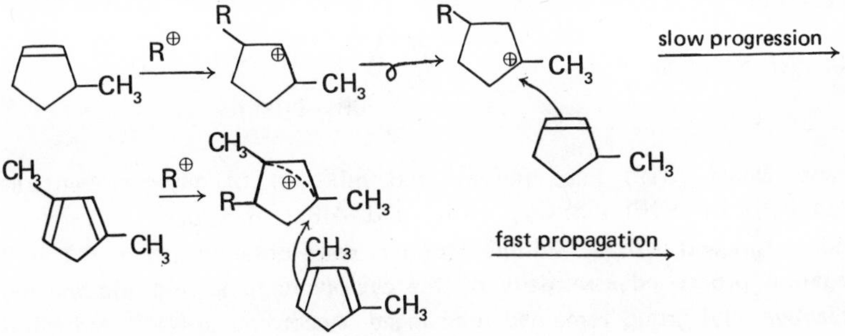

The rapid propagation of the diene is probably due to the higher stability of the substituted allylic cation as compared with the much less stable initial secondary ion in the olefin. Also the nucleophilicity of the diene is much higher than that of the olefin.

7. 2,3-Dimethylcyclopentadiene

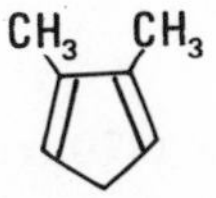

Aso and Ohara (317) briefly mention that the polymerization of 2,3-dimethyl-cyclopentadiene gives rise to an essentially pure 1,4 enchainment:

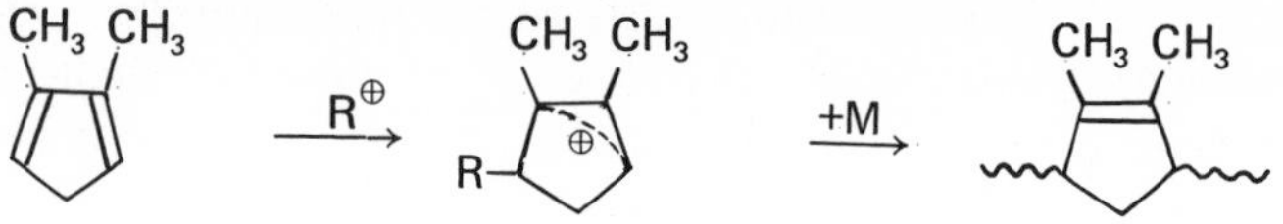

Evidently the polymerization follows the sterically least hindered route.

8. 1,2-Dimethylcyclopentadiene

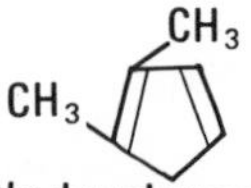

Aso and Ohara (317) mention that 1,2-dimethylcyclopentadiene can be polymerized cationically and that the microstructure of the polymer is almost a pure 3,4 enchainment:

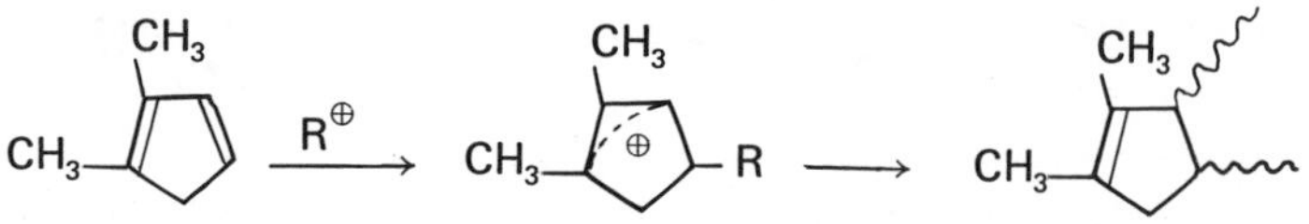

Evidently the system avoids unfavorable steric compressions and propagates by the less hindered 3,4 addition.

9. Allycyclopentadiene

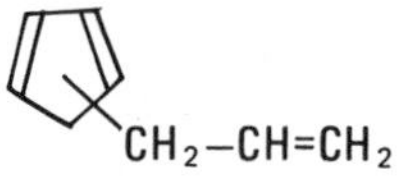

Aso and Ohara (319) have polymerized mixtures of allylcyclopentadiene isomers with $BF_3 \cdot OEt_2$, $SnCl_4$, $TiCl_4$, and $AlBr_4$ in toluene or methylene chloride, at 0 and $-78°C$. Soluble products were obtained, $[\eta] = 0.1 - 0.3$. Propagation proceeded essentially by the cyclopentadiene ring and the much less reactive allyl group remained unchanged. According to NMR and infrared evidence the structure of the polymer appears to consist of about equal amounts of 1,4 and 3,4 enchainments:

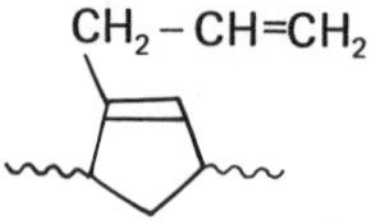

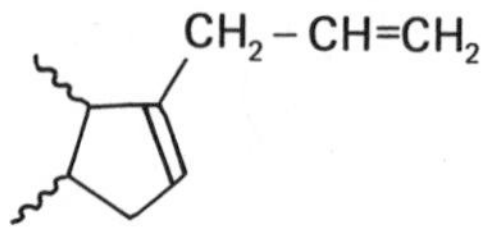

Canadian researchers (320) confirmed Aso's findings in regard to the nonreactivity of allyl groups, however, on the basis of NMR data they concluded that samples obtained with $BF_3 OEt_2$ in chloroform at room temperature ($[\eta] = 0.19$) contained ~53% 1,4 and ~47% 1,2 linkages.

10. Methallylcyclopentadiene

$$CH_3$$
$$CH_2-C=CH_2$$

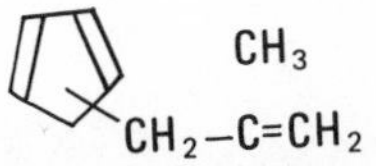

A mixture of methallylcyclopentadiene isomers has been synthesized and subsequently polymerized by $BF_3 \cdot OEt_2$ in chloroform at room temperature by Mitchell et al. (320). NMR spectroscopy indicated ~65% 1,4 and ~35% 1,2 enchainment.

11. Allylmethylcyclopentadiene

$$H_3C \qquad CH_2-CH=CH_2$$

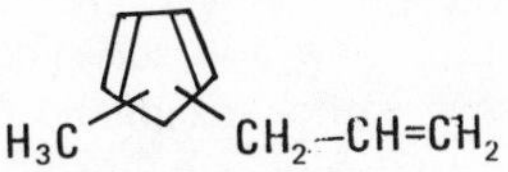

Mitchell et al. (320) were the first to synthesize allylmethylcyclopentadiene. The polymerization of the unresolved mixture of isomers was carried out in chloroform solution with $BF_3 \cdot OEt_2$ at room temperature. According to NMR spectroscopy the polymer consisted of ~69% 1,4 and ~31% 1,2 linkages:

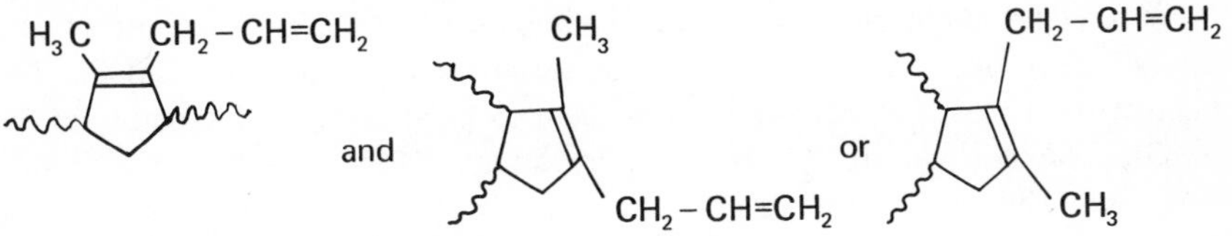

$$H_3C \quad CH_2-CH=CH_2$$

and

$$CH_3$$
$$CH_2-CH=CH_2$$

or

$$CH_2-CH=CH_2$$
$$CH_3$$

12. 6,6-Dimethylfulvene

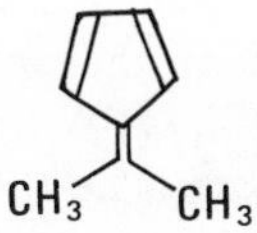

$$CH_3 \qquad CH_3$$

A brief paper by Mains and Day (321) concerns the polymerization of dimethylfulvene by a variety of Lewis acids, such as, $SnCl_4 \cdot 5H_2O$, $AlCl_3$, $SbCl_3$, and $FeCl_3$, in chloroform solution at room temperature. The products are white powdery materials, soluble in common solvents, for example, their molecular weights are in the range 1000–5000, and they have a strong tendency to absorb oxygen on standing in air. The oxidized polymers are tan and insoluble.

13. 6-Methyl-6-ethylfulvene

$$CH_3 \qquad CH_2CH_3$$

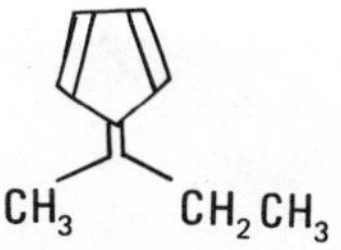

Mains and Day (321) mention that 6-methyl-6-ethylfulvene gave polymer with Lewis acids used to polymerize dimethylfulvene.

14. 1,3-Cyclohexadiene

Marvel and coworkers (322) have studied the polymerization of 1,3-cyclohexadiene with cationic initiators, for example, BF_3, PF_5, and $TiCl_4$, under a variety of conditions and obtained white polymers, the soluble fractions of which showed inherent viscosities in the range 0.04—0.19. The repeat units were mixtures of 1,2 and 1,4 enchainments.

The polymerization of 1,3-cyclohexadiene was studied by LeFebvre and Dawans (323) who also surveyed the patent literature. Among other initiator systems this author examined the cationic polymerizability with $TiCl_4$ in benzene and toluene at 0 and $30°C$. This method yielded soluble very low molecular weight, $[\eta] = 0.09$, product.

Structure analysis suggested 1,4 and 1,2 enchainments, but no firm conclusions as to the microstructure have been reached. In a companion paper Dawans (324) examined the copolymerization of cyclohexadiene with isoprene.

Imanishi et al. (325) recently investigated the structure of poly(1,3-cyclohexadiene) synthesized with $SnCl_4/CCl_3COOH(TCA)$ or $BF_3 \cdot OEt_2$ initiators in methylene dichloride or benzene diluents at $0°C$. The polymers were soluble with $[\eta] = 0.04 - 0.12$ and a softening range of $114-130°C$. The microstructure of the polymer did not seem to be affected by the nature of the initiator. According to NMR analysis, about ~20% of the unsaturation was lost. The authors speculate that chain-branching might be responsible for this. While there is evidence for 1,4 and 1,2 enchainments, no quantitative analysis as to the relative amounts of these repeat units was given.

Kinetic aspects of the polymerization of 1,3-cyclohexadiene initiated with $BF_3 \cdot OEt_2$ or $SnCl_4/TCA$ in benzene and methylene chloride diluent at $0°C$ have been investigated by Imanishi et al. (326) Polymerization in benzene was completely homogeneous, while in methylene dichloride the polymer precipitated during experiment. Both products, nontheless, were soluble in common solvents, $[\eta] = 0.04 - 0.12$.

Polymerizations with $SnCl_4/TCA$ were nonstationary that is, upon initiator addition a fast reaction commenced, although, polymerization did not go to completion and higher conversions were attained only by the repeated addition of $SnCl_4$ (but not by TCA addition). Interestingly, highest conversions were obtained with lower monomer concentrations (cf. Figure 27). This peculiar feature has not been observed when $BF_3 \cdot OEt_2$ initiator was used, however an identical phenomenon obtains with cis,cis-1,3-cyclooctadiene (see there). The authors propose that 1,3-cyclohexadiene and $SnCl_4$ form an inactive complex. With larger amounts of the diene larger quantities of $SnCl_4$ are converted into the inactive species which explains depressed final conversions at highest monomer conversions. With $BF_3 \cdot OEt_2$ the conversion—time curves did not level off as shown in Figure 27, and reached 100% conversion. Indeed, except for the very early stages, the polymerization was of stationary

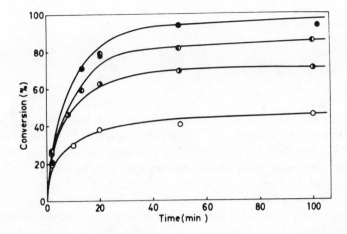

Figure 27. Time—conversion curves for the polymerization of cyclohexadiene by stannic chloride/trichloroacetic acid in methylene chloride at $0°C$. The effect of monomer concentration. $[C]_0$, 5.8 mmole/l; $[H_2O]_0$, 2.8–3.1 mmole/l; $[M]_0$ (○) 2.63 mole/l, (◑) 1.58 mole/l, (◒) 1.05 mole/l, (●) 0.53 mole/l (326).

character. The overall rate law was $R_p = k \, [M] \, [I]$, where $[M]$ and $[I]$ are monomer and initiator concentrations, respectively, $k = 0.2 \pm 0.5$ in CH_2Cl_2 at $0°C$ and $k = 0.03 \pm 0.003$ 1/mole min in benzene at $0°C$.

15. 1-Methylene-2-cyclohexene   CH_2

The cationic polymerization of 1-methylene-2-cyclohexene (neat monomer, BF_3, $-20°C$, 2 days) was first reported by Bailey and Grossens (327) who concluded that the repeat unit was mainly of 1,4 enchainment with a minor contribution of 1,2 linkages:

Mabuchi et al. (328) described the rapid polymerization of 1-methylene-2-cyclohexene by Lewis acids, for example, $BF_3 \cdot OEt_2$, $TiCl_4$, $AlEt_2Cl$, $AlCl_3$, and VCl_4, in hexane or toluene at $-78°C$. The white powdery products were largely soluble in chloroform and benzene and could be solution cast to flexible films, The polymers were amorphous, with a softening range of $80–82°C$. Infrared and NMR spectroscopy indicated that the repeat unit was essentially of 1,4 enchainment.

It is noteworthy that this monomer could not be polymerized by common free radical or anionic initiators (benzoylperoxide, azobisisobutyronitrile, butyllithium, or sodium). This behavior is in striking contrast to that of such closely related conjugated dienes as butadiene or isoprene.

16. 1-Vinylcyclohexene

$$H_2C=CH$$

The cationic polymerization and copolymerization of 1-vinylcyclohexene has been investigated by Imanishi et al. (280). The polymerizations were induced by $BF_3 \cdot OEt_2$ and $SnCl_4/CCl_3COOH(TCA)$ in toluene and methylene dichloride at $0°C$. As with other cyclic dienes, for example, cyclopentadiene, 1,3-cyclohexadiene, and 1,3-cyclooctadiene, the polymerization of 1-vinylcyclohexene proceeded by a very fast nonstationary reaction. Polymerization started immediately after initiator addition and final conversions were reached rapidly after this initial burst of polymerization. It is stated that this phenomenon was observed with all initiators examined, that is, $BF_3 \cdot OEt_2$ in toluene and $SnCl_4/$ TCA in toluene and in methylene dichloride at $0°C$. In some contrast to the 1-vinylcyclohexene system, with the other cyclic dienes under certain conditions the first nonstationary rapid polymerization is followed by a much slower stationary phase. Table 15 is a compilation of the systems investigated and the type of conversion—time plots found. With the present monomer, higher conversions can be obtained only by the introduction of larger amounts or repeated amounts of initiator. Yields are directly proportional to initiator concentration.

As in the interesting situation prevailing with 1,3-cyclohexadiene and 1,3-cyclooctadiene, an inverse relation between the initial monomer concentration $[M]_0$ and ultimate polymer yield $Y\infty$ exists with 1-vinylcyclohexene. The product $[M]_0 Y\infty$ was found to be independent of $[M]_0$ and was ~ 0.55 mole/l.

It is indeed difficult to explain this phenomenon with the hypothesis of Imanishi et al. (326) according to which fast termination and the inverse relation between the monomer concentration and ultimate conversion are due to the formation of a polymerization-inactive complex M_2C (consisting of two molecules of monomer M and one molecule of the metal halide C), which in turn was formed from a polymerization-active MC complex:

$$M + C \longrightarrow MC \underset{-M_1 \quad k_3}{\overset{+M_1 \quad k_1}{\rightleftharpoons}} M_2C$$

$$\downarrow {+TCA \atop k_2}$$

$$P\oplus \xrightarrow{\underline{+M}} \text{polymer}$$

TABLE 15 A CLASSIFICATION OF TIME VERSUS CONVERSION PLOTS OF CONJUGATED CYCLIC DIENE SYSTEMS

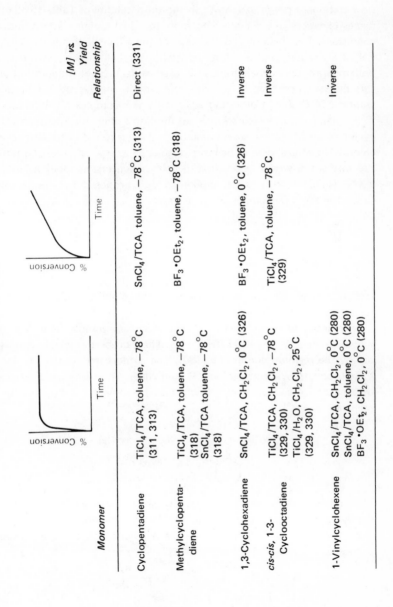

Monomer	Type of Conversion vs. Time Plots		$[M]$ vs Yield Relationship
Cyclopentadiene	TiCl$_4$/TCA, toluene, -78°C (311, 313)	SnCl$_4$/TCA, toluene, -78°C (313)	Direct (331)
Methylcyclopenta-diene	TiCl$_4$/TCA, toluene, -78°C (318) SnCl$_4$/TCA toluene, -78°C (318)	BF$_3$·OEt$_2$, toluene, -78°C (318)	
1,3-Cyclohexadiene	SnCl$_4$/TCA, CH$_2$Cl$_2$, 0°C (326)	BF$_3$·OEt$_2$, toluene, 0°C (326)	Inverse
cis-cis, 1-3-Cyclooctadiene	TiCl$_4$/TCA, CH$_2$Cl$_2$, -78°C (329, 330) TiCl$_4$/H$_2$O, CH$_2$Cl$_2$, 25°C (329, 330)	TiCl$_4$/TCA, toluene, -78°C (329)	Inverse
1-Vinylcyclohexene	SnCl$_4$/TCA, CH$_2$Cl$_2$, 0°C (280) SnCl$_4$/TCA, toluene, 0°C (280) BF$_3$·OEt$_2$, CH$_2$Cl$_2$, 0°C (280)		Inverse

Imanishi et al. proposed this scheme (326) to explain the cationic polymerization behavior of 1,3-cyclohexadiene and 1,3-cyclooctadiene monomers which both show first a fast nonstationary polymerization phase followed by a second slow stationary phase. As shown in the third column of Table 15 all conjugated dienes except 1-vinylcyclohexene show this dual kinetic pattern under specific conditions. For example, in the case of the 1,3-cyclooctadiene—TiCl$_4$/TCA in CH$_2$Cl$_2$ system only the fast, nonstationary behavior is observed; however, both phases occur when toluene is used as the diluent. Imanishi et al. explain this by assuming that in toluene (but not in methyl chloride) the inactive complex M$_2$C slowly decomposes to the active complex MC plus monomer M. This totally unsupported *ad hoc* assumption cannot be employed with 1-vinyl-cyclohexene because a slow phase has not been found in toluene or in methyl chloride. It would seem to be more reasonalbe to argue that fast termination and the inverse monomer concentration versus conversion relationship is due to allylic (suicide) termination similar to that proposed by Mondal and Young (330) for the 1,3-cyclooctadiene system:

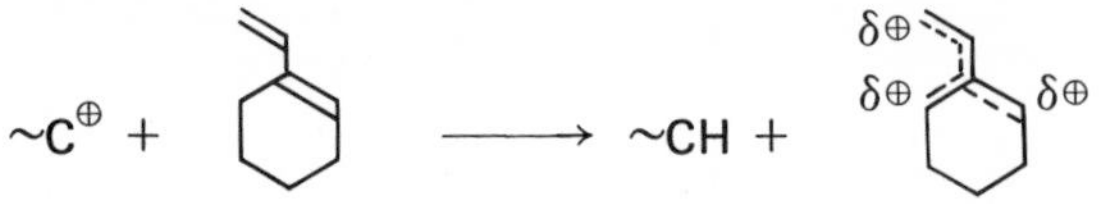

and to assume that the newly formed highly stabilized allylic ion is unable to sustain even a slow reinitiation.

Cationically polymerized 1-vinylcyclohexene is a soluble, white, low molecular weight product $\overline{M}_n \sim 1200$ (DP $\sim$ 11). These properties remain almost unaffected by the reaction conditions (280). The authors investigated the structure of their products by bromination, IR, and NMR. The spectra indicated a preponderance of 1,4 enchainment accompanied by some 1,2 linkages in the polymer prepared by SnCl$_4$/TCA:

The polymer produced by BF$_3 \cdot$OEt$_2$ shows evidence for the presence of methyl groups that, according to the authors, cannot be explained as coming from the initiator complex. Instead a nebulous isomerization mechanism was proposed:

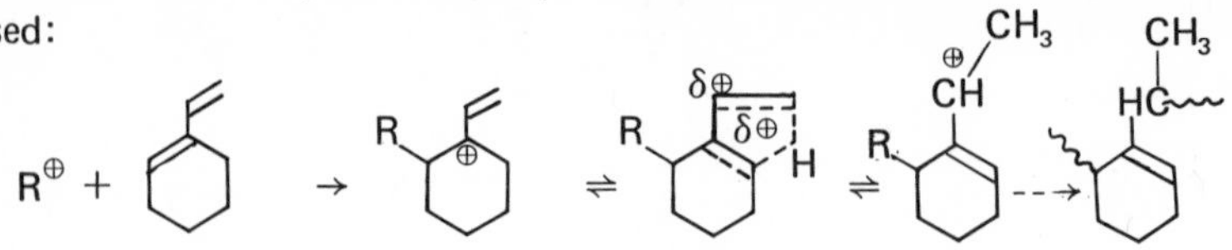

It is much simpler to assume a rapid, acid catalyzed protonation-proton elimination sequence giving rise to an energetically favorable diolefin which is then incorporated into the polymer:

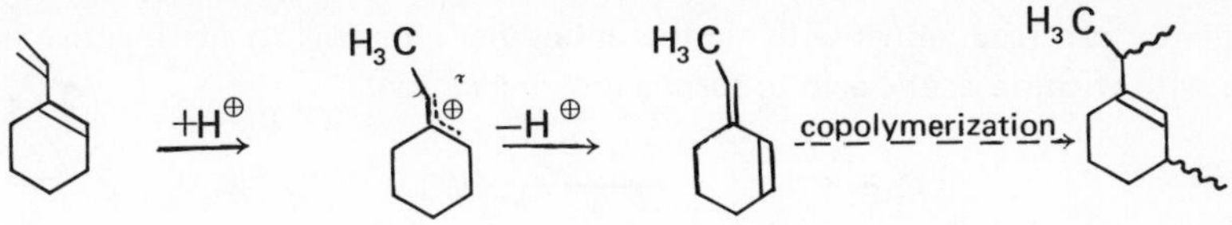

17. 1,2-Dimethylenecyclohexane

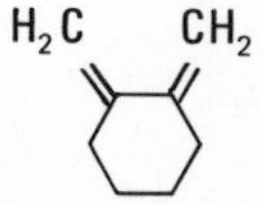

1,2-Dimethylenecyclohexane was polymerized in the bulk with $BF_3 \cdot OEt_2$ at Dry Ice temperature for 3 days. Under these conditions 10% conversion gave a white, highly crystalline polymer, m.p. = $164-165°C$, with the proposed repeat structure (332):

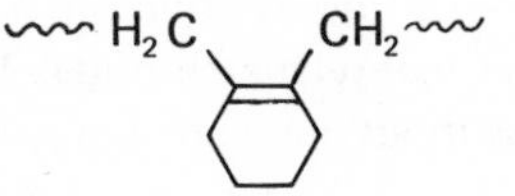

The polymer was identical in all respects with the material obtained by emulsion polymerization of the same monomer (m.p. = $164.5-165°C$). When the monomer was treated with gaseous BF_3 at room temperature, a liquid oligomer was obtained.

18. cis,cis-1,3-Cyclooctadiene

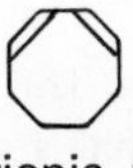

In the course of their studies on the cationic polymerization behavior of cyclic dienes, Imanishi et al. (329) also investigated the cis,cis-1,3-cyclooct-adiene/$TiCl_4$/CCl_3COOH/CH_2Cl_2 or $C_6H_5 \cdot CH_3$ system at $-78°C$. In methylene chloride diluent a burst of polymerization ensued on initiator addition, however, the reaction rapidly stopped and leveled off at conversion levels determined by the amount of initiator added. There was a straight relation between the amount of monomer converted and the initiator concentration, however, molecular weights were independent of initiator concentration, $[\eta] = 0.05$, $\overline{M}_n \sim 1600-1800$, at all conversions. In toluene, the situation was somewhat different: after initiator addition and the first burst of polymerization, the rate did not level off but progressed steadily albeit more slowly to higher conversions. Interestingly an inverse relationship between the yields (conversion) and initiator concentration was found. This situation is very similar to that with 1,3-cyclohexadiene (see there). To explain this phenomenon the authors assume a two-step mechanism: first monomer M and metal halide C give a complex MC, which subsequently can react either with another monomer molecule to give inactive species or with trichloroacetic acid to form a growing cation:

$$M + C \xrightarrow{\quad fast \quad} MC$$

The first, fast polymerization phase is supposed to be initiated by the initially formed MC complex, while the second, slow period (in toluene) is due to a slow decomposition $M_2C \longrightarrow M + MC$ followed by a reaction with trichloroacetic acid $MC + TCA \longrightarrow P^{\oplus}$. The fact that the second, slow polymerization occurs only in toluene is explained by assuming that the decomposition of the M_2C complex is unfavorable in methylene chloride. These peculiar assumptions are difficult to take seriously without any corroborating evidence.

The results of Imanishi et al. (329) have also been criticized by Mondal and Young (330) who investigated kinetic aspects of the $TiCl_4/H_2O$ initiated polymerization of 1,3-cyclooctadiene. These authors confirmed the finding that the polymerization consists of two phases—a rapid, nonstationary phase whose initial rate follows the law:

$$Rate = k \ [TiCl_4] \ [H_2O] \ [Monomer]$$

and a slow, stationary phase. The ultimate yields are directly proportional to the concentration of active chains but inversely proportional to the monomer concentration. The second, stationary phase is much slower. The effect of monomer (and $TiCl_4$) concentration on the yield obtained by both groups (329, 330) is shown in Figure 28. Mondal and Young (330) explained this phenomenon by postulating allylic termination with unreacted monomer:

The concept of allylic (suicide) termination has been treated in some depth

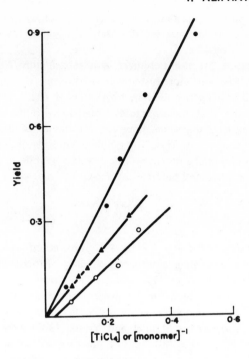

Figure 28. Dependence of yield upon [TiCl$_4$] and [monomer]. Vertical axis: yield; horizontal axis: [TiCl$_4$] (mole/l) and [monomer]$^{-1}$ (l/mole^{-1}). (●) Temperature, -78°C, [TiCl$_4$] = 0.227 mole/l [H$_2$O] = 0.003 mole/l. (○) Temperature, -78°C [monomer] = 6.7 mole/l. [H$_2$O] = 0.002 mole/l. (▲) Data of reference 329 [TiCl$_4$] = [TCA] = 0.0336 mole/l, (temperature, -78°C) (330).

in conjunction with isobutylene (see there). The key equation derived by making the above simple assumption is:

$$\text{Yield} = \frac{k_p}{k_{mt}} \cdot \frac{[P^{\oplus}]_0}{[M]_0}$$

where k_p and k_{mt} are the rate constants for propagation and chain transfer with monomer and $[P^{\oplus}]_0$ and $[M]_0$ are the initial concentrations of active chain ends and monomer, respectively. This expression smoothly explains the phenomenon of the inverse relation between yields and monomer concentration.

The slow, stationary polymerization that follows the first burst is initiated by the stable allylic cations formed during termination (or rather slowdown) of the first phase. It is reasonable to expect that reinitiation of the second phase (either by direct initiation by the allyl cation or by proton elimination) will be slow because it involves the destruction of highly resonance stabilized secondary allylic cations.

The microstructures of the polymers synthesized with $TiCl_4$, $TiCl_4/TCA$, $SnCl_4/TCA$, and BF_3 initiator systems in methylene chloride and toluene solvents at $-78°C$ have been studied by Imanishi et al. (329). Infrared and NMR spectroscopy showed that the structures obtained with the various initiator systems were essentially the same, however, the nature of the solvent importantly affected enchainment. Thus in methylene chloride, polymerization gives almost exclusively 1,4 enchainment and no branching while in toluene there is 1,4 enchainment and branchy polymer is formed:

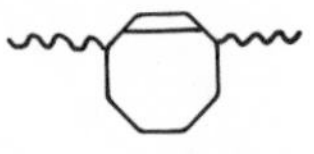

Whether the branch sites are 1,2 or 1,4 units could not be determined.

All the polymers were white, amorphous powders of low intrinsic viscosities , $[\eta]$ = 0.10, softening range $172-184°C$. The linear products obtained in toluene solvent were soluble in aromatic and chlorinated hydrocarbons, however the branchy materials were insoluble, indicating crosslinking.

The products prepared by Mondal and Young (330) were also white, low molecular weight ($\overline{M}_n > 10,000$), soluble powders, softening points $\sim 200°C$, and easily oxidizable in air. Spectroscopic analysis indicated some branches.

19. p-Xylylene

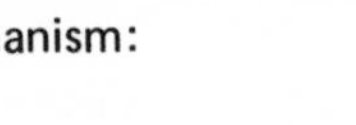

This extremely reactive conjugated multiolefin is rapidly polymerized with small amounts of BF_3, $SbCl_5$, $AlCl_3$, $TiCl_4$, H_2SO_4, and CCl_3COOH in an inert solvent, for example, hexane, at $-78°C$ (333). The following reaction sequence was proposed for the polymerization mechanism:

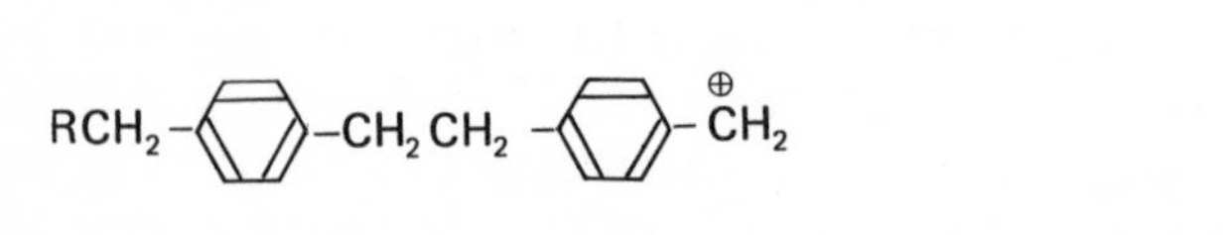

Very few details concerning the polymerization conditions are given. The molecular weights were probably very low since compression molded translucent buttons were extremely brittle.

20. 1-Methyl-4-Isopropyl-1,3-Cyclohexadiene (α-Phellandrene) and

21. 1-Methyl-4-Isopropyl-2-Cyclohexene (β-Phelladrene)

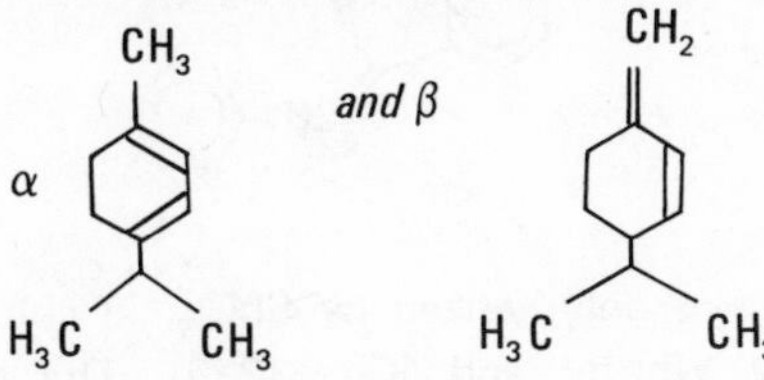

The polymerization of phellandrene (probably the α-isomer, although the exact identity or composition of isomer mixture has not been given) by trichloroacetic acid to products of DP = 9—10 has been mentioned (613a).

22. Benzofulvene

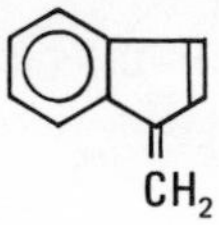

Maréchal mentions (334) the polymerization of benzofulvene with $TiCl_4$ in CH_2Cl_2 for 5 min at $-72°C$. Under these conditions a soluble yellow powdery polymer was obtained with 80% conversion, $[\eta] = 0.2$, softening point 300°C, yellowing (probably, thermal degradation) at ~265°C. Bromine analysis indicated about one double bond per repeat unit. Beyond this, structure analysis has not been reported.

23. Methylbenzofulvene

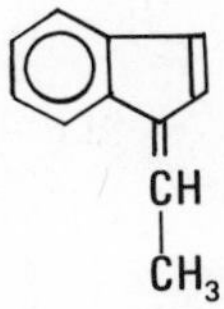

Maréchal (334) mentions the polymerization of this monomer with $TiCl_4$ in CH_2Cl_2 at $-72°C$, with a 100% yield of a colorless solid, softening point = 250°C, yellow discoloration at ~230°C, $[\eta] = 0.14$. There is one double bond per repeat unit. No further information is available.

24. Ethylbenzofulvene

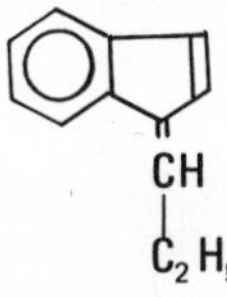

Maréchal (334) discusses the polymerization of this monomer using $TiCl_4$ in CH_2Cl_2 for 5 min at $-72°C$ with a 94% yield of a colorless polymer, softening point = 250°C, no discoloration upon heating to this temperature, $[\eta]$ = 0.16. One double bond per repeat unit has been reported.

25. Benzalindene

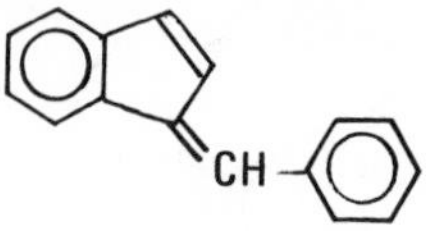

This monomer was polymerized by SbCl$_5$ in chloroform solution at ambient temperature by Whitby and Katz (335). The precipitated product was a yellow powder, m.p. = 252–255, DP = 6.

26. Cinnamalindene

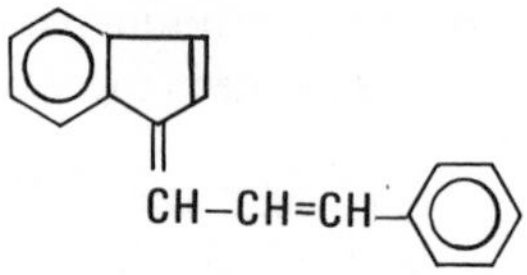

Whitby and Katz (335) polymerized this monomer with SbCl$_5$ and SnCl$_4$. The reaction between SbCl$_5$ and cinnamalindene was so vigorous that it led to charring, hence SnCl$_4$ was preferred. The polymerization was carried out in chloroform, and the solid product was purified by reprecipitation to give a yellow powder, m.p. = 238–242°C, DP = 4.

27. Cinnamalfluorene

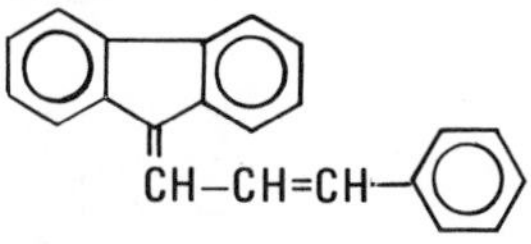

This monomer was polymerized by Whitby and Katz (335) who used SbCl$_5$ in chloroform at ambient temperatures and obtained a solid of DP = 8. The soluble polymer was fractionated (from benzene into methanol) and the melting points of the fractions were determined. For the highest fraction, DP = 12 and m.p. = 365–368°C and for the lowest fraction, DP = 4 and m.p. = 286–290°C. According to these authors TiCl$_4$, SnCl$_4$, AlCl$_3$, AlBr$_3$, and so on are less active polymerization initiators than SbCl$_5$.

28. 1-Isopropylidene-3a,4,7,7a-tetrahydroindene

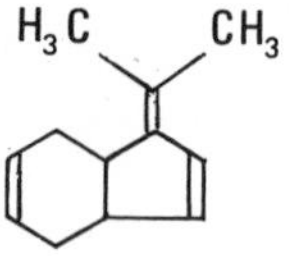

Cesca et al. (336) synthesized and investigated the cationic polymerization of

1-isopropylidene-3a,4,7,7a-tetrahydroindene. While free radical and anionic initiators gave only oligomers, a wide variety of Friedel—Crafts acids, for example, AlEtCl$_2$, BF$_3$·OEt$_2$, SnCl$_4$, TiCl$_4$, and, in particular, the AlEt$_2$Cl / tBuCl system, readily produced high polymer. Most of the work was carried out with premixed AlEt$_2$Cl/tBuCl 1:3 systems in   toluene solvent.

If the aim of the synthesis is the production of high molecular weight polymer in high yield, this technique has two disadvantages. First, premixing the alkylaluminum halide with the alkyl halide results in reduced initiator efficiency because the AlEt$_2$Cl + tBuCl reaction gives tBuEt (neohexane) + AlEtCl$_2$, removes the essential initiating species, and, particularly at high AlEt$_2$Cl/tBuCl ratios, produces more highly chlorinated aluminum compounds, even AlCl$_3$, with diminished initiating activity. Indeed the authors mention that the AlEt$_2$Cl/tBuCl system gives homogeneous solutions in chlorinated or aromatic solvents but is insoluble in aliphatic hydrocarbons. This may be due to the *in situ* formation of sparingly soluble AlCl$_3$ or AlEtCl$_2$·AlCl$_3$. Secondly the use of toluene as solvent is disadvantageous as it may result in greatly diminished molecular weights because of alkylative transfer reactions. Thus not too surprisingly the use of premixed AlEt$_2$Cl/tBuCl systems in toluene gave high conversions only at comparatively low monomer/AlEt$_2$Cl ratios (~100) and modest molecular weights even at $-78°$C, M_n = 9350. Similarly the premixing of various Brønsted acids, for example, HCl, H$_2$O, CCl$_3$OH, and CH$_3$COOH, with AlEt$_2$Cl produced very low conversions and very low molecular weight products.

All the products were white, amorphous, soluble powders, with a softening range of 150—170°C. The molecular weights of the methanol insoluble polymers were generally low and the highest value reported was $\overline{M}_n$ = 15,700, [η] = 0.176; molecular weight dispersities were $\overline{M}_w/\overline{M}_n$ ~ 1.3—2.4.

Infrared and NMR spectroscopy and chemical evidence indicate that the major repeat unit of the polymer arises by a 1,4-type enchainment:

29. *1-Isopropylidenedicyclopentadiene*

Cesca et al. (337) polymerized this conjugated diene by a variety of Lewis acid initiator systems, for example, BF$_3$·OEt$_2$, TiCl$_4$, EtAlCl$_2$, Et$_2$AlCl, and Et$_3$Al/ tBuCl in methylene chloride, n-heptane, and toluene diluents in the range 0 to

−78°C. Under most conditions completely soluble products were obtained. This finding is remarkable in view of the highly active sites of unsaturation in the monomer. The molecular weights are quite high (frequently $\overline{M}_n > 50,000$). The authors fractionated polymer samples, determined $\overline{M}_n$ and $[\eta]$ values, and calculated the a exponent in the Mark-Houwink equation. On the basis of the high value of this exponent, $a = 0.74$, the authors propose an essentially linear high polymer. Extensive spectroscopic investigations substantiate this proposition and the idea that the repeat unit of the polymer arises by a 1,4-type enchainment:

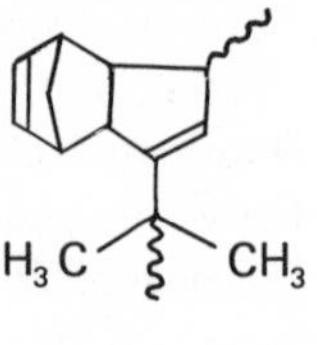

This monomer was copolymerized with 1-isopropylidene-3a,4,7,7a-tetrahydroindene using $BF_3 \cdot OEt_2$ in methylene chloride at −78°C. The reactivity ratios indicate that the former monomer is more reactive ($r = 2.29 \pm 0.68$) than the latter ($r = 1.05 \pm 0.6$), which is attributed to the additional strain caused by the methylene bridge in the former molecule.

30. Hydroxybenzylbenzalindene

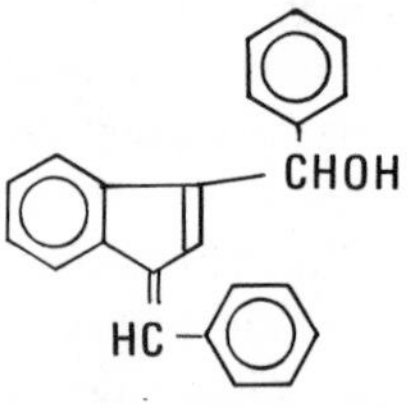

Whitby and Katz (335) polymerized this monomer with $SbCl_5$ and $SnCl_4$. The $SbCl_5$ induced polymerization in chloroform at ambient temperature gave after repeated precipitation (from chloroform into alcohol) a deep yellow colored powder, m.p. = 297−300°C, average DP = 6. The polymerization with $SnCl_4$ proceeded more slowly than that with $SbCl_5$, m.p. = 195−200°.

F. SMALL RING COMPOUNDS

1. Cyclopropane

The polymerization of cyclopropane by cationic initiators was first studied by Tipper and Walker (338). The work was performed not much after the studies of Fontana and Kidder (30, 31) on propylene appeared (see under Propylene) and was strongly influenced by the latter. Thus Tipper and Walker used equimolar $AlBr_3/HBr$ mixtures as initiators in n-heptane solvent in the range +22 to −78°C. In contrast to Fontana and Kidder, Tipper and Walker carried out experiments under rigorously anhydrous high vacuum conditions. Measurable rates were obtained for dilatometric experiments in the range −10 to −78°C. The overall activation energy calculated from the straight Arrhenius rate plots was 6±1 kcal/mole. The degrees of polymerization were increasing from DP ~5.8 at +25°C to ~15.7 at −78°C. The overall activation energy of the polymerization calculated from log DPs obtained at low temperatures was −1.6 kcal/mole. No reaction took place in nitroethane diluent (AlBr$_3$ ·CH$_3$NO$_2$) and the addition of water, methanol, or trichloroacetic acid to the cyclopropane /AlBr$_3$ /n-heptane system completely inhibited the polymerization in the absence or presence of HBr. Iodine retarded the polymerization.

The structure of the oligomer was analyzed by infrared spectroscopy and it was stated that "no appreciable branching was apparent in the polycyclopropane." In view of the structure work on polypropylene (see propylene), this is an unexpected conclusion. Unfortunately the spectra have not been published and further independent confirmation with modern high resolution spectroscopy is needed before this statement can be accepted. Interestingly, the alkylation of benzene with cyclopropane in the presence of AlCl$_3$ at low temperatures affords n-propylbenzene as the only monoalkylated product (339, 340). At higher temperatures (65°C) with sulfuric acid isopropylbenzene was obtained (339).

In line with contemporary thinking are the results and conclusions of Pinazzi et al. (341). These authors polymerized cyclopropane in sealed tubes with gaseous BF$_3$ in methylene chloride at 80°C for 4 hr. They obtained in 42% yield an oligomer of molecular weight 500, or DP 11. Infrared and NMR spectroscopy indicated a linear polypropylene structure for the oligomers. The polymerization was postulated to involve intramolecular hydride migration.

2. Ethylcyclopropane   ▷−CH$_2$−CH$_3$

Pines et al. (342) mention the conjunct polymerization of ethylcyclopropane in the presence of AlCl$_3$ plus HCl at 25°C.

3. n-Propylcyclopropane   ▷−CH$_2$CH$_2$CH$_3$

Pines et al. (342) mention the facile conjunct polymerization of n-propylcyclopropane with AlBr$_3$ plus HBr or with AlCl$_3$ at 24°C.

4. Isopropylcyclopropane
$$\triangleright - \underset{\underset{CH_3}{|}}{\overset{\overset{CH_3}{|}}{CH}}$$

Isopropylcyclopropane was polymerized with AlBr$_3$ in the range −10 to −50°C to a pale yellow liquid with low degrees of polymerization (DP $\sim$ 3–4) (343). Infrared and NMR spectroscopic analysis indicated the following repeat unit:

$$-CH_2-CH_2-\underset{\underset{H_3C-CH-CH_3}{|}}{CH}-$$

This interesting and unexpected finding suggests that the polymerization of at least this cyclopropane derivative does not proceed by conventional carbenium ion intermediates. A carbenium ion mechanism would include the following rearrangement–propagation:

$$-CH_2-CH_2-\underset{\underset{H}{\underset{|}{\overset{|}{H_3C-C-CH_3}}}}{\overset{\oplus}{CH}} \qquad -CH_2-CH_2-CH_2-\underset{\underset{CH_3}{|}}{\overset{\overset{CH_3}{|}}{C\oplus}}$$

and would lead to *gem*-dimethyl substituents in the polymer.

5. *1,1-Dimethylcyclopropane*

1,1-Dimethylcyclopropane is readily polymerized in ethyl chloride diluent or in bulk with $AlBr_3$ or $AlCl_3$ at $-50°C$ to a low molecular weight ($\sim$1000) product (342, 344). No reaction occurs below $-50°C$ or with $BF_3\cdot OEt_2$. On the ba-sis of infrared evidence the repeat unit of the oligomer was proposed to be (344):

$$-CH_2-CH_2-\underset{\underset{CH_3}{|}}{\overset{\overset{CH_3}{|}}{C}}-$$

that is, identical to the repeat structure of the polymer obtained by the cationic polymerization of 3-methyl-1-butene (see there). Japanese investigators (345) found that among $AlBr_3$, $AlCl_3$, $SnCl_4$, $TiCl_4$ and $BF_3\cdot OEt_2$, only $AlBr_3$ was active in methylene chloride at $0°C$. Effective polymerization occurred ($>$90% conversion in $\sim$30 min) in *n*-hexane as well, however, the degree of polymerization of the viscous liquid obtained was quite low (DP $\sim$ 5).

6. *1,2-Dimethylcyclopropane*

Pines et al. (342) mention the conjunct polymerization of 1,2-dimethylcyclo-propane in the presence of $Al(OH)Cl_2$ at $24°C$.

7. *Bicyclo [3.1.0] hexane*

8. *Bicyclo [4.1.0] heptane*

9. *Bicyclo [5.1.0] octane*

10. *Bicyclo [6.1.0] nonane*

11. *Bicyclo [10.1.0] tridecane*

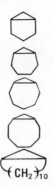

These bicyclo[n.1.0] alkanes, where n = 3,4,5,6, and 10 are listed together as there is only one group of investigators who studied their polymerization behavior under cationic conditions (316, 341). Thus Pinazzi et al. (341) polymerized the above compounds with a variety of Lewis acids, such as BF_3, $BF_3 \cdot OEt_2$, $AlBr_3$, $SnCl_4$, and $TiCl_4$, in methylene chloride diluent in the temperature range 20–80°C. Polymerization did not occur at the low temperatures (<0°), usually employed in cationic polymerizations. Generally polymerizations proceeded slowly to moderate or low yields and gave mostly trimers. The highest molecular weight product, a decamer, was obtained with the compound n = 4. The structure of the oligomers was analyzed by IR and NMR techniques. According to these investigators cationic attack occurs at the cyclopropane ring which probably opens in a concerted fashion to yield a propagating cyclic carbenium ion:

12. *1-Methylbicyclo [4.1.0] heptane*

13. *2-Methylbicyclo [4.1.0] heptane*

14. *3-Methylbicyclo [4.1.0] heptane*

The polymerization of 1-methyl-, 2-methyl-, and 3-methylbicycloheptanes with $SnCl_4$, $TiCl_4$, and $AlBr_3$ in methylene chloride solvent was investigated at relatively high temperatures (80°C with $AlBr_3$) by Pinazzi et al. (346). While very little details are given, it is mentioned that NMR spectroscopy suggests, two $-CH_3$ groups per repeat unit. On the basis of this information the polymerization may involve the opening of the cyclopropane ring in a manner similar to that postulated with bicyclo[4.1.0] heptane and bicyclo[n.1.0] alkanes (341) (see there).

15. *1-Methylbicyclo[5.1.0]octane*

16. *1-Methylbicyclo[6.1.0]nonane*

17. *1-Methylbicyclo[10.1.0]tridecane*

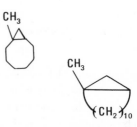

These macrocyclic monomers have been polymerized by $VOCl_3$, $BF_3 \cdot OEt_2$, $TiCl_4$, and $AlCl_3$ in methylene chloride at $80°C$ (216, 347). Product molecular weights were similar to those obtained with the unsubstituted bicyclic structures, for example, $3 < DP < 5$.

Pinazzi et al. propose the following structure for the repeat unit:

where n = 5,6, and 10.

18. *Spiropentane*

Spiropentane was oligomerized at low rates to soluble products (DP $\sim$ 10—15) with a variety of Lewis acids, for example, $AlCl_3$, WCl_6, $ZrCl_4$, $MoCl_5$, $BF_3 \cdot OEt_2$, $TiCl_4$, and $SnCl_4$, in methylene chloride diluent at 20—80°C (216, 348, 349). According to results of IR and NMR spectroscopic investigation, the structure of the oligomer for the most part resembles that of cyclized polyisoprene.

19. *2,4-Spiroheptane*

20. *2,5-Spirooctane*

The polymerization of these two spiro compounds was attempted with $AlBr_3$

(84). At 35–40°C an exothermic reaction takes place. The structures of the products have not been determined, but their infrared spectra do not resemble those of poly(vinylcyclopentane) or poly(vinylcyclohexane).

21. 2,6-Spirononane

22. 2,7-Spirodecane

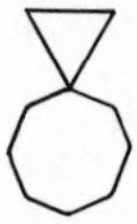

23. 2,11-Spirotetradecane

$(CH_2)_{11}$

These spiro compounds were polymerized by Pinazzi et al. (216) "in a cationic medium of methylene chloride at relatively high temperatures in the presence of cationic catalysts." The oligomers allegedly had the repeat unit:

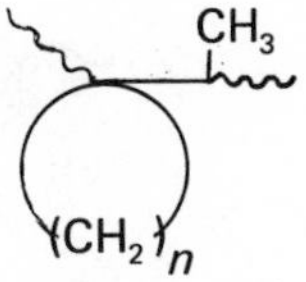

CH_3

$(CH_2)_n$ where n = 6,7, or 10

24. Dicyclopropyl

Pinazzi et al. (216) mention that "in cationic medium polymerization [of dicyclopropyl] occurs by ring opening of the two cyclopropane structures and the formation of polymers structured as":

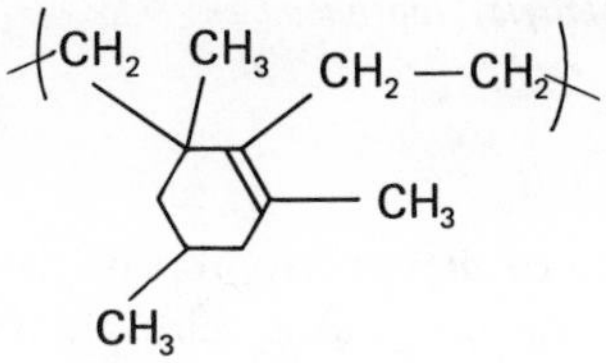

25. Phenylcyclopropane

Murahashi et al. (350) investigated the cationic polymerization of phenylcyclopropane with BF_3, $AlBr_3$, and $AlCl_3$ in methylene chloride in the range 25 to −78°C. No product was obtained at −78°C and low oligomers formed at 0 and $-20°C$. The highest polymer $(\overline{M}_n$ 1400) was obtained with $AlCl_3$ at room temperature after 31 hr, however, the product contained only 35.1% methanol insoluble fraction. Spectroscopic analysis (IR, NMR, and UV), chemical analysis, and oxidation studies indicated a linear structure:

Propagation is visualized to be a repetitive alkylation of the phenyl ring by cyclopropyl cations.

Contrary to these finding Aoki et al. (345) who also reacted phenylcyclo - propane with $AlBr_3$ in n-hexane at $0°C$ and obtained a viscous liquid in 90% conversion, interpret their infrared spectra to indicate that the propagation step takes place as follows:

26. *1,1-Dichloro-2,2-dimethylcyclopropane*

27. *1,1-Dichloro-2,2,3-Trimethylcyclopropane*

28. *1-Chloro-1-bromo-2,2-dimethylcyclopropane*

The cationic polymerization of these cyclopropane derivatives is mentioned by Pinazzi et al. (216) who state that Lewis acid initiators yield oligomers with the structures such as:

One mole of HCl is lost per monomer unit during polymerization.

29. *1,1-Dichloro-2-phenylcyclopropane*

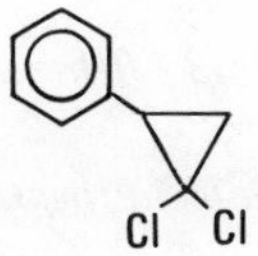

30. *1,1-Dichloro-2-phenyl-2-methylcyclopropane*

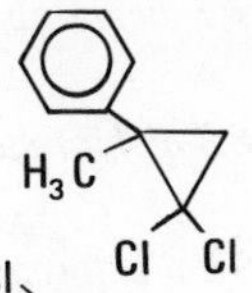

31. *1,1-Dichloro-2-p-chlorophenyl-2-methylcyclopropane*

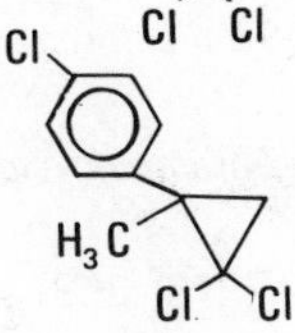

Pinazzi et al. (216) mention the oligomerization of these compounds to structures with, for example, the following repeat unit:

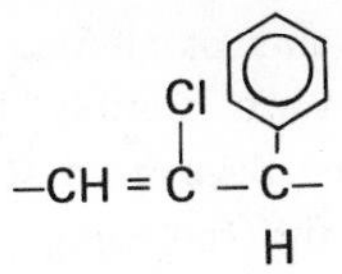

$$-CH = \underset{\underset{H}{|}}{C} - \overset{\overset{Cl}{|}}{C} -$$

32. *1-Methyl-1-vinyl-2,2-dichlorocyclopropane*

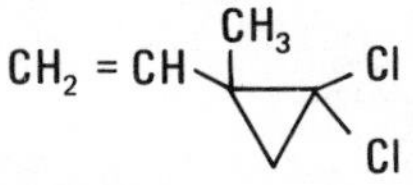

Pinazzi and coworkers studied the polymerization of 1-methyl-1-vinyl-2,2-di-chlorocyclopropane with Lewis acids such as WCl_6, $TiCl_4$, and $SnCl_4$ in methylene chloride solvent at 20 and 80°C (216, 348, 351). Liquid oligomers ($\bar{M}_n$ 700–1000 or DP 4.5–6.5) were obtained. Structure analysis by NMR spectroscopy suggest for the predominant repeat units:

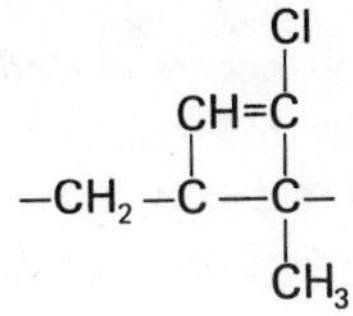

and

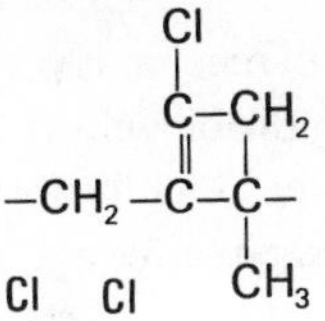

33. *7,7-Dichlorobicyclo[4.1.0]heptane*

34. *8,8-Dichlorobicyclo[5.1.0]octane*

$n = 4,5,6$

35. *9,9-Dichlorobicyclo[5.1.0]nonane*

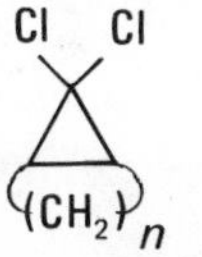

36. *1-Methyl-6,6-dichlorobicyclo [3.1.0] hexane*

37. *1-Methyl-7,7-dichlorobicyclo [4.1.0] heptane*

38. *1-Methyl-8,8-dichlorobicyclo [5.1.0] octane*

39. *1-Methyl-9,9-dichlorobicyclo [6.1.0] nonane*

40. *7-Chloro-7-bromobicyclo [4.1.0] heptane*

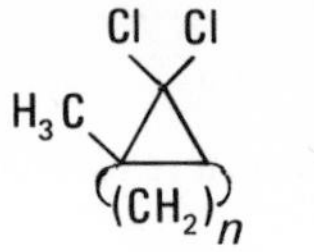

$n = 3,4,5,6$

Pinazzi et al. (216, 352) synthesized this series of dichlorobicyclo[n.1.0]alkanes and 1-methyl-dichlorobicyclo[n.1.0] alkanes by reacting the corresponding cyclo-alkenes with dichloro carbene and polymerizing them with $TiCl_4$, $AlCl_3$, and $SnCl_4$ in methylene chloride solvent in sealed tubes at temperatures above 40°C. Conversions were modest (10–20%) and no polymerization occurred below 40°C. The molecular weights of all the products were in the oligomer range, 500–1500, and they were soluble in common organic solvents.

The structures of the oligomers have been studied by NMR and IR spectro-scopy on the basis of which the following polymerization mechanism was proposed:

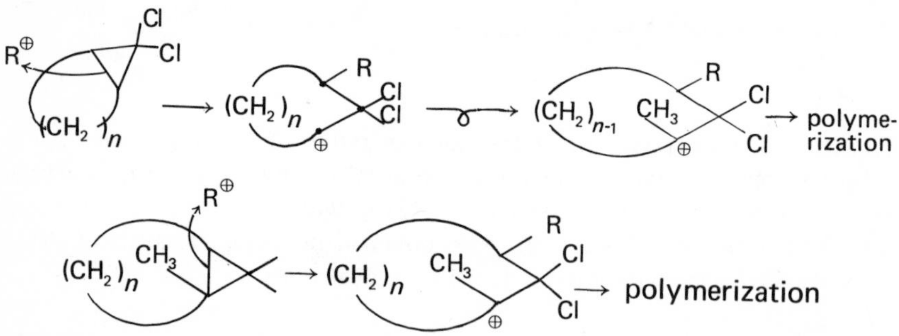

According to this mechanism, ring contraction occurs during the oligomeriza-tion in the dichlorobicyclo[n.1.0] alkane series and the two repeat structures obtained from the dichlorobicyclo[n.1.0] alkanes and the 1-methyl-dichloro-bicyclo[n.1.0] alkane series are, for all practical purposes, identical:

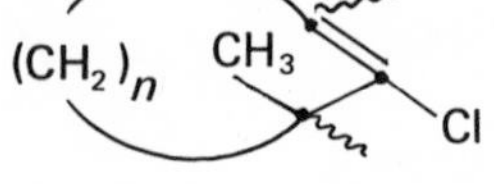

In view of the scanty experimental data available (spectra only described, not shown) and the complexity of the structures and propositions involved, these conclusions should be taken with a grain of salt.

41. *13,13-Dichlorobicyclo[10.1.0]tridecane*

42. *13,13-Dibromobicyclo[10.1.0]tridecane*

43. *1-Methyl-13,13-dichlorobicyclo[10.1.0]tridecane*

44. *1-Methyl-13,13-dibromobicyclo[10.1.0]tridecane*

Pinazzi and coworkers (216, 353) synthesized these macrocyclic compounds by reacting the appropriate cyclododecenes with dihalocarbenes and examined their polymerization behavior with $AlCl_3$, $TiCl_4$, and $SnCl_4$ in methylene chloride solvent. While no experimental details are given, it is stated that appreciable rates are obtained above 50°C and ~30% conversion occurs at 80°C after 24 hr. The average degree of polymerization is 10.

On the basis of rather scanty spectroscopic and chemical evidence the authors propose a 1,3 enchainment for the first two monomers, whereas for the latter two (methyl substituted) compounds they envision a conjugated diene intermediate and 1,4 enchainment:

I and II   →   X = Cl or Br

III and IV   →   →   X = Cl or Br

To substantiate the intermediate formation of the diene, 2-methyl-3-chloro-cyclotridecadiene-1,3 was prepared and polymerized. Spectroscopic inspection indicated that the structures of the polymers obtained from this diene and those from the bicyclic compounds were identical (353).

45. *Bicyclo[6.1.0]nonene-4*

According to Pinazzi et al. (216) the cationic polymerization of this monomer occurs by transannular reaction and yields the following repeat structure:

46. Methylcyclobutane

Pines et al. (342) mention the conjunct polymerization of methylcyclobutane in the presence of $AlBr_3$/HBr or $AlCl_3$/HCl at near ambient temperatures.

47. Ethylcyclobutane

Pines et al. (342) mention the conjunct polymerization of ethylcyclobutane with $AlCl_3$/HCl or $Al(OH)Cl_2$ at room temperature.

48. Isopropylcyclobutane

Pines et al. (342) mention the conjunct polymerization of isopropylcyclobutane by $AlCl_3$/HCl or $AlBr_3$ at 25°C.

G. BICYCLIC OLEFINS

1. α-Pinene

Compared to β-pinene, the polymerization chemistry of α-pinene is rather dull. As mentioned earlier, a good summary of work with α- and β-pinene carried out prior to 1950 can be found in reference 221. Roberts and Day (221) studied the oligomerization of α-pinene with a series of Friedel-Crafts halides, for example, $AlBr_3$, $AlCl_3$, $ZrCl_4$, and BF_3, in toluene at 40–45°C. The softening point of the products was in the range 67–85°C. Further experiments were carried out with $AlCl_3$ which showed maximum yields of solid oligomer at 40°C. The softening points rose from ~65 to ~105°C with a corresponding rise in number average molecular weights from ~600 to ~800 (DP up to ~6). The softening point–molecular weight relation of α-pinene and limonene are similar. Furthermore, some of the physical characteristics (e.g., density and refractive index) of α-pinene oligomers are also very similar to those of limonene oligomers. These and other data were interpreted to suggest that the oligomers obtained from α-pinene were in fact limonene oligomers because prior to oligomerization α-pinene is isomerized to limonene. Terpinolene and the terpines contribute negligibly to the solids. The question of the structure of limonene oligomers was considered and it was concluded that a simple vinyl polymerization cannot explain the results of ozonalysis which indicated less than one double bond per oligomer unit.

Irradiation induced oligomerization of α-pinene at room temperature gave H_2, some isomers, and a trimer (354).

Finnish authors reported (355) that α-pinene does not copolymerize with isobutylene or styrene in the presence of BF_3 or $AlCl_3$ in CH_2Cl_2 in the range −50 to +30°C.

2. β-Pinene

Roberts and Day (221) were the first to systematically investigate the polymerization behavior of α- and β-pinene. A good brief summary of earlier work (prior to 1950) with the pinenes, constituents of turpentine, can be found in their paper. These authors studied the polymerization of β-pinene in toluene solvent at 40–45°C with a variety of Lewis acids, for example, $AlBr_3$, $AlCl_3$, $AlCl_3$·OEt_2, $ZrCl_4$, BF_3, BF_3·OEt_2, and $SnCl_4$; even the mild Friedel-Crafts acids $BiCl_3$, $SbCl_3$, and $ZnCl_2$ were found to have some polymerization activity. The average molecular weight of the solid β-pinene polymers was ~1500 (DP ~ 5). Maximum solid polymer yield was obtained at −10°C. The effect of molecular weight on the softening point of the resins was determined. The softening point rises from ~65 to ~140°C with a parallel increase in the molecular weight from ~1000 to ~2000. Beyond a molecular weight of ~2000 the softening points do not change. Chemical evidence (ozonolysis, refractive indices, etc.) led the authors to conclude that the repeat structure of the β-pinene polymer is:

They also proposed reasonable carbenium ion isomerization mechanism to account for this structure:

β-Pinene can be cationically polymerized to relatively high molecular weights due to a number of favorable factors. First of all, it has a very reactive *exo*-methylene double bond. Secondly, the tertiary cation that is formed upon cationation can escape the "steric trap" by opening the strained four-membered ring. The opening of this ring importantly contributes to the necessary driving force of the polymerization. *exo*-Methylene cyclohexane does not give high molecular weight polymer because propagation is impeded on account of steric compression in the transition state. Thirdly, the opening of the four-membered ring results in the formation of a new tertiary cation.

Further research by others, such as by Marvel and coworkers (356, 356a), confirmed the conclusions of Roberts and Day. Marvel et al. (356) briefly investigated the polymerization of β-pinene with Ziegler-Natta initiators, in particular with iBu$_3$Al/TiCl$_4$ 1:1 systems in n-heptane, and obtained respectable yields over the range +25 to −68°C. Mainly by comparing the infrared absorption spectrum of conventional commercial poly(β-pinene) (Piccolyte S-115) with that of their own products, the authors concluded that the structures of these materials were essentially identical. A further conclusion that may be drawn is that Marvel's initiator contained active cationic sites and that in fact he produced an acidic Ziegler-Natta initiator that initiated a conventional cationic polymerization (he obtained ~70% yield at −68°C). Most of his products were of very low molecular weight, η_{inh} = 0.02–0.1.

Huet and Maréchal (357) have polymerized β-pinene by a variety of Lewis acids, for example, TiCl$_4$, AlBr$_3$, SnCl$_4$, and BF$_3$·OEt$_2$, in methylene chloride in the range +20 to −75°C and in ethyl chloride in the range −75 to −130°C. By mixing various solvents such as vinyl chloride, CHClF$_2$, and CClF$_3$ with ethyl chloride, the polymerization temperature could be lowered, for example, to −150°C, however, the molecular weights did not seem to increase with decreasing temperatures below ~−75°C. The highest inherent viscosity, 0.13, was obtained with a polymer prepared with BF$_3$ in a mixture of EtCl/CHClF$_2$ at −75°C. These authors showed that α- and β-pinene do not copolymerize and that β-pinene does not give a copolymer with styrene when the mole fraction of β-pinene in the charge is larger than 0.5, however, copolymerization does occur when the mole fraction is less than 0.2.

An examination of yield and inherent viscosity data showed that the yield (percent) of methanol-insoluble polymer and η_{inh} increase simultaneously. At a particular BF$_3$·OEt$_2$ concentration (e.g., 0.4 mole/l, the η_{inh} versus time plot showed a maximum (η_{inh} = 0.11 at 30 min). This was interpreted to indicate the simultaneous occurrence of two phenomena: The ascending part of the viscosity curve was attributed to postpolymerization branching rather than to linear chain growth because the final polymer yield was obtained in less than 5 min, that is, much sooner than the maximum viscosity was reached. The descending branch of the curve was interpreted to indicate polymer degradation by excess initiator.

While the first proposition is acceptable, this author cannot subscribe to the second one, that explaining the descending part of the η_{inh} curve. In the absence of better data the descending branch of the plot is ascribed by this author to experimental variation (only one data point necessitated the drawing of the descending part in the curve in Figure C of ref. 357).

Spectroscopic analysis in regard to the polymer repeat structure was in agreement with the conclusions of Roberts and Day.

There are two reports by Finnish workers (355, 358) describing the polymerization, or rather copolymerization, of β-pinene. One of these papers describes the polymerization of β-pinene in the presence of isobutylene or styrene with BF_3 or $AlCl_3$ in methylene chloride solvent in the range -10 to $-50°C$, however, according to the authors, copolymerization did not occur (355). In the second paper, the authors claim that β-pinene can be copolymerized with styrene or α-methylstyrene in m-xylene solvent (358) and they attribute this difference in copolymerization behavior to the nature of the solvents. More evidence is needed before these reports can be accepted.

Indeed, research in this writer's laboratory (359) has proven that β-pinene and isobutylene can be readily copolymerized by a number of Lewis acids, such as BF_3, $AlCl_3$, and $AlEtCl_2$. The reactivity ratios depend on the nature of the Lewis acid and other experimental conditions. For example, with $AlEtCl_2$ in C_2H_5Cl at $-100°C$ $r_{IB} = 0.6$ and $r_{\beta P} = 1.0$. Thus β-pinene is a more reactive cationic monomer than isobutylene. The copolymers, which may contain any proportion of the two monomers, are interesting rubbers (high isobutylene content) or plastics (high β-pinene content). Under suitable conditions azeotropic copolymerization can be achieved.

It is not surprising that isobutylene and β-pinene can be polymerized to random copolymers, considering that both compounds readily homopolymerize by cationic techniques and there is a great similarity in their structures. In regard to their structures, the double bond systems in isobutylene and β-pinene are almost identical:

and the propagating species are also very similar:

The reactivity ratios of the β-pinene isobutylene comonomer pair obtained in methyl chloride solvent at low temperatures are strongly affected by the nature of the Lewis acid and the copolymerization temperature as shown below:

Lewis Acid	Temperature (°C)	r_{IB}	$r_{\beta P}$
$AlEtCl_2$	−50	0.3	3.4
$AlEtCl_2$	−78	0.5	1.9
$AlEtCl_2$	−100	0.6	1.8
$AlEtCl_2$	−110	1.0	1.0
BF_3	−100	6.0	0.7
$AlCl_3$	−100	0.8	1.2

Since all these experiments have been carried out in the polar methyl chloride solvent at temperatures below −78°C, the dielectric constant (ϵ) of the system was certainly much higher than 10, and probably closer to 17. [The ϵ of ethyl chloride at −72°C is 16.5 (360)]. Also, the homopolymerization of β-pinene and isobutylene, and consequently their copolymerization, proceed by a conventional cationic (carbenium ion) propagation and it would be inconceivable to assume a pseudocationic mechanism.

These facts then refute Plesch's hypothesis (361) that "in solvents of ϵ greater than about 10 ion pairs are probably largely irrelevant to the polymerization under a wide range of conditions." Other data that are also in disagreement with Plesch's hypothesis have been obtained in this writer's laboratories (362)*.

The literature on the radiation induced polymerization of α- and β-pinene is of interest not only as a separate field of challenging endeavor, but also because it contains two reports that are significant for the field of cationic polymerization as a whole. The first report by Bates et al. (362a) concerns the effect of water on the polymerization of β-pinene. Their main findings were that polymer yield was extremely sensitive to the amount of water in the irradiation system and that highest yields (G values) were obtained in the driest experiment. Their thoughts on the effect of water on cationic polymerization may be summarized by one simple equation:

$$\sim CH_2-\overset{\overset{\displaystyle CH_3}{|}}{\underset{\underset{\displaystyle R}{|}}{C}}{}^{\oplus}(H_2O)_n \longrightarrow \sim CH{=}\overset{\overset{\displaystyle CH_3}{|}}{\underset{\underset{\displaystyle R}{|}}{C}} + H_3\overset{\oplus}{O}(H_2O)_{n-1}$$

$$\longrightarrow \sim CH_2-\overset{\overset{\displaystyle CH_3}{|}}{\underset{\underset{\displaystyle R}{|}}{C}}{-}OH + H_3\overset{\oplus}{O}(H_2O)_{n-2}$$

*The product ratios obtained with the 2,2,4-trimethyl-1-pentene/Lewis acid/tBuCl/CH_3Cl or CH_2Cl_2 system in the temperature range −40 to −90°C are strongly and characteristically dependent on the nature of the Lewis acid, for example, Me_3Al, Et_2AlCl, and Me_2AlCl, employed.

According to this concept the destruction of the propagating carbenium ion, that is, termination, is determined by the degree of solvation of the hydrodium ion $H_3O^{\oplus}(H_2O)_n$ which in turn is controlled by the amount of water in the system. According to this view the propagating carbocation rapidly attracts by coulombic forces dispersed clusters of water molecules in the low dielectric system, and as soon as a sufficient number of water molecules have surrounded the growing cation to form an energetically much favored solvated hydronium ion, termination takes place by either of the above paths.

This concept and the calculations pertaining to the exothermicity of these reactions under various conditions represent an excellent guide in helping to assess the influence of various amounts of water not only on radiation induced polymerizations but also on cationic reactions in general. Also, for the first time, Bates and coworkers explained the effect of water on termination. Two years later Williams and coworkers (331) published a more comprehensive investigation on the subject in which they confirmed and amplified earlier results, however, conceptually no novel information was described.

The second significant report concerning an allegedly crystalline, high molecular weight (DP $\sim$ 2700, $\overline{M}_n \sim$ 37,000) poly(β-pinene), unfortunately, cannot be examined and has not been followed up. This report appears as a private communication by Leese in the reference section of Pinner's chapter (363). According to this source poly(β-pinene) was obtained by radiation induced polymerization and was fractionated into a chloroform soluble and insoluble fraction. The soluble fraction had a DP $\sim$ 1360 and m.p. = 150–179°C, whereas the insoluble fraction melted at 208–235°C. The latter fraction was soluble in hot chlorobenzene and appeared to have twice the DP of the soluble portion (i.e., DP $\sim$ 2700). The polymer was highly crystalline and the crystals showed birefringence, however the sample prepared by $AlCl_3$ was nonbirefringent. The absence of birefringence was attributed to structural irregularities in the latter polymer, but the basic structures of these materials were thought to be identical. Unfortunately no more information is available on this subject.*

Not only is the molecular weight of this poly(β-pinene) at least an order of magnitude higher than those obtained by chemical catalysis (this, by the way, might not be too surprising since irradiation induced polymerization usually yields highest molecular weight polymers, for example, isobutylene, styrene, cyclopentadiene, and isobutylvinylether), but this material would be the first crystalline β-pinene polymer described and a further investigation of its physical properties would be of theoretical and practical significance.

Finally it should be noted that β-pinene is a constituent of a large number of commercially available resins used in a large number of applications.

---

*It is worth mentioning that crystalline (by x-rays) poly(β-pinene) has been obtained in this writer's laboratory (359) by heating pure β-pinene for extended periods of time under a blanket of nitrogen.

3. Norbornene (Bicyclo[2.2.1]hept-2-ene)

Sartori et al. (364) investigated the polymerization of norbornene with a cationic Ziegler-Natta catalyst, for example, $iBu_3Al/TiCl_4$ = 1:2 in n-heptane at $0°C$, and obtained a crystalline (by x-rays) white powder, intrinsic viscosity 0.1 dl/g (in tetraline at $135°C$). The complete absence of evidence of unsaturation in the NMR spectra and an analysis by infrared spectroscopy led the authors to the conclusion that the polymer consists entirely of the repeat unit:

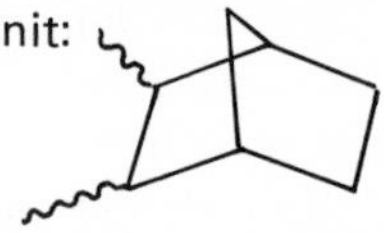

In the course of their studies on the polymerization of norbornene with Ziegler-Natta initiators, Saegusa and coworkers (365) mention the cationic polymerization of this monomer with $AlCl_3$, $TiCl_4$, and $BF_3·OEt_2$ and the synthesis of low molecular weight polymer which the authors assumed was of the above structure. Another group of Japanese scientists (366) employed $MoCl_5$ which produced high polymer. Infrared analysis suggested a ring-opening polymerization.

Kennedy and Makowski (367) were also interested in the polymerization behavior of norbornene and obtained in the presence of $AlEtCl_2$ in ethyl chloride at −78 and −100°C white, soluble solids. While the molecular weights were rather low ($\bar{M}_n$ = 1470 and 1940, respectively), the softening points were quite high, 235 and 260°C, respectively. In contrast to the samples obtained by Sartori et al. (364) these polymers were found to be amorphous by x-rays. On the basis of infrared spectroscopy, Kennedy et al. came to the same conclusion as Sartori et al. in regard to the recurring unit of the polymer. However, in spite of the agreement of these workers concerning the structure of cationically synthesized polynorbornene, this writer has reservations in accepting earlier statements [including his own (367)] in this regard. This reservation is based on the demonstrated exceptional facility for various isomerizations of the norbornyl cation (368). It is suggested that the structure of polynorbornene synthesized by conventional cationic initiators (possibly excluding cationic coordination initiators of the Ziegler-Natta type) is a mixture of various recurring units that have been formed by isomerization of the norbornyl skeleton prior to propagation (74). A possible contributing structure may be the following *syn, exo*-norbornane derivative:

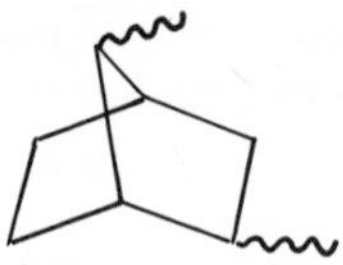

Unfortunately it would be very difficult to prove (or disprove) this suggestion with present day analytical techniques.

4. 5-Methylnorbornene (5-Methylbicyclo[2,2,1]-hept-2-ene)

Takada et al. (369) mention the oligomerization of 5-methylbicycloheptene by typical cationic agencies, for example, $BF_3 \cdot OEt_2$ and $AlBr_3$, to very low yields of colorless or yellowish powders of very low molecular weights.

5. 2-Methylenenorbornane (2-Methylenebicyclo[2.2.1]heptane)

2-Methylenenorbornane is a very reactive but nonpolymerizable cationic monomer (367) which gives only viscous oils (oligomers) with $AlEtCl_2$ neat or in ethyl chloride diluent in the range −30 to −100°C. Competition experiments show that this olefin is much more reactive than norbornene which, nonetheless, readily gives solid polymer with $AlEtCl_2$.

The reason for the nonpolymerizability of 2-methylenenorbornane is the severe steric compression in the transition state of propagation ("buried carbenium ion"):

and the fact that this structure is reluctant to isomerize to a sterically less crowded carbenium ion.

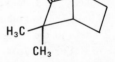

6. Camphene (2-Methylene-3,3-dimethylbicyclo[2.2.1]heptane)

Camphene was first polymerized almost a century ago when Landolph (370) noted that polymers are obtained on heating with BF_3 in a sealed tube at 250°C. Takada et al. (369) briefly mention that the camphene cannot be polymerized by cationic initiators, that is, it gives only small amounts of oligomers when contacted with $BF_3 \cdot OEt_2$, $TiCl_4$, $AlBr_3$, or $SbBr_3$.

7. 2-Vinylnorbornane (2-Vinylbicyclo[2.2.1]heptane)

Imoto and coworkers wanted to study the influence on the polymerization behavior of exo and endo vinyl groups attached to the bicyclo[2.2.1] heptane system (371). However, attempts to prepare the *endo*-2-vinylnorbornane were unsuccessful and six attempted syntheses resulted in the more stable exo derivative. Polymerizations of this derivative were carried out in sealed tubes under nitrogen atmosphere. Cationic polymerizations with $BF_3 \cdot OEt_2$ gave after 24 hr at room temperature ~55% yield of brown, benzene soluble greases. It is not clear whether or not a solvent was used in these experiments. (Ziegler-Natta initiators gave white solids.) Infrared spectroscopy indicated that the vinyl groups disappeared during polymerization.

Somewhat later Kennedy and Makowski (372) synthesized a mixture of exo-(35.9%) and endo-2-vinylnorbornane (63.4%) and polymerized it with $EtAlCl_2$ in ethyl chloride in the range +10 to $-100°C$. The reaction at $+10°C$ gave after 43 min a white soluble powder in 17.5% yield, softening range 95–133°C, $\overline{M}_n = 885$, whereas that at $-78°C$ gave after 3 days 6.1% yield of a product that softened at 152°C, $\overline{M}_n = 1442$. Infrared spectra suggested that the polymer was of a conventional vinyl enchainment:

$$-CH_2-CH-$$

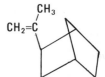

8. *Isoproenylnorbornane (2-(Bicyclo[2.2.1]heptane)propene)*

Kennedy and Makowski (372) synthesized this monomer (exo/endo = 50.8:44.7%) and mention that no methanol insoluble polymer formed even after 2 weeks in the presence of $EtAlCl_2$ in ethyl chloride diluent in the range +25 to $-78°C$. Evidently steric compression prohibits propagation of this olefin.

9. *Methylenenorbornene (2-Methylenebicyclo[2.2.1]hept-5-ene)*

Sartori et al. (373) investigated the cationic polymerization of methylenenorbornene by the use of $EtAlCl_2$ and $AlBr_3$ at $-78°C$ and VCl_4 at $-20°C$ in n-heptane solvent. Soluble crystalline (by x-rays) polymers were obtained, $[\eta] \sim 0.3$, softening range 150–160°C. On the basis of infrared spectroscopy the authors propose a nortricyclic recurring unit:

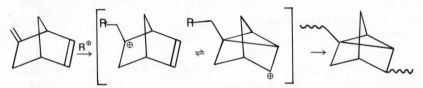

The polymerization mechanism has been discussed in some detail by Kennedy and Makowski (74, 367, 372) who propose the following sequence:

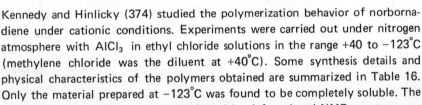

Evidently, the initially formed tertiary carbenium ion is unable to propagate because of prohibitively large steric compression. However, isomerization to the less crowded secondary ion provides an escape from this "steric trap" and propagation to high polymer may proceed.

10. Norbornadiene (Bicyclo [2.2.1] hepta-2,5-diene)

Kennedy and Hinlicky (374) studied the polymerization behavior of norbornadiene under cationic conditions. Experiments were carried out under nitrogen atmosphere with $AlCl_3$ in ethyl chloride solutions in the range +40 to $-123°C$ (methylene chloride was the diluent at $+40°C$). Some synthesis details and physical characteristics of the polymers obtained are summarized in Table 16. Only the material prepared at $-123°C$ was found to be completely soluble. The structure of the polymer was investigated by infrared and NMR spectroscopy and on the basis of this analysis it was concluded that the polymer essentially consisted of 2.6-disubstituted nortricyclene repeat units:

It was suggested that crosslinked (insoluble) material forms at high temperatures because of "2,3-type" reactions:

TABLE 16 POLYMERIZATION OF NORBORNADIENE WITH AlCl$_3$ (374)

Synthesis temperature (°C)	−123	−78	0	+40
Solvent	C$_2$H$_5$Cl	C$_2$H$_5$Cl	C$_3$H$_5$Cl	CH$_2$Cl$_2$
[M] (mole/l)	1.39	1.39	1.39	0.69
Time (min)	37	25	29	4
Yield (%)	17.2	42.0	71.0	30.0
Benzene soluble fraction (%)	100	72.5	34.5	59
$\overline{M}_n$ of soluble fraction	5520	8680	3680	2980
	9850[a]			
Melting behavior[b]		Decomposition before melting		
Crystallinity (x-ray)	Amorphous			
T_g (by torsion braid analysis) (°C) 320°				

[a]Prepared at −127°C.
[b]No melting up to +285°C.

The physical transitions of a linear polynorbornadiene, $\overline{M}_n$ = 9850, have been studied by torsion braid analysis (375). The thermomechanical spectrum of the polymer, determined at less than 1 cps in the range −180 to +500°C, revealed the glass transition temperature to be 320°C, which is the highest known T_g for a hydrocarbon addition polymer.

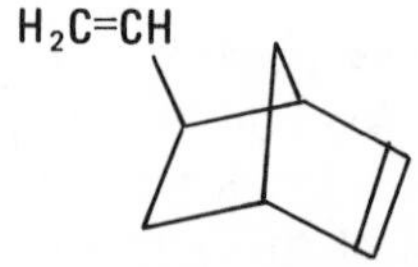

11. 2-Vinylnorbornene (2-Vinylbicyclo[2.2.1]hept-2-ene)

2-Vinylnorbornene was synthesized and its polymerization studied by Kennedy and Makowski (367). These workers used EtAlCl$_2$ in ethyl chloride diluent in the range −30 to −135°C. No polymer was obtained at −100°C and below, but reaction readily occurred at −78°C and above and white powdery products were obtained. The product obtained at −78°C was largely (~78%) soluble in toluene, $\overline{M}_n$ ~ 4020. At −30°C predominantly insoluble product was isolated. Analysis of the infrared spectra indicated the recurring unit (374):

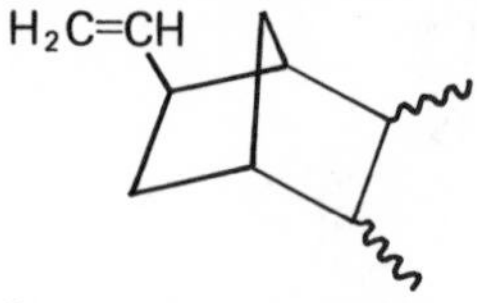

This may be an oversimplification as the polymer probably also contains other repeat units that are formed by the isomerization of the norbornyl skeleton, for example

The formation of insoluble product at higher temperatures indicates crosslinked material. Evidently under these conditions the relatively unreactive vinyl groups also enter the reaction.

12. 2-Isopropenylnorbornene (2-Isopropenylbicyclo [2.2.1]-hept-5-ene)

Kennedy and Makowski (367) investigated the cationic polymerization of 2-isopropenylnorbornene with $EtAlCl_2$ in ethyl chloride diluent in the range of −30 to −100°C. Polymerization occurred readily, and insoluble products were obtained. Evidently, in this molecule the reactivities of the two kinds of double bonds are comparable so that insoluble, crosslinked materials are formed even at −100°C.

13. 5-Methylbicyclo [2,2,2]-oct-2-ene

In the course of their study on the Ziegler-Natta polymerization of various bicyclic monomers, Takada et al. (369) mention that 5-methylbicyclo[2,2,2]-octene-2 gives only traces of polymer with $BF_3 \cdot OEt_2$ at 0 and −78°C.

14. 1,2-Dihydro-endo-dicyclopentadiene

Cesca et al. (376) polymerized 1,3-dihydro-endo-dicyclopentadiene with $EtAlCl_2/tBuCl$ (1:1) initiator in methylene chloride at −20 and −78°C and obtained ~50% yields of white products with $\bar{M}_n$ = 1670 and 2150, respectively, softening points 265–280°C. Infrared spectra showed no evidence for unsaturation. The 1308 cm^{-1} band indicated the presence of norbornane structure. According to the authors the polymerization involved the double bond opening of the norbornene ring system:

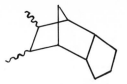

In view of the demonstrated facile rearrangements in the norbornyl cation system, it is conceivable that the structure of these polymers also includes repeat units arising by a prior Wagner-Meerwein rearrangement (see dicyclopentadiene).

Japanese investigators (377) who studied the polymerization of this olefin with Ziegler-Natta initiators mention that $Et_3Al/TiCl_4$ (Al/Ti = 2:3) produces an amorphous, low molecular weight product with repeat units formed by the opening of the norbornene unsaturation. This is not too surprising considering the cationic nature of the Al/Ti ratio employed. With high Al/Ti ratios ring-opening occurs and crystalline polymers are formed.

15. 9,10-Dihydro-endo-dicyclopentadiene

In the course of their studies on the polymerization of dicyclopentadiene, Cesca et al. (576) also investigated the polymerization of 9,10-dihydro-endo-dicyclopentadiene. Polymerizations were carried out with $EtAlCl_2/tBuCl$ initiator in methylene chloride solvent at $-78°C$. Only very low molecular weight oligomers were obtained ($\overline{M}_n \sim 540$, softening point 85–95°C). Infrared and NMR spectroscopy seems to indicate a polymerization step involving hydride migration:

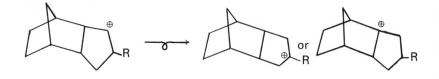

16 and 17. Dicyclopentadiene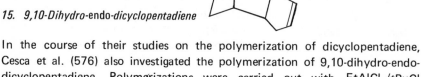

There are two dicyclopentadienes: the endo and the exo isomer:

endo-DCP   *exo*-DCP

The *endo*-dicyclopentadiene is commercially available. The exo isomer is thermodynamically more stable and if a pathway is available, for instance by a suitable carbenium ion rearrangement, isomerization to the exo isomer takes place.

Corner et al. (378) studied the polymerization of *exo*- and *endo*-dicyclopenta-dienes in neat systems with $BF_3 \cdot OEt_2$ initiator at room temperature. Polymeri-zation occurred readily and solid polymers were obtained: $\bar{M}_n$ = 1450 from the *exo*-DCP; $\bar{M}_n$ = 820 from the *endo*-DCP. The results of infrared spectroscopy suggest that the repeat unit of the polymer obtained from the *exo*-DCP is:

whereas that prepared from the *endo*-DCP, was formed by a Meerwein-Wagner rearrangement:

The driving force of the rearrangement in this case is evidently provided by the *endo* → *exo* isomerization. This seems to be the first report that has shown spectroscopic proof for the rearrangement postulated in conjunction with the cationic polymerization of the parent compound norbornene (374).

Cesca et al. (376) have also studied the polymerization of *endo*-DCP. These authors employed a variety of cationic initiators ($AlCl_3$, $EtAlCl_2$, $EtAlCl_2/tBuCl$, BF_3, $TiCl_4$, $SnCl_4$, $Et_2AlCl/tBuCl$, etc.) mostly in methylene chloride diluent in the range +10 to −78°C. Polymerizations occurred readily and amorphous (by x-rays) white powders in the molecular weight range 1300–4450 were obtained. The softening points were between ~180 and >320°C, depending on $\bar{M}_n$. All the products were soluble. Infrared and NMR spectroscopy suggested "complicated" norbornane structures indicating that both double bonds in the monomer are involved in the overall reaction. The authors think that the polymer is largely linear (high value of the α exponent in the Mark-Houwink relation) but branches may also be present (376). On the basis of some infrared evidence the authors speculate that in the case where the cyclopentene double bond is attacked (in contrast to the more reactive norbornene double bond) a nortricyclene structure might form via an intramolecular hydride shift:

Takada et al. (379) briefly mention the polymerization of dicyclopentadiene with TiCl₄ to a low molecular weight product consisting largely of repeat units formed by opening the double bond in the norbornene ring structure.

18. Di-endo-methylene hexahydronaphthalene

According to Foster and Hepworth (380) the cationic polymerization ($BF_3 \cdot OEt_2$, neat system, at room temperature under dry N_2, yield $\sim$ 50%) of di-*endo*-methylenehexahydronaphthalene gives a soluble and completely saturated polymer of $\bar{M}_n$ = 1050. On the basis of an infrared band at 3042 cm⁻¹ the authors propose the formation of a half-cage recurring unit by a transannular polymerization:

II. AROMATIC OLEFINS

A. STYRENE AND DERIVATIVES

1. Styrene (Vinylbenzene)

$$H_2C=CH$$

Styrene, a fairly reactive cationic monomer, has a long and distinguished history in the annals of cationic polymerizations. Because of its ease of polymerization and controllable and conveniently measurable rates and molecular weights, this monomer has been the subject of numerous scientific studies, particularly kinetic investigations.

Since the overall physical properties of cationically synthesized polystyrene were found to be rather unattractive for large scale commercial exploitation, this monomer was and still remains merely an attractive candidate for laboratory kinetic exercises and mechanism studies. The paramount deficiency of cationically synthesized polystyrenes from the commercial–technological point of view is their comparatively low molecular weight and therefore their brittleness. Thus very little polymer characterization work has been done with these polymers. In contrast, a variety of free radical induced polymerizations readily and cheaply produce high molecular weight polystyrenes required in most applications.

Table 17 is a summary of essentially all the scientific investigations that concern the cationic polymerization of styrene.

The cationic polymerizability of styrene was described in the last century (see Chapter 3). The second historical entry of interest in regard to styrene is a publication by Williams (381–383). This author and his associates were the first to study the reaction mechanism and the kinetics of a cationic polymerization system. It is of interest that the system Williams selected, that is, the styrene/SnCl₄ system, turned out to be an excellent choice as evidenced by the large number of research workers (see Table 17) who also investigated this monomer/coinitiator combination. Today Williams' original system is certainly among the best elucidated polymerization reactions on record.

Mathieson's review published in 1963 (459) is an excellent summary of kinetic aspects of the polymerization chemistry of styrene up to 1962 and serves as a point of departure for the present discussion. The compilation in Table 17 also includes the early studies discussed by Mathieson.

A glance at this table suffices to indicate that over the years investigators polymerized styrene with practically all the known cationic initiator systems: Brønsted acids, Lewis acids, alkylmetal systems, carbenium ion salts, iodine, clays, and so on. A somewhat more detailed look at the table reveals that almost without exception all studies involve rate studies, effects of reaction variables on yields, investigations carried out under highest purity conditions to study the effect of initiators, and so on, in other words, kinetic studies. No group of workers became interested in the systematic investigation of the physical properties of cationically polymerized styrene, obviously because of their unattractive physical properties. Few studies show detailed molecular weight or molecular weight distribution or branching studies and practically none deals with problems of tacticity or structural irregularities. A notable exception is the very recent work of Matsuzaki et al. (460) who examined with ^{13}C NMR

TABLE 17 CATIONIC POLYMERIZATION OF STYRENE

Initiator System	Solvent	Temperature (°C)	Remarks	References
Bronsted Acids				
HCl	$(ClCH_2)_2$	25		384
HBr	$(ClCH_2)_2$	25		384
H_2SO_4	$(ClCH_2)_2$	25		384–390
H_2SO_4	$(ClCH_2)_2$	30	Nonsteady state polymerization	391, 392
H_2SO_4	$CHCl_3$	Room temp.	Spectra studies	393
H_2SO_4	C_6H_6	<−50 (?) 20–40	Heterophase system	394
$HClO_4$	CH_2Cl_2	Room temp.	Spectroscopy	395
$HClO_4$	CH_2Cl_2	15–19	No evidence for $C^\oplus$ during polymerization	396
$HClO_4$	CH_2Cl_2	0	"Pseudocationic" polymerization	397
$HClO_4$	$(CH_2)_2$; EtCl; $EtNO_2$;$CCl_4/(ClCH_2)_2/$ $EtNO_2$	−30, 0, 25	Reproducible kinetics in presence of small amounts of water	398
$HClO_4$	CH_2Cl_2	−80	Kinetics by stopped flow technique	399
$HClO_4$	CH_2Cl_2		Reinterpretation of molecular weights in ref. 401	400
$HClO_4$	CH_2Cl_2	0, −60, −97	Two stage polymerization	401, 615
$HClO_4$	CH_2Cl_2	0 to −77	n-Bu_4NClO_4 and $LiClO_4$ reduce rates	402
$HClO_4$	$(ClCH_2)_2$	25, 0–30	"Transfer dominated" polymerization with fast initiation and no termination	403
$HClO_4$	$(ClCH_2)_2$; $CCl_4/(ClCH_2)_2$	25–30	Fast initiation; kinetics	404
$HClO_4$	CH_2Cl_2	Room temp.	Spectra studies	393
$HClO_4$	CH_2Cl_2	<−50(?) −19 to +19	Calorimetric study	405
$HClO_4$	CH_2Cl_2; $(ClCH_2)_2$	−29 to room temp.	Kinetics of pseudocationic polymerization	486

		Temp (°C)		Ref.
CCl_3COOH/H_2O	$(ClCH_2)_2$; $EtNO_2$; $EtBr$; $MeNO_2$; $C_6H_6/MeNO_2$; bulk	0, 25, 40	Kinetics	487–492
CF_3COOH	Neat	−15, 0, 22	Effect of medium	387
$HFSO_3$	$(ClCH_2)_2$	25	Semiquantitative information	384
$HClSO_3$	$(ClCH_2)_2$	25	Semiquantitative information	384
Lewis acids of the Friedel-Crafts halide type				
$AlCl_3/H_2O$	Neat	25	Spectra and coinitiation studies	493, 461
$AlCl_3/H_2O$	CCl_4	0, 25, 35	Initial kinetic studies	493–496
BF_3/H_2O	CCl_4	0, 25	A brief kinetic report	497
$BF_3 \cdot OEt_2(H_2O?)$	C_6H_6; $(ClCH_2)_2$; $C_6H_6/(ClCH_2)_2$	30	Initial kinetic studies	398, 411, 430, 431
	C_6H_6	30	Rate studies	499
	C_6H_6	30	Inhibition studies with Et_2O, THF, Et_2O	500
	C_6H_6	30	H_2O is promoter	468
	CH_2Cl_2; $(ClCH_2)_2$	25	H_2O is promoter	467
	CH_2Cl_2	0	The effect of salts nBu_4NX where X = ClO_4, BF_3, I	437
	C_6H_6; CCl_4; $C_6H_5NO_2$; $(ClCH_2)_2$, $(Cl_2CH)_2$	20	Rate acceleration with $(CH)_2C=C(CN)_2$ in $(ClCH_2)_2$ and $(Cl_2CH)_2$	470
$BF_3 \cdot THF/H_2O$	Toluene, CCl_4, CH_2Cl_2; C_6H_6	−70 to +40	Structure by ^{13}C NMR	460
		30	Rate studies	499
$BF_3 \cdot C_2H_5OH(H_2O?)$	C_6H_6	30	A brief experiment	501
$BF_3 \cdot C_2H_5OH(H_2O?)$	$C_6H_6/(ClCH_2)_2$; C_6H_6; $(ClCH_2)_2$	30	Initial kinetic studies	498

TABLE 17 (continued)

Initiator System	Solvent	Temperature (°C)	Remarks	Reference
$TiCl_4/H_2O$	CH_2Cl_2	−30 to −90	Superseded by reference 403	406
$TiCl_4/H_2O$	Toluene; hexane; $(ClCH_2)_2$; EtBr; iPrBr; tBuBr	25, −62	Preliminary studies, now partly obsolete	407–410
$TiCl_4/H_2O$	C_6H_6; $C_6H_6/(ClCH_2)_2$	30		411
$TiCl_4/H_2O$	CH_2Cl_2	−25, −60, −90	Extremely dry conditions	412
$TiCl_4/CCl_3COOH$	Toluene; hexane	−62, 25		407, 409, 410
$SnCl_4$ (H_2O?)	$CCl_4/C_6H_5NO_2$	−19, 0, 25, 27.5	Chain transfer studies	413–424
$SnCl_4/H_2O$	CCl_4	0, 25, 30	Now classical studies	381–383, 414, 425–428
$SnCl_4/H_2O$	C_6H_6	25, 30		411, 425, 429–432
$SnCl_4/H_2O$	Cyclohexane	0, 25, 30		416, 428, 433, 434
$SnCl_4/(H_2O$?)	$(ClCH_2)_2$	20, 30	Rate studies	435
$SnCl_4(H_2O$?)	$(ClCH_2)_2$	25		427, 429, 433, 434, 436
$SnCl_4(H_2O$?)	CH_2Cl_2	0	The effect of nBu$_4$NX where X = ClO$_4$, BF$_4$, I	437
$SnCl_4(H_2O$?)	EtCl	30		438, 439
$SnCl_4(H_2O$?)	$CHCl_3$;C_6H_5Cl	30		428
$SnCl_4(H_2O$?)	C_6H_5Br	25		433, 434
$SnCl_4/H_2O$	$C_6H_5NO_2$	25		427, 433, 434, 440, 441

Catalyst	Solvent	Temp.	Notes	Ref.
SnCl$_4$/H$_2$O	EtNO$_2$	25		433, 434
SnCl$_4$(H$_2$O?)	CCl$_4$/(ClCH$_2$)$_2$	25		426, 427
SnCl$_4$/H$_2$O	Toluene,1,1,2,2-tetrachloroethane	35	The effect of water on rates	674
SnCl$_4$/H$_2$O	Hexane, toluene, CHCl$_3$, CH$_2$Cl$_2$, MeNO$_2$	0	Hammett plot, effect of solvent on rates and molecular weight	686
SnCl$_4$(H$_2$O?)	C$_6$H$_6$/(ClCH$_2$)$_2$	30		411, 502
SnCl$_4$/H$_2$O	C$_6$H$_6$/C$_6$H$_5$NO$_2$	25		433, 434
SnCl$_4$/CCl$_3$COOH	C$_6$H$_6$	30		430,431 503,504
SnCl$_4$/CCl$_3$COOH	Cyclohexane	30		428
SnCl$_4$/CCl$_3$COOH(+?)	CCl$_4$; CHCl$_3$; (ClCH$_2$)$_2$; C$_6$H$_5$Cl	30		428
SnCl$_4$/CCl$_3$COOH (H$_2$O?)	CH$_2$Cl$_2$	0	The effect of nBu$_4$NX where X = ClO$_4$, BF$_4$, I	437
SnCl$_4$/iPrCl or tBuCl	(ClCH$_2$)$_2$	25		436, 427
SnCl$_4$/iPrCl or tBuCl	C$_6$H$_5$NO$_2$	25		440, 427
SnCl$_4$/HCl	CCl$_4$; C$_6$H$_5$NO$_2$	25		505, 506
SnCl$_4$/HBr	CCl$_4$; C$_6$H$_5$NO$_2$	25		507
SnBr$_4$/HCl	CCl$_4$; C$_6$H$_5$NO$_2$	25		507
SnBr$_4$/HBr	(ClCH$_2$)$_2$	25		385
SnBr$_4$(H$_2$O?)	(ClCH$_2$)$_2$; C$_6$H$_6$	25		385, 508
SbCl$_5$/H$_2$O	CCl$_4$; EtNO$_2$;cyclohexane	25		383, 508, 509
FeCl$_3$/H$_2$O	C$_6$H$_6$; C$_6$H$_6$/(ClCH$_2$)$_2$	30		411
PF$_5$	CH$_2$Cl$_2$	−78, 80, 30	Structure by ^{13}C NMR: 61–58% racemic	460
ReCl$_5$/ClCH$_2$COOC$_2$H$_5$ (H$_2$O?)	C$_6$H$_6$; C$_6$H$_6$/(ClCH$_2$)$_2$	0, 10, 20	Kinetic studies	510–512

TABLE 17 (continued)

Initiator System	Solvent	Temperature (°C)	Remarks	References
Alkylaluminum Systems				
$AlEtCl_2(H_2O?)$	Toluene	Room temp.	Qualitative experiment; polymerization in open system	251
$AlEtCl_2/HCl$ or H_2O	Toluene	Room temp.	Coinitiation experiments; under anhydrous conditions polymerization only with HCl or H_2O	251
$AlEtCl_2/tBuCl$	Toluene or CH_2Cl_2?	−78	Qualitative experiments with Hammett indicators: highly active initiator	442
$AlEt_2Cl/H_2O$	C_6H_6	20	Protogen required for initiation	40
$AlEt_2Cl/C_6H_5COOH$	C_6H_6	20	Protogen required for initiation	40
$AlEt_2Cl/LiCl \cdot xH_2O$ or $MgCl_2 \cdot xH_2O$	C_6H_6	20	Protogen required for initiation	40
$AlEt_2Cl/tBuCl$	Toluene or CH_2Cl_2	−78	Qualitative experiments with Hammett indicators: highly active initiator system	442
$AlEt_2Cl/tBuCl$	C_6H_5Cl	−50, +20	Flash polymerization, yields inversely proportional to temperature	443
$AlEt_2Cl/tBuCl$	Heptane, C_6H_5Cl, EtCl	−78, −50, −20 and +11	Effect of solvent and temperature on yield and molecular weight	443
$AlEt_2Cl/tBuCl$	Heptane, Me-cyclohexane, C_6H_5Cl	−50, +20	Effect of solvent and temperature on yield	443
$AlEt_2Cl/C{=}C{-}C{-}C$ with Cl; $AlEt_2Cl/RCl$ where R = nBu, iPr, secBu, allyl, crotyl, tBu, benzyl, diphenyl, methyl, trityl, etc.	CH_3Cl	−50	Coinitiation by RCl determined by stability and availability of $R^{\oplus}$; GPC data, very high molecular weights	443, 444
$AlEt_3/H_2O$ (1:1 and 2:3)	Toluene or CH_2Cl_2?	−78	Qualitative experiments with Hammett indicators, highly active initiator system	442

Initiator system	Solvent	Temp (°C)	Remarks	Ref.
AlEt₃/H₂O (various ratios)	CH₂Cl₂		AlEt₃/H₂O premixed; polymerization only when ratio 1.0–0.5; no polymerization when ratio >1.0 or <9.5; $\overline{M}_v \sim 200{,}000$	445
AlEt₃/H₂O (1:1)	Hexane, toluene, CH₂Cl₂, EtBr, Me₂O	−78	Qualitative experiments on effect of solvent	445
AlEt₃/H₂O(~1.1) plus CH₃OCH₂Cl	CH₂Cl₂	−78	CH₃OCH₂Cl addition to inactive AlEt₃/H₂O gives active initiator	445
AlEt₃/H₂O/CH₃OCH₂Cl (1:4 to 0.66:0.1)	CH₂Cl₂	−78	CH₃OCH₂Cl activator	445
AlEt₃/H₂O/CH₃OCH₂Cl (1:1:0.1)	Hexane, toluene, CH₂Cl₂, EtBr, Me₂O	−78	Qualitative experiments on effect of solvent	445
AlEt₃/H₂O(~1.1) plus CH₃COCl	CH₂Cl₂	−78	CH₃COCl addition to inactive AlEt₃/H₂O gives active initiator	445
AlEt₃/H₂O/CH₃OCH₂Cl (1:0.9:0.1)	Toluene	−78	Fresh system active, aging (1 hr at 110°C) destroys activity	445
AlEt₃/H₂O/CH₃COCl (1:0.5:10)	CH₂Cl₂	−78	CH₃CO end-group in polymer $\overline{M}_v$ = 17,000	445
AlEt₃/H₂O(~1.1) plus tBuCl	Toluene	−78	tBuCl addition to inactive AlEt₃/H₂O gives active initiator	445
AlEt₃/H₂O/tBuCl (1:2 to 0.66:0.2)	Toluene	−78	tBuCl activator	445
AlEt₃/tBuCl	Toluene or CH₂Cl₂?	−78	Qualitative experiments with Hammett indicators	442
AlEt₃/tBuCl (1:0.25)	Toluene	−78	Highly active initiator; [η] = 0.18	445
AlEt₃/CH₃OCH₂Cl (1:0.2)	CH₂Cl₂	−78	Highly active initiator; [η] = 0.22	445
AliBu₃/tBuCl	Toluene or CH₂Cl₂	−78	Qualitative experiments with Hammett indicators highly active initiator	442

235

TABLE 17 (continued)

Initiator System	Solvent	Temperature (°C)	Remarks	References
Salts				
CH_3COClO_4	CH_2Cl_2	0	GPC shows two populations	446
$CH_3COClO_4(H_2O?)$	CH_2Cl_2	0	The effect of nBu$_4$NX where X = ClO$_4$, BF$_4$, I	437
$(C_6H_5)_3CCl$ SnCl$_5$	$(ClCH_2)_2$	30	Initiation studies $k_i = 11.6 \times 10^{-2}$ ℓ/mole min	447
$C_6H_5CH(CH_3)Br/$ AgClO$_4$	CH_2Cl_2	0	Addition of RBr to St + AgClO$_4$ initiates polymerization	405
RX/AgClO$_4$ where R = allyl, iPr, tBu, benzyl, trityl, etc. and X = Cl, Br, I	Neat	0	Addition of RX to St + AgClO$_4$ initiates polymerization. Effect of R and X studied	448
$C_6H_5CH_2Br/BF_3COOAg$	Neat	0	A brief experiment	448
$C_6H_5CH_2Br/AgNO_3$	Neat	0	A brief experiment	448
$C_6H_5CH_2Br/AgBF_4$	Neat	0	A brief experiment	448
$(C_6H_5)_3C^{\oplus}SbCl_6^{\ominus}$	CH_2Cl_2	+20 to −78	Mentioned only	303
$C_7H_7^{\oplus}$ BF$_4^{\ominus}$, and SbCl$_6^{\ominus}$	CH_2Cl_2	?	Mentioned only	675
Miscellaneous Materials				
I$_2$	Neat	0, 25		449
I$_2$	$CCl_4(ClCH_2)_2$	30	Kinetics	450
I$_2$	$(ClCH_2)_2$	20, 30, 40	Rate studies	435, 451
I$_2$	$(ClCH_2)_2$	30	Solvent effects	432
Cation exchange resin	Neat	0	A brief experiment	453
Acid clay cracking catalyst	Neat	0–25	Qualitative observations	454
13X Zeolite	Neat	30	Kinetics, very high molecular weights	685
Various acids	CH_3NO_2	30	Electrodialysis	383

All types of acids	SO_2	0, 25		455, 456
γ $Al_2O_3 \cdot AlEt_3$	Toluene or CH_2Cl_2?	−78	Qualitative experiments with Hammett indicator; active initiator	442
γ Al_2O_3	Toluene or CH_2Cl_2?	−78	Qualitative experiments with Hammett indicator; active initiator	442
$ZnEt_2/H_2O$ (1:1)	Toluene or CH_2Cl_2?	−78	Qualitative experiments with Hammett indicator; active initiator	442
$ZnEt_2/H_2O/tBuCl$ (1:0.9:0.1)	CH_2Cl_2	−78	Active system	445
$ZnEt_2/tBuCl$	Toluene or CH_2Cl_2?	−78	Qualitative experiments with Hammett indicator	442
$ZnEt_2/H_2O/CH_3OCH_2Cl$ (1:0.9:0.1)	CH_2Cl_2	−78	Active system	445
CCl_3COCl/H_2O	Neat	+20 to 50	Low molecular weights	457
$Me_3SiCl/HgCl_2$	$(CH_2Cl)_2$	0–30	To introduce Si as head-group	458

spectroscopy the stereoregularity of polystyrenes prepared by various mechanisms. According to these authors the syndiotacticity of polystyrene prepared with $BF_3 \cdot OEt_2$ in toluene increases with decreasing temperatures, that is, the racemic dyads rise from 0.50 at $+40^{\circ}C$ to 0.59 at $-40^{\circ}C$.

It appears that the cationic homopolymerization of styrene has been, is, and will remain in the foreseeable future an interesting system for kinetic or model studies, but polystyrene produced by cationic techniques will not become a technologically commercially important polymer.

A discussion of some of the more significant fundamental studies carried out since 1962 in the field of cationic styrene polymerization follows.

Two significant new areas have emerged since Mathieson's review: (1) styrene polymerizations induced with alkylaluminum compounds and (2) pseudocationic polymerizations.

This field has also been greatly enriched by the systematic and continuous investigation of the Kyoto group mainly under Professor Higashimura which has produced a large amount of information on all facets of the kinetics and mechanistic details of cationic styrene polymerization. Recent findings of this group are summarized below. In addition to these continued and systematic studies a number of interesting new ideas have appeared, for example, cationic styrene polymerizations in the presence of electron acceptors or in the presence of electric fields. These reports are also discussed.

Alkylaluminum Initiator Systems.   A series of interesting investigations on the polymerization mechanism of styrene have been carried out with various alkylaluminum based initiators. The first report in the scientific literature that styrene can be polymerized by triethylaluminum in conjunction with suitable protogens such as H_2O or HCl is by Sinn et al. (251). These workers investigated the polymerization of styrene and other olefins with typical Ziegler-Natta initiators and in the course of these studies noted that, in particular, the polymerization of styrene can be readily induced by introducing small quantities of HCl or H_2O to quiescent styrene/$AlEtCl_2$ mixtures in toluene. These experiments had to be carried out in high vacuum equipment as the purification of styrene and toluene with the $LiAlH_4$ and highly purified nitrogen was insufficient to remove the last traces of protogenic impurities. The "stopping" experiment was carried out by first filling the evacuated and baked-out glass apparatus with N_2 dried through $AlEt_3$; then the styrene was dried by stirring with $AlEt_3$ (the $AlEt_3$ was pretreated by stirring at $100^{\circ}C$ with KCl) and distilled into a flask containing $AlEtCl_2$. No reaction occurred when the styrene/$AlEtCl_2$ mixture was stirred even at room temperature. The monomer could be distilled off the alkylaluminum compound without apparent change. However, when a trace of water was admitted to the quiescent styrene/$AlEtCl_2$ mixture, the system turned orange and polymerization ensued. Styrene/$AlEt_3$/toluene mixtures that could not be initiated by small amounts of water, could be induced to polymerize by admitting gaseous HCl. No quantitative statements as to the necessary amount of protogens could be made. It was emphasized that LiAlH$_4$-dried monomer contains sufficient amounts of cationogens ("cocatalyst") to initiate the polymerization with AlRCl$_2$ or AlR$_2$Cl, however, monomer distilled off AlR$_3$ is unable to induce such polymerizations (251). These authors were mainly interested in elucidating mechanistic details of polymerizations initiated by Ziegler-Natta systems and therefore did not follow up the above observations nor investigate the products.

Next, Russian researchers (40) described styrene polymerizations initiated with AlEt$_2$Cl plus various chemicals such as H$_2$O, CH$_3$COOH, C$_6$H$_5$·COOH, and hydrated salts, that is, LiCl·xH$_2$O or MgCl$_2$·xH$_2$O in benzene solvent at 20°C. Diethylaluminum chloride alone was found to be inactive. While AlEt$_2$Cl/protogens were active initiator systems, AlEt$_3$ plus the same protogens did not induce polymerizations. In ethyl chloride diluent, however, the AlEt$_2$Cl/protogen initiated polymerization occurred rapidly which led the authors to conclude that the initiating ion pair is generated from C$_2$H$_5$Cl by the process:

$$AlR_2Cl + C_2H_5Cl \;\rightleftharpoons\; AlR_2Cl_2^{\ominus} C_2H_5^{\oplus}$$

While conceptually this formalism is attractive and initiation by this mechanism probably occurs with other more suitable alkyl halides (i.e., alkyl halides that give rise to stable carbenium ions), in this particular case with ethyl chloride, in the opinion of this writer initiation was most likely due to the HCl or H$_2$O impurity in the solvent. The Russian authors did not mention any purification of their diluent, a chemical that commonly contains more than trace amounts of HCl stemming from its manufacturing process. The present author has found that with KOH treated ethyl chloride the AlEt$_2$Cl/EtCl system is almost completely inactive for the polymerization of styrene even in the presence of large amounts of ethyl chloride, that is, when it is used as the reaction medium.

A more substantial study of various AlEt$_3$ based systems has been reported by Saegusa et al. (445). Among other things, these authors polymerized styrene with various AlEt$_3$/H$_2$O mixtures in CH$_2$Cl$_2$ at −78°C. The initiator solution was prepared by premixing the ingredients. It was established that the AlEt$_3$/H$_2$O ratio is important in determining the success of the polymerization. Active systems were obtained only when the AlEt$_3$/H$_2$O ratio was between 1.0 and 0.66 and no polymerization occurred beyond these limits; indeed optimum yields and molecular weights were obtained within the narrower limits AlEt$_3$/H$_2$O 1.0–0.5. The molecular weights obtained in this study are remarkable; for example, a sample obtained with AlEt$_3$/H$_2$O = 1.0, at 17.3% conversion, gave $[\eta]$ = 0.80 corresponding to a $\bar{M}_v$ = 190,000, and another with AlEt$_3$/H$_2$O = 0.66, at 13.5% conversion gave $[\eta]$ = 0.86 corresponding to $\bar{M}_v$ = 210,000

(Table 1 in reference 445). The role of water was visualized as follows:

$$(AlEt_3)_2 + H_2O \ \rightarrow \ Et_2Al-O-AlEt_2 \ (+2EtH)$$

$$Et_2Al-O-AlEt_2 + H_2O \ \rightarrow \ H^{\oplus} \ [Et_2Al-O-AlEt_2 \cdot OH]^{\ominus}$$

It is implied, although not stated, that initiation involves protonation with this hypothetical acid.

In further experiments the authors added various other chemicals such as tBuCl, CH_3COCl, and CH_3OCH_2Cl to the $AlEt_3/H_2O$ mixture. The $AlEt_3/H_2O$ ratio in these experiments was purposely chosen to be beyond the active initiator range of 1.1. Polymerizations occurred upon the addition of small amounts of the above agents to monomer/$AlEt_3/H_2O$ mixtures in CH_2Cl_2 at $-78°C$, however, in these instances product molecular weights were much lower than in those without the third component (see above). Finally, in experiments run in the absence of water it was found that $AlEt_3/CH_3OCH_2Cl$ (in methylene chloride) and $AlEt_3/t$BuCl (in toluene) at $-78°C$ are effective initiating systems. Initiation was attributed to ion pairs, for example,

$$AlEt_3 + CH_3OCH_2Cl \ \rightleftharpoons \ [CH_3OCH_2]^{\oplus} [AlEt_3Cl]^{\ominus}$$

$$[CH_3OCH_2]^{\oplus} [AlEt_3Cl]^{\ominus} + CH_2{=}\underset{\underset{C_6H_5}{|}}{CH} \ \longrightarrow \ CH_3OCH_2CH_2\overset{\oplus}{\underset{\underset{C_6H_5}{|}}{CH}} [AlEt_3Cl]^{\ominus}$$

or in the ternary system $AlEt_3/H_2O/CH_3OCH_2Cl$:

$$Et_2Al-O-AlEt_2 + CH_3OCH_2Cl \ \rightleftharpoons \ [CH_3OCH_2]^{\oplus} [Et_2Al-O-AlEt_2 \cdot Cl]^{\ominus}$$

$$[CH_3OCH_2]^{\oplus} [Et_2Al-O-AlEt_2 \cdot Cl]^{\ominus} + CH_2{=}\underset{\underset{C_6H_5}{|}}{CH} \ \longrightarrow$$

$$CH_3OCH_2CH_2\overset{\oplus}{\underset{\underset{C_6H_5}{|}}{CH}} [Et_2Al-O-AlEt_2 Cl]^{\ominus}$$

Direct evidence for this hypothesis came from end-group studies. Thus according to infrared spectroscopy, a polymer (MW $\sim$ 17,000) prepared with the $AlEt_3/H_2O/CH_3COCl$ (1.0/0.5/10) initiating system contained CH_3CO-end-groups.

And last but not least Saegusa et al. (445) carried out experiments with aged initiating systems and recognized that aging destroys initiating ability. Thus polymerization occurred when CH_3OCH_2Cl was added to a quiescent styrene/

$AlEt_3/H_2O$ ($AlEt_3/H_2O = 1.1$) mixture in toluene at $-78°C$; however, polymerization was absent when an aged (1 hr at $110°C$) $AlEt_3/H_2O/CH_3OCH_2Cl$ (1/0.9/0.1) mixture was added to styrene under the same conditions. This observation was interpreted to mean that the active initiating principle was the unstable ion pair:

$$Et_2Al-O-AlEt_2 + CH_3OCH_2Cl \rightleftharpoons [CH_3OCH_2]^{\oplus} [Et_2Al-O-AlEt_2 \cdot Cl]^{\ominus}$$

The present author concurs with this interpretation and suggests that aging probably gives rise to CH_3OCH_2Et and compounds containing $>Al-Cl$ bonds; this mixture is inactive on two counts: first because of the absence of initiating carbenium ions and second because of the in situ generation of an ether, a powerful chain stopping nucleophile.

The same Japanese authors (442) also characterized the acidity of $AlEt_3/H_2O$ (1:1, 2:3, and 2:1) mixtures by using various Hammett indicators. According to these admittedly qualitative studies the 1:1 $AlEt_3/H_2O$ system is a stronger acid than $AlCl_3$. Species such as

$$Et_2Al-O-(AlO)_{\overline{n}}\,AlEt_2$$
$$\underset{Et}{|}$$

were suggested to account for the acid sites. The value of n and the degree of aggregation of these species may determine the acid strength. Independent experiments suggested that the acid sites are successively destroyed by the introduction of increasing amounts of pyridine.

Still in the area of alkylaluminum based initiators, Kennedy (443, 444) investigated aspects of the polymerization of styrene with a variety of $AlEt_2Cl/RCl$ systems where R was nBu, iPr, secBu, allyl, methallyl, iBu, tBu, crotyl, benzyl, trityl, and so on. The experiments were carried out by adding the RCl component to quiescent styrene/$AlEt_2Cl$ systems stirred in methyl chloride at $-50°C$. With certain alkyl halides, such as tBuCl, immediate flash polymerization occurred. After 5 min the reactions were terminated with cold CH_3OH and the polymer was recovered by evaporating the liquids and drying in vacuo. Some of the data contained in references 443 and 444 are summarized in Table 18. [The factor 10^{-3} in the molecular weight columns was unfortunately omitted in the original reference (444).]

Evidently, certain $AlEt_2Cl/RCl$ combinations are efficient initiators for the polymerization of styrene and give rise to very high molecular weight products. It is of interest that experiments with $AlEt_2Cl$ can be conveniently carried out in "open systems," that is under a blanket of purified nitrogen, and cumbersome high vacuum conditions are unnecessary to demonstrate the need for the initiating cationic species RX. With $AlEtCl_2$ or $AlCl_3$ this cannot be done.

Initiation studies with these aluminum containing Lewis acids must be carried out in high vacuum. As has been demonstrated by Sinn et al. (251) and Jordan and Treolar (461), $AlEtCl_2$ and $AlCl_3$ require trace amounts of protogenic impurities (moisture, etc.) for initiation. The reason for this difference in the behavior of these Lewis acids, that is, between $AlEt_3$ and $AlEt_2Cl$ on the one hand and $AlEtCl_2$ and $AlCl_3$ on the other hand, is obscure.

A brief glance at the fourth column in Table 18 reveals an enormous range of initiating efficiencies among the various $AlEt_2Cl/RCl$ combinations. A more detailed examination suggests that alkyl halides with low R–Cl heterolytic bond dissociation energies are efficient initiators while those with high bond strengths are poor ones. Or, alkyl halides giving rise to certain stable carbenium ions are active initiators. However, excessive carbenium ion stability may be detrimental to initiating ability. Thus triphenylmethyl chloride, which among the compounds examined most readily gives the cation, is a poor initiator, probably because the trityl carbenium ion is too stable relative to the propagating styryl ion.

The formation of the initial ion pair probably occurs according to:

$$RCl + AlEt_2Cl \;\rightleftharpoons\; R^{\oplus} [AlEt_2Cl_2]^{\ominus}$$

The high initiating activity of isobutyl chloride is probably due to the rearrangement:

$$iBuCl + AlEt_2Cl \;\longrightarrow\; iBu^{\oplus} AlEt_2Cl_2^{\ominus} \xrightarrow{\;\sigma\;} tBu^{\oplus} AlEt_2Cl_2^{\ominus}$$

whereas the unexpectedly low activity of benzyl chloride has been attributed to self-alkylation, a process that occurs with greatest velocity even at $-120°C$:

$$C_6H_5CH_2Cl + AlEt_2Cl \;\longrightarrow\; C_6H_5CH_2^{\oplus} AlEt_2Cl_2^{\ominus} \xrightarrow{-HCl} \; +C_6H_4CH_2+_n$$

A quantitative examination of the $AlEt_2Cl/RCl$ ratios on initiating activity shows that alkyl halides that generate relatively unstable carbenium ions ($nBu^{\oplus}$, $iPr^{\oplus}$, and $secBu^{\oplus}$) give polymer only when this ratio is $\ll 1$, that is, in the presence of a large excess of RCl, whereas those giving stable ions ($tBu^{\oplus}$, crotyl$^{\oplus}$, etc.) readily initiate even when their concentration is very small. With allyl chloride or methallyl chloride, which give fairly stable cations, an increase in the $AlEt_2Cl/RCl$ ratio from 0.07 to $\sim$3 also increases the initiating efficiency by about three orders of magnitude. Initiating efficiencies in excess of 10^6 are obtained with alkyl halides yielding substituted allyl cations, the t-butyl ion, and cations stabilized by phenyl groups (except the trityl ion).

These observations can be explained by the basic concept of carbenium ion generation from RCl in conjunction with $AlEt_2Cl$ as shown by the above reaction. To account for the low but measurable initiating activity of nBuCl,

TABLE 18 POLYMERIZATION OF STYRENE WITH THE AlEt$_2$Cl/RCl System (444)

RCl	AlEt$_2$Cl/RCl	Conversion (%)	grams PSt/mole RCl	$\bar{M}_w^a \times 10^{-3}$	$\bar{M}_n^a \times 10^{-3}$	$\bar{M}_w/\bar{M}_n$
CH$_3$CH$_2$CH$_2$CH$_2$Cl	0.08	1.0	25	568	311	1.8
CH$_3$CHCH$_3$ (Cl)	0.07	0.65	13	593	265	2.0
CH$_3$CHCH$_2$CH$_3$ (Cl)	0.08	1.0	25	572	282	2.0
CH$_2$=CH–CH$_2$	0.073	0.8	17	–	–	–
	0.15	7.3	340	570	282	2.0
	0.30	4.4	400	551	308	1.8
	3.0	3.1	2,860	–	–	–
CH$_2$=C–CH$_2$ / CH$_3$ (Cl)	0.073	0.26	6	462	175	2.6
	0.18	0.95	52	335	149	2.2
	0.37	0.82	93	412	197	2.1
	3.7	2.00	2,300	454	218	2.1
CH$_2$=CH–CHCH$_3$ (Cl)	45	93.1	>1,250,000	386	54	7.2
	80	87	>2,120,000	–	–	–
	96	86.2	>2,540,000	520	159	3.3
	114	66.3	2,320,000	479	152	3.1
	214	25.1	1,730,000	604	244	2.5
CH$_3$CH=CHCH$_3$ (Cl)	3.6	98.5	>10,900	360	45.5	7.9
	123	85.8	>3,140,000	503	95.7	5.3
	234	88.3	>6,200,000	483	136	3.5
	752	10.0	2,220,000	576	287	2.0

TABLE 18 (continued)

Compound					
CH₃CCH₃ (Cl) / CH₃	100	85.8	> 2,600,000	566	169	3.3
	129	9.2	360,000	550	270	2.0
	268	5.2	420,000	–	–	–
	750	1.6	375,000	576	266	2.2
CH₃CHCH₂ (Cl) / CH₃	0.08	95.0	> 2,270	501	96	5.2
	0.23	43.0	3,050	605	314	1.9
	0.47	12.6	1,840	602	243	2.5
	2.0	1.12	690	–	–	–
	4.1	0.45	560	568	336	1.69
C₆H₅CH₂ (Cl)	4.4	9.4	12,500	587	310	1.89
C₆H₅CHCH₃ (Cl)	4.1	100	> 126,000	–	–	–
	136	74	3,050,000	516	190	2.7
	278	34.9	2,930,000	626	288	2.2
(C₆H₅)₂CH (Cl)	6.8	100	> 208,000	368	52.6	7.0
	220	62.3	4,150,000	599	229	2.6
	440	27.8	3,700,000	531	230	2.3
(C₆H₅)₃C (Cl)	11	–	–	–	–	–

ªThe factor 10^{-3} was unfortunately omitted in the heading in reference 444.

244

iPrCl, and secBuCl and, in particular, for the fact that initiation occurs only in the presence of a large excess of these halides, it was proposed that the true initiating species is produced in two steps. First the more acidic $AlEtCl_2$ (or perhaps even $AlCl_3$) is produced and in turn is able to initiate even with traces of carbenium ion generators or protogenic impurities, for example:

$$i\text{PrCl} + \text{AlEt}_2\text{Cl} \rightleftharpoons i\text{Pr}^{\oplus}\text{AlEt}_2\text{Cl}_2^{\ominus} \rightleftharpoons i\text{PrEt} + \text{AlEtCl}_2$$

$$\text{AlEtCl}_2 + i\text{PrCl} \rightleftharpoons i\text{Pr}^{\oplus}\text{AlEtCl}_3^{\ominus} \rightleftharpoons i\text{PrEt} + \text{AlCl}_3$$

The following simplified scheme summarizes the main events during polymerization:

$$\text{RCl} + \text{AlEt}_2\text{Cl} \rightleftharpoons \text{R}^{\oplus}\text{AlEt}_2\text{Cl}_2^{\ominus} \xrightarrow{n\text{St}} \text{RSt}_n^{\oplus}\text{AlEt}_2\text{Cl}_2^{\ominus}$$

$-\text{RE}t$ \qquad $-\text{RSt}_n\text{Et}$

$- - - - \text{AlEtCl}_2$

$$\text{RCl} + \text{AlEtCl}_2 \rightleftharpoons \text{R}^{\oplus}\text{AlEtCl}_3^{\ominus} \xrightarrow{n\text{St}} \text{RSt}_n^{\oplus}\text{AlEtCl}_3^{\ominus}$$

$-\text{RE}t$ \qquad $-\text{RSt}_n\text{Et}$

$- - - - \text{AlCl}_3$

$$\text{RCl} + \text{AlCl}_3 \rightleftharpoons \text{R}^{\oplus}\text{AlCl}_4^{\ominus} \xrightarrow{n\text{St}} \text{RSt}_n^{\oplus}\text{AlCl}_4^{\ominus} \longrightarrow \text{RSt}_n\text{Cl} + \text{AlCl}_3$$

According to this mechanism the first cation is formed in conjunction with the $\text{AlEt}_2\text{Cl}_2^{\ominus}$ counterion. At this junction initiation can occur by alkylating styrene or else the ion pair can stabilize itself by generating REt and $AlEtCl_2$. Indeed in the absence of monomer this reaction is probably the only one that occurs (462). The propagating ion may terminate after each addition step by the same process (ethylation), giving rise to an ethyl capped polystyrene molecule $RSt_n Et$ and, again, $AlEtCl_2$. The ethylaluminum dichloride is, of course, a potent coinitiator in its own right and in conjunction with the RCl initiator may induce the further polymerization of styrene. The overall mechanism may subsequently progress according to the second or even third equation shown in the above scheme.

Alkyl halides yielding stable cations most likely initiate according to the first reaction. The relatively weakly acidic $AlEt_2Cl$ is nonetheless a sufficiently strong chlorine acceptor to generate carbenium ions from these species. With nBuCl, iPrCl, or secBuCl, $AlEt_2Cl$ may not be acidic enough to break the R–Cl bond and cation generation is delayed until the *in situ* formation of sufficient quantities of the more aggressive $AlEtCl_2$ or $AlCl_3$. However, even with these very strong Lewis acids the ion concentrations remain low and only low polymer conversions are obtained.

There is strong independent, although admittedly circumstantial, evidence that indicates the existence of an initiator destroying competitive reaction $R^\oplus + AlEt_2Cl_2^\ominus \rightarrow REt + AlEtCl_2$. First, no polymerization will take place if tBuCl and $AlEt_2Cl$ are premixed and the mixture is added to styrene. This reaction yields neohexane, tBuEt, plus $AlEtCl_2$, and the last compound cannot initiate the polymerization in the absence of RCl. Second, the experiments of Saegusa et al. (408) with aged $AlEt_3/H_2O/CH_3OCH_2Cl$ show a similar phenomenon. Third, Cesca et al. (687) found that premixed $AlEt_2Cl/t$BuCl 3:1 systems were less effective than charges in which the tBuCl component was added last. Fourth, Kennedy showed (443) that on successive additions of tBuCl to styrene/$AlEt_2Cl$ mixtures, polymerization takes place but initiating activity disappears before the $AlEt_2Cl/t$BuCl ratio reaches stoichiometric equivalence. Evidently, concurrent with initiation, competing coinitiator consuming side reaction also occurs.

If polymerizations do occur in styrene/$AlEt_2Cl$/RCl systems, they are flash reactions and final conversions are reached within seconds after RCl addition. In independent runs Kennedy examined the effect of temperature, solvent polarity, and the nature of RCl on styrene consumption (443). Figure 29 summarizes some of the results obtained with

$$\underset{\displaystyle CH_3-\overset{\displaystyle |}{\underset{}{CH}}-CH=CH_2}{\overset{\displaystyle Cl}{}}$$

as RCl at $20°C$. Rates are faster in polar solvents than in nonpolar ones (see rates at $20°C$ in various solvents). Surprisingly, however, rates are faster at lower temperatures in polar solvent (chlorobenzene) than in nonpolar diluents. These observations can be explained assuming faster termination in nonpolar than in polar media where solvation might cause charge separation. Since the polarity of the medium increases with decreasing temperatures, increased solvation occurs and retards termination under these conditions.

The molecular weights produced by suitable $AlEt_2Cl$/RCl systems are remarkably high. For example, weight average molecular weights in the range 400–600 $\times 10^3$ and number average molecular weights of 50–300 $\times 10^3$ can be readily obtained. Indeed, it appears that the molecular weights of these polystyrenes are the highest on record among cationically synthesized polystyrenes.

The results of Kagiya et al. (448) obtained with $AgClO_4$/RCl initiating systems have a direct bearing on the initiation research carried out with aluminum containing Lewis acids, in particular, with AlR_2X and AlR_3. These experiments were carried out by dissolving silver perchlorate in styrene and adding to this

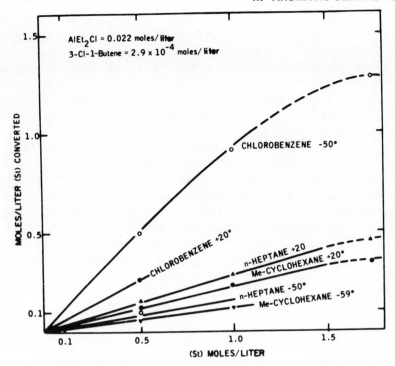

Figure 29. The effect of initial styrene concentration on the amount of styrene converted using the $AlEt_2Cl$/3-chloro-1-butene system in various solvents at -50 and $+20°C$ (443).

solution stoichiometric amounts of various organic halides in styrene at $0°C$. This technique for initiation was first reported by Plesch (405) with the $AgClO_4$/ 1-phenylethyl bromide system. Kagiya et al. built upon this base. Rapid polymerization occurred only when the silver salt and the organic halide interacted as evidenced by the precipitation of silver halide (polystyrene is soluble in its own monomer). The reaction was followed by withdrawal of aliquots at desired time intervals and precipitation into methanol.

The pertinent initiation and propagation steps can be visualized as follows:

$$RX + AgClO_4 \rightarrow R^{\oplus}ClO_4^{\ominus} + Ag\ X\downarrow \xrightarrow{St} RSt^{\oplus}ClO_4^{\ominus}$$

$$RSt^{\oplus}ClO_4^{\ominus} + \xrightarrow{nSt} RSt_n^{\oplus}ClO_4^{\ominus}$$

With tBuBr and tBuI linear conversion–time plots were obtained. The straight lines extrapolated back to the origin indicating rapid initiation and stationary polymerization. According to the authors no termination occurs, although conversions up to only 20% are shown (Figure 2 of reference 448). In contrast, with RX = tBuCl, $(C_6H_5)_3CCl$, $(C_6H_5)_2CHCl$, $C_6H_5CH_2Br$, $C_6H_5CH(CH_3)Br$, and iPrI the conversion–time plots were more or less convex to the abscissa indicating rate acceleration and no termination. Conversion–time[2] plots gave straight lines and the slopes were regarded as a measure of initiating activity: $(C_6H_5)_2CHCl \gg C_6H_5CH_2Br > C_6H_5CH(CH_3)Cl > (CH_3)_3CCl > (CH_3)_2CHCl > (C_6H_5)_3CCl$. Comparing the initiating activity of the halides, the Japanese authors found the sequence chloride < bromide < iodide.

These and some other data were analyzed to mean that, in the first approximation, initiating activity of organic halides is inversely proportional to their ionic (heterolytic) dissociation energy.

In this context the work of Minoura and Toshima (458) merits consideration. These investigators initiated the polymerization of styrene in 1,2-dichloroethane by chlorosilane compounds in conjunction with metal halides, for example, $FeCl_2$, $CdCl_2$, $HgCl$, and $HgCl_2$, in the temperature range 0–30°C. Unfortunately the mixing sequence of the reagents was not specified in the experimental section. However it is clear from the available data that polymerization only proceeds when both the chlorosilane and the metal halide are present in the system. Infrared spectroscopy indicated the incorporation of silicon into the polymer, and UV spectra provided evidence for the existence of $Me_3SiCl \cdot FeCl_3$ complexes. On the basis of these and other data the authors propose initiation by

$$Me_3SiCl + HgCl_2 \rightarrow Me_3\overset{\delta\oplus}{Si}Cl - Hg\overset{\delta\ominus}{Cl}_2 + monomer \rightarrow Me_3\underset{\underset{\underset{C_6H_5}{|}}{\overset{|}{CH_2=CH}}}{\overset{\uparrow}{SiCl}}\!\!-\!\!-\!\!HgCl_2 \rightarrow$$

$$\underset{\underset{C_6H_5}{|}}{\underset{\overset{\oplus}{CH_2-CH}}{\overset{\overset{\ominus}{HgCl_3}}{\underset{Me_3Si}{|}}}}$$

which would explain the presence of silicon in the polymer.

Pseudocationic Polymerization. In the course of their investigation on the polymerization of styrene with $HClO_4$ in methylene chloride at low temperatures, Gandini and Plesch (396) noted that the reaction involves three stages: Stage I is a period of very fast polymerization during which electrical conductivity and an orange–red color are present. This is followed by a much slower process, Stage II, characterized by the absence of conductivity and color and this in turn is followed by Stage III during which the conductivity and yellow color reappear. These and a series of other observations could not be explained by conventional concepts of a cationic polymerization which involve more or less free ionic species. According to Plesch (463), during Stage II the propagating species may be perchlorate esters, $-CH_2CH(C_6H_5)OClO_3$, or perhaps perchlorate esters in conjunction with free ions.

The exploration of the mechanism of the still controversial pseudocationic polymerizations occupies several research groups in England (463), Ireland (463a), Italy (464), and Australia (465). The application of the Hammett $\sigma\rho$ concept could be of use in elucidating the mechanism of pseudocationic polymerizations. If ions are not involved in the mechanism, ρ should be the same for electron releasing as well as electron withdrawing substituents.

Investigations by the Kyoto Group. Professors Okamura and Higashimura and their students have greatly contributed to our understanding of cationic styrene polymerization. This group has concentrated on the detailed elucidation of rates and polymerization mechanism. A competent review that summarizes the work of these researchers up to 1969 is in reference 466. The reader is referred to this source in which available quantitative information in regard to the elementary steps, that is, initiation, propagation, transfer, and termination, is assembled and discussed.

Recently, Masuda and Higashimura (437) investigated the effect of various salts on the polymerization of styrene. For example, in one set of experiments they added small amounts of nBu_4NClO_4 to styrene in CH_2Cl_2 at $0°C$ and initiated the polymerization with $SnCl_4(H_2O?)$, $SnCl_4 \cdot CCl_3COOH$, or $BF_3 \cdot OEt_2$ and discovered that the rate increased considerably in the presence of the salt. A possibility for this phenomenon proposed by the authors is that the counteranion $ClO_4^{\ominus}$ added in the form of salt exchanges with the original counteranion $G^{\ominus}$ introduced with the initiator, and that the more stable $ClO_4^{\ominus}$ counterion induces a faster rate than the original anion:

$$\sim\!\!C^{\oplus}\text{-----}G^{\ominus} + nBu_4NClO_4 \quad \rightleftharpoons \quad \sim\!\!\sim\!\!C^{\oplus}\text{-----}ClO_4^{\ominus} + nBu_4NG$$

Salts like nBu_4NBF_4 and nBu_4NI, however, decreased the rate which was attributed to slow polymerizations in the presence of $BF_4^{\ominus}$ and $I^{\ominus}$ counteranions. The fact that large amounts of nBu_4NClO_4 also decreased the rate remains unexplained. The molecular weights of the products obtained in the presence of nBu_4NClO_4 were considerably lower than those obtained in the absence of this salt (limiting viscosity dropped from 0.3 to <0.1). This again was viewed as evidence for counteranion exchange.

Another interesting recent report by this group (446) concerns the molecular weights and molecular weight distributions (MWD) of polystyrenes. First styrene was polymerized with CH_3COClO_4 in CH_2Cl_2 at $0°C$ to various conversions and the MWD was found to be bimodal. The low molecular weight peak increased with increasing conversions. Next, polymerizations were carried out in the presence of increasing amounts of nBu_4NClO_4 which resulted in the reduction and, with larger quantities, complete elimination of the higher molecular weight peak. The polarity of the medium also affected MWD: in the less polar CH_2Cl_2/C_6H_6 3:1 mixture the high molecular peak was sharply reduced. On the other hand $BF_3 \cdot OEt_2$ and $SnCl_4 \cdot CCl_3COOH$ produce monomodal distributions and the peak is at a higher molecular weight than that obtained with CH_3COClO_4. These results are shown in Table 19.

The bimodality of the MWD obtained with CH_3COClO_4 is strong evidence for the existence of two independently propagating species of different activities.

Another piece of interesting information that has emerged recently is that $BF_3 \cdot OEt_2$ is at best a sluggish initiator system for the polymerization of styrene and that H_2O greatly enhances the rate. Background information of this problem can be found in a paper by Giusti et al. (467). This group of workers most carefully investigated the styrene/$BF_3 \cdot OEt_2/H_2O$ system in $ClCH_2CH_2Cl$ and CH_2Cl_2 solvents at $25°C$ and found that H_2O was an effective rate accelerator. While polymerizations progressed in the absence of H_2O, higher rates were obtained in the presence of certain moderate concentrations of water ($[H_2O]/[BF_3 \cdot OEt_2]$ less than 1.0, maximum rate 0.5). Similar results have recently been obtained by Higashimura et al. (468). These investigators confirmed the findings of Giusti et al. that BF_3OEt_2 is an initiator of very low activity and that the overall rate can be increased by a factor of ∼10 in the presence of modest quantities of H_2O. Further, the rate increase was found to be caused not by higher k_ps (this effect would account for about 50% rate increase), but rather by the enhanced concentration of propagating ions. The active ion concentration was determined by an ingenious method using 2-bromothiophene terminator.

Miscellaneous Findings and Leads Since 1963.   Plesch and his coworkers investigated the polymerization of styrene under carefully controlled conditions. Plesch and coworkers (468) summarized their work with the styrene/$TiCl_4/CH_2Cl_2/H_2O$ system collected over many years by two independent operators, Longworth and Panton, and consider previously published data on this system as obsolete or superseded. Unfortunately, in spite of the large amount of careful work, the authors are frustrated by complex phenomena and have to conclude: "It is . . . disappointing that the chemical information which [this work has] yielded is so meager. Perhaps the most significant point to emerge is that with the same catalyst, cocatalyst, and solvent, the polymerizations of styrene and isobutene are utterly different in almost every respect. . . . Kinetic studies are not the most suitable for investigating these systems. . . ." Indeed, it is disappointing that even in a "simple" system such as this, we still do not understand the effect of water on the polymerization, do not have a meaningful rate law, do not have a clear idea of the effect of water concentration on the molecular weights, do not understand the nature of the propagating species, and so on.

TABLE 19 MOLECULAR WEIGHTS $\bar{M}_n$, $\bar{M}_w$, and $\bar{M}_w/\bar{M}_n$ OF POLYSTYRENES OBTAINED WITH AcClO$_4$, IN THE ABSENCE AND PRESENCE OF nBu$_4$NClO$_4$ (446)[a]

nBu$_4$NClO$_4$ mmole/l	Overall Values			Low Molecular Weight Peak				High Molecular Weight Peak			
	$\bar{M}_n \times 10^{-3}$	$\bar{M}_w \times 10^{-3}$	$\bar{M}_w/\bar{M}_n$	$\bar{M}_n \times 10^{-3}$	$\bar{M}_w \times 10^{-3}$	$\bar{M}_w/\bar{M}_n$	Fraction	$\bar{M}_n \times 10^{-3}$	$\bar{M}_w \times 10^{-3}$	$\bar{M}_w/\bar{M}_n$	Fraction
–	3.78	13.6	3.60	2.08	3.10	1.49	0.875	15.8	23.4	1.49	0.125
0.05	2.49	6.19	2.49	2.12	3.12	1.47	0.974	16.2	21.3	1.31	0.026
0.1	2.28	4.92	2.16	2.08	3.09	1.49	0.985	16.4	20.1	1.23	0.011
1.0	2.07	3.09	1.49	2.07	3.09	1.49	1.000	0	0	0	0.00

[a] [M] = 1.0 mole/l, [AcClO$_4$] = 0.37 mole/l n CH$_2$Cl$_2$, at 0°C; conversion 35–42%.

The authors obtained some interesting molecular weight data. The molecular weight (DPs calculated from intrinsic viscosity data) versus monomer concentration plots at three temperatures gave straight lines through the origin indicating that chainbreaking with monomer does not occur. Their plot, shown in Figure 30, also indicates the synthesis of high polymers up to $\bar{M}_n \sim 50,000$ (Figure 1 in reference 469). This product may be one of the highest molecular weight polystyrenes prepared by conventional cationic techniques on record until that date. The molecular weights of polystyrenes obtained by chemical and γ-ray initiated polymerizations are compared later in this section.

An interesting publication in this field of polymer science is the report by Panayotov et al. (470) who polymerized styrene with $BF_3 \cdot OEt_2$ in the presence of electron acceptors such as tetracyanoethylene (TCE), chloranil (CA), and 1,3,5-trinitrobenzene (TNB). A short investigation indicates that in the presence

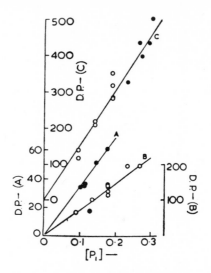

Figure 30. The dependence of the DP on the monomer concentration (○) W. R. L., (●) C. J. P. (A) $-25°C$, (B) $-63°C$, (C) $-91°C$ [H_2O] = 0, i.e., residual after 5 hr pumping (469).

of electron acceptors the rate of polymerization is strongly increased in chlorinated hydrocarbons (carbon tetrachloride, tetrachloroethane, and dichloroethane), but remains largely unaffected in benzene and nitrobenzene solvents. The rate enhancing effect of TCE shows a maximum at a certain TCE/BF_3OEt_2 ratio. Finally the UV spectrum of a TCE/styrene system changes (two strong peaks disappear) upon $BF_3 \cdot OEt_2$ addition.

To explain these observations the Bulgarian authors propose that the initially formed donor–acceptor complex M·EA between monomer M and electron acceptor EA is destroyed by the addition of $BF_3 \cdot OEt_2$ because the electron acceptor in the system seeks out the counteranion $G^\ominus$ as the better donor molecule:

$$\sim\sim\oplus G^\ominus \xrightarrow{\text{M·EA}} \sim\sim M^\oplus [G \cdot EA]^\ominus$$

This complexation of the counterion enhances the separation of ions and thus accelerates the rate. This effect operates only in "inert" solvents; it does not occur in donor solvents (benzene) or in acceptor media (nitrobenzene). The maximum rate at a certain $EA/BF_3 \cdot OEt_2$ ratio is due to a saturation effect: the rate increases with increasing $EA/BF_3 \cdot OEt_2$ until counteranion complexation is maximum; beyond this point the EA seeks out the monomer whose reactivity decreases because of complexation.

Systematic studies dealing with the cationic polymerizations under electric field carried out by Ise (471) provide improved insight into the cationic polymerization mechanism of styrene and other monomers. Ise regards the polymerizing system as a weak electrolyte solution in which ion pairs and free ions coexist. Under a strong electric field weak electrolyte solutions exhibit increased conductivity due to field facilitated ionization of ion pairs (the so called second Wien effect). According to Ise and his school this effect can also be observed with suitable polymerizing systems. Thus if field facilitated ionization occurs, one obtains higher polymerization rates and possibly higher molecular weights in the presence of an electric field than in its absence. The increase in the concentration of free ions explains the higher polymerization rates and molecular weights.

Experimentally, these workers first established proper measuring conditions to avoid the many possible pitfalls involved in quantitative electrochemical operations. Initial experiments showed that the rate of many cationic homopolymerizations, for example α-methylstyrene/iodine and isobutylvinyl ether/iodine, increased under an electric field. According to further kinetic research this effect was caused by an increase in the rate constant (and not to a change in mechanism), and the increased rate constant reflected an enhanced entropy contribution and not a change in the activation energy. The application of electric field did not always increase the molecular weights.

Ise's most interesting results were obtained by investigating the effect of various solvents on the rate of styrene polymerization with $BF_3 \cdot OEt_2$ in the presence of electric field (472). In a series of experiments the dielectric milieu of the polymerization system was changed from 2.4 to 35 by the use of solvent mixtures of toluene, 1,2-dichloroethane, and nitrobenzene; Figure 31 shows the findings.

The ordinate R_{pE}/R_{po} is the ratio of rates determined in the presence and absence of electric field, respectively. Mathematically:

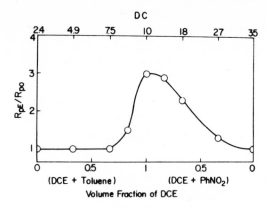

Figure 31. Field effects and solvent composition: styrene/BF_3OEt_2, $25°C$, $E = 0.25$ V/cm, (472).

$$\frac{R_{pE}}{R_{po}} \approx \frac{\alpha_E/(1-\alpha_E)k_p'' + k_p'}{\alpha_o/(1-\alpha_o)k_p'' + k_p'}$$

where α_E and α_o are the degrees of dissociation in the presence and absence of electric field, and k_p' and k_p'' are the rate constants for ion pair and free ion propagation, respectively. According to the figure, the rate does not change until a sufficient amount of 1,2-dichloroethane (DCE) is mixed with toluene to raise the dielectric constant ϵ to ~7.5. From this point on the rate ratio increases until it reaches a maximum value in pure DCE at a medium dielectric constant value of ~10. By admixing highly polar nitrobenzene to DCE, the rate ratio decreases until it again reaches the base line in pure nitrobenzene. In other words, the rate ratios are 1 in very low and very high dielectric media, but rise significantly (up to threefold) at medium ϵ values. The author's ingenious explanation can be summarized along these lines: at large or small α, the $R_{pE}/R_{po} = 1$. However, in toluene, at low ϵ values, tight ion pairs are present and the application of an external electric field is insufficient to bring about dissociation and an increase in free ion concentration. In nitrobenzene, at high ϵ values, almost completely dissociated free ions exist so that the application of the electric field does not much affect the free ion concentration and, again, no rate enhancement is observed. At large or small α, R_{pE}/R_{po} ~ 1. However, at medium ϵ values, in DCE, the combined effect of solvation by DCE plus the external field is able to enhance free ion concentration so that the rate ratio increases in the presence of an electric field: in the range of intermediate values of α, $R_{pE}/R_{po} > 1$. In addition to $BF_3 \cdot OEt_2$, similar phenomena were obtained with $SnCl_4/CCl_3COOH$ and $TiCl_4/CCl_3COOH$ initiator systems (473).

Aspects of Ise's work have recently been questioned (474), however, the original authors rebutted the criticism (475).

The electroinitiated polymerization of styrene was briefly described by Breitenbach and Srna (476). Low yields of a respectable molecular weight of polystyrene were produced around the anode when a $(Et_4N)^{\oplus} BF_4^{\ominus}$/styrene system was electrolyzed in nitrobenzene solution with 0.35 mA for 60 min.

Research on the γ-ray induced polymerization of styrene should be noted. The understanding of γ-ray initiated polymerizations of vinyl monomers in general and that of styrene in particular has significantly advanced during the last decade. One of the important conclusions of a major review in 1963 (477) on polymerization by radiation was the recognition that vinyl polymerizations in the liquid phase may occur not only by a free radical mechanism but may simultaneously proceed by an ionic process. At the present there remains no doubt that the γ-ray induced polymerization of styrene and many other monomers ($\sim$50), may proceed under suitable conditions by a cationic mechanism. Briefly, careful work carried out under increasingly pure conditions by a number of groups showed that in the absence of moisture at low temperatures the rate of polymerization gradually changes from 0.5 order, indicating termination by a free radical combination process, close to first order process, suggesting unimolecular termination characteristic for a cationic mechanism. Indeed, a larger number of monomers exhibit this behavior, namely, they polymerize by an ionic mechanism at low temperatures ($\sim$$-78^{\circ}$C) but by a conventional free radical process at higher temperature levels ($\sim$$0^{\circ}$C). The polymerization behavior of styrene has been thoroughly investigated in this regard (477a).

Major reviews of this field of chemistry include recent references (478–481). Several researchers investigated the irradiation induced polymerization of styrene in methylene chloride solvent at -78°C (482–485) and there is general agreement that the mechanism is one of free cationic propagation.

It is of interest to compare polystyrene molecular weight data obtained in γ-ray induced polymerization in CH_2Cl_2 at -78°C with those collected from polymerizations initiated by $TiCl_4$ also in CH_2Cl_2 solvent at -78°C. According to Williams (478) DP is $\sim$500 ($\bar{M}_n \sim$ 50,000) in irradiation experiments, that is, when transfer to monomer is the molecular weight determining parameter. Interestingly this value is very close to that obtained by chemical initiation [$TiCl_4$/H_2O(?)] in the same solvent at similar temperatures (469). This could mean either that both systems propagate by free carbocations or that the counterion in the $TiCl_4$ system (perhaps the $TiCl_4OH^{\ominus}$), although present, does not get involved in the molecular weight determining transfer step. In other words, the presence or absence of counterion does not influence the molecular weight of polystyrenes obtained in these systems. This is in sharp contrast to the findings obtained with the isobutylene/γ-rays and isobutylene/BF_3 or $AlCl_3$ systems, where irradiation induced free ion polymerization produced polyisobutylenes of very much higher molecular weight than those obtained with chemical (BF_3, $AlCl_3$, or $AlEtCl_2$) initiation (136). For the isobutylene system it was proposed that the counterion assists chain transfer to monomer (or more precisely, the counterion impedes propagation to a greater extent than it does the competing process of monomer transfer). It is conceivable that with isobutylene two monomer transfer processes coexist: one, a relatively slower, spontaneous monomer transfer process that operates with free ions and another, a relatively faster "assisted" monomer transfer that operates in the presence of suitable counterions:

Isobutylene-type monomer transfer transition state

In this transition state the spreading of the charge is aided by the presence of a counterion. The exothermicity of this transfer process is very low and may be due to the slightly more favorable solvation of the small t-butyl cation as compared to that of the polymer cation (136). That the molecular weights of polystyrene obtained in γ-ray and $TiCl_4/H_2O$(?) induced polymerizations are quite comparable could be explained by assuming that in these systems only one kind of monomer transfer is operating, that this process is independent of the counterion, and that the transition state of this process leads to indane end-groups:

The driving force for this process is the formation of a new bond by intramolecular alkylation.

Last but not least the results by Barson et al. (685) deserve comment. These authors polymerized styrene by 13X synthetic zeolite at 30°C and obtained extremely high molecular weight products ($\overline{M}_n = 1$–3.8×10^6) and unusually narrow molecular weight distributions ($\overline{M}_w/\overline{M}_n \sim 1.2$–$1.5$). The cationic polymerization mechanism was visualized to be one in which initiation occurs by protonation. The proton is provided by adsorbed water impurities in the zeolite. Propagation also takes place at the zeolite surface. Termination or transfer is expected to occur by dissociation from the zeolite.

Evidently in the 13X zeolite induced polymerization of styrene to very high molecular weights the counteranion remains embedded in the solid phase. Similar situations have been discussed in conjunction with isobutylene where complex solid or colloidal counteranions give rise to very high molecular weight products.

2. p-Methylstyrene

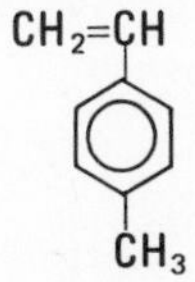

$$CH_2=CH$$

$$CH_3$$

The polymerization kinetics of p-methylstyrene using iodine initiator in ethylene chloride at $30°C$ was studied by Higashimura and coworkers (513, 514). The propagation rate constant was 5.7 l/mole min (styrene = 0.22 l/mole min). These authors also copolymerized p-methylstyrene with styrene, p-chloro-styrene, and p-methoxystyrene. By determining the rate constant of the cross propagation step they were able to assign a relative reactivity scale to these monomers:

p-methoxystyrene > p-methylstyrene > styrene > p-chlorostyrene

Higashimura et al. (515) also studied the effect of solvent on the polymeriza-tion and copolymerization of p-methylstyrene. 1,3-Dichloroethane, chloroform, carbon tetrachloride, and mixtures of these were used as solvents and iodine was the initiator. The rate constant was several orders of magnitude larger for polymerizations carried out in $(CH_2Cl)_2$ than in CCl_4.

Heublein and Dawczynski (686) studied details of the homopolymerization of p-methylstyrene by $SnCl_4/H_2O$ and $0°C$. The overall rate constants determined for various solvents were as follows: 1.9 in hexane ($\bar{M}_n = 1760$), 4.70 in toluene ($\bar{M}_n = 1870$), 10.50 in chloroform ($\bar{M}_n = 1660$), 17.40 in methylene chloride ($\bar{M}_n = 2150$), and 59.10 in nitroethane ($\bar{M}_n = 2440$).

According to Krentsel and coworkers (516, 517, 613) poly(p-methylstyrene) synthesized by low temperature cationic technique, for example $BF_3 \cdot OEt_2$ in toluene or ethyl chloride in the range 0 to $-130°C$ contains "anomalous structures":

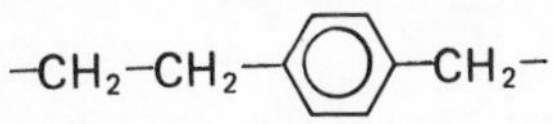

$$-CH_2-CH_2-\langle O \rangle-CH_2-$$

Great efforts have been made by Kennedy et al. (518) to reproduce this unusual claim but in spite of lengthy experimentation no evidence for the above structure could be found. Rather, polymerizations initiated with perchloric acid, $BF_3 \cdot OEt_2$, $AlBr_3$, and $AlEtCl_2$ under a variety of conditions lead to the conclusion that cationically synthesized poly(*p*-methylstyrene) is of a conventional 1,2 structure.

Higashimura and coworkers (519) studied the copolymerization behavior of *p*-methylstyrene with *p*-methyl-β-styrene (see under β-methylstyrene) and Overberger et al. studied that with *p*-chlorostyrene (520).

$$CH_2=CH$$

3. p-*Ethylstyrene*

$$CH_2CH_3$$

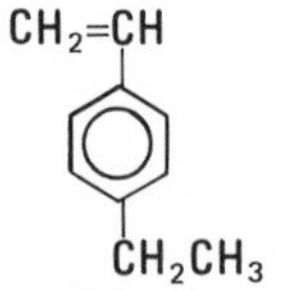

Very little cationic polymerization research has been carried out with this olefin. The only reference in the scientific literature of which this writer is aware is by Overberger et al. (536) who copolymerized *p*-ethylstyrene (and a host of other styrene derivatives) with *p*-chlorostyrene in the course of their systematic study of reactivities of styrene and substituted styrenes. According to their results the reactivity ratios of *p*-ethylstyrene and *p*-chlorostyrene are 4.1 ± 0.7 and 0.29 ± 0.4, respectively, very similar to the values obtained with *p*-methylstyrene/*p*-chlorostyrene (4.5 ± 0.7 and 0.22 ± 0.05), indicating little if any steric interference by the extra $-CH_2-$ group in the para position of the title monomer.

4. p-*Dodecylstyrene* $CH_2=CH-\bigcirc-C_{12}H_{25}$

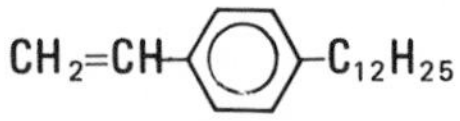

Heublein and Dawczynski (686) briefly examined the polymerization of this monomer by using $SnCl_4/H_2O$ in CH_2Cl_2 at $0°C$. Interestingly, the overall rate of polymerization was almost four times as high as that of the *p*-methyl derivative and the molecular weight of the polymer ($\bar{M}_n = 4700$) was almost double than that of poly(*p*-methylstyrene) under the same conditions.

$$CH_2=CH$$

5. p-*Isopropylstyrene*

$$H_3C-CH-CH_3$$

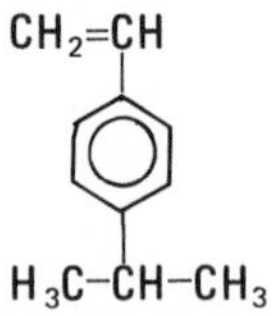

Kennedy (74) mentions in a review article that *p*-isopropylstyrene rapidly polymerizes with $AlCl_3$ in ethyl chloride even at as low a temperature as $-168°C$ to a conventional 1,2 structure and that no isomerization (hydride migration)

takes place. The polymer is a white amorphous (by x-rays) solid, soluble in benzene and starts to soften at ~112°C. Similar conclusions have been reached by Disselhoff (521) who used $AlCl_3$, $SnCl_4$, $TiCl_4$, BF_3, and H_2SO_4 initiator systems and by Krentsel and coworkers (517) who probably used $BF_3 \cdot OEt_2$.

6. p-tert-*Butylstyrene* $CH_2=CH-\langle\bigcirc\rangle-\overset{\overset{\displaystyle CH_3}{|}}{\underset{\underset{\displaystyle CH_3}{|}}{C}}-CH_3$

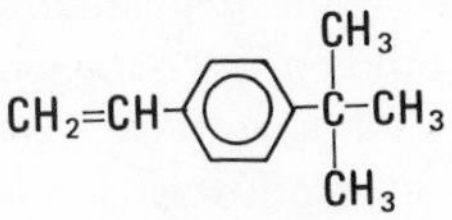

Heublein and Dawczynski (686) briefly examined the polymerization behavior of this monomer using $SnCl_4/H_2O$ in CH_2Cl_2 at 0°C. The overall rate of polymerization was found to be about five times that of styrene, however, the molecular weight of the polymer ($\overline{M}_n$ = 7600) was about half that of polystyrene.

7. o-*Methylstyrene*

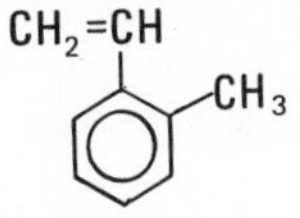

$$CH_2=CH$$
$$\underset{\bigcirc}{\overset{|}{}}-CH_3$$

As Magagnini et al. (522) have shown, the cationic polymerization of o-methylstyrene (H_2SO_4, $AlBr_3$, $BF_3 \cdot OEt_2$ in EtCl and CH_2Cl_2, −10 to −126°C) yields a conventional 1,2 enchainment and no hydride migration takes place even at lowest temperatures.

8. o-*Isopropylstyrene*

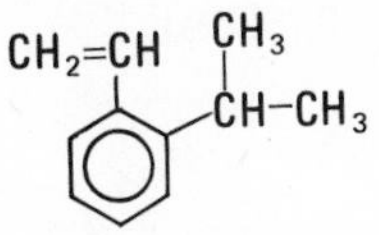

$$CH_2=CH \quad CH_3$$
$$\underset{\bigcirc}{\overset{|}{}}-CH-CH_3$$

o-Isopropylstyrene was polymerized cationically first by Aso et al. (523) using $AlCl_3$ at low temperatures. These authors proposed that hydride migration occurs during polymerization. In contrast, Kennedy et al. (524) who reinvestigated the cationic polymerization of this monomer concluded from spectroscopic evidence and by comparison with free radical initiated polymerization product that the cationic polymer ($AlCl_3$ and $AlBr_3$ in EtCl and CH_2Cl_2) is of a conventional 1,2 enchainment.

9. 2,4-*Dimethylstyrene*

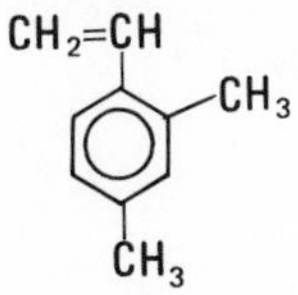

$$CH_2=CH$$
$$\underset{\underset{CH_3}{\bigcirc}}{\overset{|}{}}-CH_3$$

Anton and Maréchal (525) studied the homopolymerization of 2,4-dimethylstyrene with a variety of initiators, for example, $TiCl_4$, $SnCl_4$, BF_3OEt_2, $AlBr_3$, and H_2SO_4, in methylene chloride and 1,2-dichloroethane solvents in the range -70 to $+70°C$. Respectable molecular weights were obtained with BF_3OEt_2 and $AlBr_3$ at $-70°C$ ($\bar{M}_n = 74 \times 10^3$ and 46×10^3, respectively). The softening range of these polymers was 195–198°C and 185–188°C. In another paper by the same authors (526) $SnCl_4$ was used in the temperature range $+30$ to $-70°C$, however, only low intrinsic viscosity product was obtained, $[\eta] = 0.15$–0.40.

Copolymerization studies by the same authors indicate (525) that 2,4-dimethylstyrene is about 1.5 times as reactive as styrene against a styrene or 2,4-dimethylstyrene carbocation.

10. *2-Methyl-5-isopropylstyrene (2-Vinyl-p-cymene)*

Lalande et al. (527) mention the polymerization of this monomer with $AlCl_3$ in *n*-heptane at 20°C (yield, 16%; $\bar{M}_n = 1000$–5000).

11. *2-Isopropyl-5-methylstyrene (3-Vinyl-p-cymene)*

Lalande et al. (527) mention the polymerization of this monomer with $AlCl_3$ in *n*-hexane at 20°C (yield, 12%; $\bar{M}_n = 1000$–5000).

12. *Vinylmesitylene*

There is a brief report in the literature (528) that vinylmesitylene can be polymerized to an extremely brittle low molecular weight polymer by $SnCl_4$ at room temperature.

13. *2,6-Dimethyl-4-t-butylstyrene*

The cationic (and radical) polymerization of this monomer was briefly investigated by Korshak and Matveeva (529) who report that polymerization was successful with gaseous BF_3 (1 wt. %) in CCl_4 or $AlCl_3$ in nitrobenzene (10% solution) at $-15°C$. The colorless glassy product fuses in the $180–190°C$ range; the molecular weights (viscosities in benzene solution) were 13,500 with $AlCl_3$ and 29,900 with BF_3, however, no other details are given.

Interestingly, this monomer could not be polymerized by free radical initiators (by heating, with benzoyl peroxide, or direct sun light) (529).

14. p-*Fluorostyrene*

Aliev (530) studied the copolymerization of p-fluorostyrene with isobutylene with the purpose of preparing new lubricants or oil thickeners. Copolymerizations were carried out in CH_3Cl, with $AlCl_3$ under various conditions. The highest molecular weight 21,000 was obtained at $-100°C$.

15. p-*Chlorostyrene*

The homopolymerization of this olefin (using CCl_3COOH) to low molecular weight products, DP = 2–15, has been mentioned (613a). In a more substantial study Brown and Pepper (531) used $HClO_4$ and $(CH_2Cl)_2$ solvents at $25°C$. The number average molecular weight was $\sim$1380. They found the polymerization rate first order in monomer concentration and calculated the rate constant for propagation k_p = 290 l/mole min. This rate constant was about a factor of 3.5 lower than that for styrene. Higashimura and coworkers (513) came to similar conclusions. The latter authors investigated the homopolymerization of p-chlorostyrene and other styrene derivatives by the use of iodine in $(CH_2Cl)_2$ at $30°C$ and obtained k_p = 0.075 l/mole sec for p-chlorostyrene and k_p = 0.22 ℓ/mole sec for styrene. According to these data also, p-chlorostyrene is about one-third as reactive as styrene. According to Heublein and Dawczynski (686), who polymerized p-chlorostyrene by $SnCl_4/H_2O$ in CH_2Cl_2 at $0°C$, the overall rate constant of polymerization of this monomer is about one-fifth that of styrene. The $\bar{M}_n$ obtained by these authors was 14,700. The lower reactivity of p chlorostyrene as compared to the parent compound is expected considering the position of the electron withdrawing chlorine substituent in the molecule.

While there is a dearth of information on the homopolymerization of p-chlorostyrene, the copolymerization of this monomer with a variety of olefins has been studied in detail by many investigators. The first report on the copolymerization with styrene seems to be that in 1948 by Alfrey and Wechsler (532). Since that time the copolymerization of p-chlorostyrene has been studied with the following monomers: isobutylene (291, 533), styrene (513, 531, 534, 535), α-methylstyrene (536–538), α-ethylstyrene (520a), p-methylstyrene (473, 520), p-ethylstyrene (538), m-methoxystyrene (520), trans-β-methylstyrene (520), cis-β-methylstyrene (520), trans-α,β-dimethylstyrene (520a), p-bromostyrene (520), and cyclo-1,3-octadiene (539). The reactivity ratios of all these monomer pairs are available and inspection substantiates the conclusion reached above that p-chlorostyrene is a rather sluggish cationic monomer.

16. o-*Chlorostyrene*

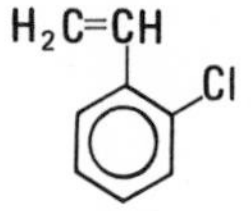

Alfrey et al. (540) described the copolymerization of this monomer with p-methoxy-β-methylstyrene (anethole) using $SnCl_4$ in CCl_4 at $0°C$. The reactivity ratios were: $r_{oClSt} = 0.03 \pm 0.005$ and $r_{pMMSt} = 18 \pm 3$.

17. m-*Chlorostyrene*

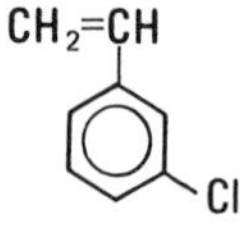

Overberger et al. (520) mention the copolymerization of styrene with m-chlorostyrene using $SnCl_4$ in CCl_4 at $0°C$. The reactivity ratios $r_{St} = 3.3 \pm 0.4$ and $r_{mClSt} = 0.3 \pm 0.05$ indicate a relatively sluggish cationic monomer.

18. 3,4-*Dichlorostyrene*

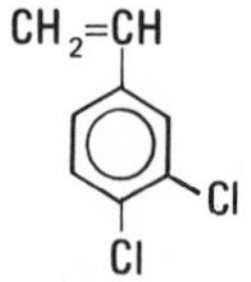

Florin (541) studied the copolymerization of this monomer with styrene using $AlCl_3$, $AlBr_3$, $SnCl_4$, $TiCl_4$, and $BF_3 \cdot OEt_2$ in CCl_4 solution as well as $AlCl_3$ and H_2SO_4 in $C_6H_5NO_2$ solution at $0°C$ and determined the corresponding reactivity ratios. For example, with $AlBr_3$ $r_{3,4DSt} = 0 \pm 0.2$ and $r_{St} = 6.8 \pm 0.8$ and with $SnCl_4$ $r_{3,4DSt} = 0.48 \pm 0.08$ and $r_{St} = 3.1 \pm 0.1$. It appears that 3,4-dichlorostyrene is relatively more reactive than 2,5-dichlorostyrene.

$$CH_2=CH$$

19. *2,5-Dichlorostyrene*

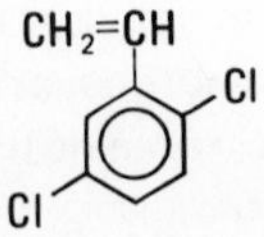

Alfrey and Wechsler (532) mention in a footnote that the reactivity of this monomer is even lower than that of p-chlorostyrene and one year later in 1949 Florin (542) substantiated this statement by an experiment in which he copolymerized styrene with 2,5-dichlorostyrene with $AlCl_3$ and $SnCl_4$ in C_2H_5Cl at $0°C$ and found $r_{St} = 14.8 \pm 2$ and $r_{2,5ClSt} = 0.34 \pm 0.2$.

$$CH_2=CH$$

20. p-*Bromostyrene*

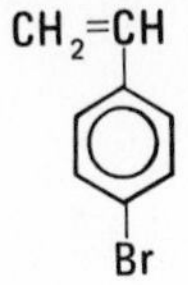

The homopolymerization of this monomer has been briefly examined by Heublein and Dawczynski (686) who used $SnCl_4/H_2O$ at $0°C$. The overall rate of polymerization was somewhat lower than that of p-chlorostyrene, but somewhat higher than the p-iodo derivative. The $\overline{M}_n$ was 15,000.

Overberger et al. (520) mention the copolymerization of this monomer with p-chlorostyrene with $SnCl_4$ in CCl_4 at $0°C$. According to the reactivity ratios ($r_{Cl} = 1.0 \pm 0.1$ and $r_{Br} = 1.0 \pm 1.0$) these two monomers have identical reactivities and probably give perfectly random copolymers.

21. p-*Cyanostyrene* $CH_2=CH$—⟨◯⟩—CN

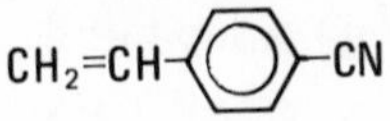

Heublein and Dawczynski (686) briefly investigated the polymerization behavior of this monomer using $SnCl_4/H_2O$ in CH_2Cl_2 at $0°C$. A very slow overall polymerization, but a respectable molecular weight product ($\overline{M}_n = 19,100$) was obtained. It is remarkable that p-cyanostyrene is polymerizable under these conditions at all considering this monomer contains a highly nucleophilic $-CN$ group.

$$CH_2=CH$$

22. p-*Hydroxystyrene*

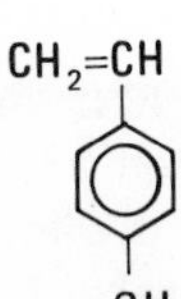

Kato (543) polymerized p-hydroxystyrene in methylene chloride with $BF_3·OEt_2$

in the range +12 to $-22°C$ and obtained in rapid reactions high (92–96%) conversions of fairly high molecular weight amorphous polymer, $[\eta]$ = 0.3–0.5. Much higher intrinsic viscosity product was obtained by cationic technique than by free radical polymerizations. The structure of the polymer was studied by infrared and ultraviolet spectroscopy. It was established that the products were of a conventional vinyl enchainment.

The author has claimed but did not prove the copolymerization of p-hydroxystyrene with ethyl vinyl ether with $BF_3 \cdot OEt_2$ at $-22°C$.

Kato (543) also reports a brief study on the solid state polymerization of this monomer with $BF_3 \cdot OEt_2$ at solid Dry Ice temperature. A dispersion of monomer/n-hexane/$BF_3 \cdot OEt_2$ at $-78°C$ gave 43% conversion of an amorphous polymer having a conventional structure for which $[\eta]$ = 0.157.

23. o-*Hydroxystyrene*

Kato and Kamagawa (543, 544) studied the polymerization of o-hydroxystyrene with $BF_3 \cdot OEt_2$ in ethylene chloride and methylene chloride solvents in the range −15 to $-30°C$. In his later paper (543) this author reports three experiments (in the first publication there is only one experiment) in which high conversions of very low molecular weight polymer were obtained $[\eta] \sim 0.06$, $\bar{M}_n \sim$ 3900. Polymerization rates of this monomer appeared to be slower than that of the para isomer but faster than the meta derivative. Structure analysis by IR and UV indicates that the polymerization and polyalkylation proceed simultaneously. Evidently not only the vinyl group but also the phenol nucleus is involved in this polyreaction which yields highly branched products.

24. m-*Hydroxystyrene*

Kato (543) reports the polymerization of m-hydroxystyrene with $BF_3 \cdot OEt_2$ in methylene chloride diluent in the range +3 to $-40°C$. As compared to the para and ortho isomers, low rates of polymerization and low molecular weight products, $[\eta] \sim 0.04$, were obtained. Structure studies by IR and UV spectrocopy indicate branched molecules which presumably formed by simultaneous polyaddition by the vinyl group and polyalkylation of the phenol moiety.

25. p-*Methoxystyrene*

Staudinger and Dreher (545) were among the early investigators who studied the cationic polymerization of this monomer. These authors used $SnCl_4$ in benzene at $0°C$ and obtained a polymer that after precipitation with methanol exhibited a softening range of 129–136°C.

Higashimura and coworkers (546) polymerized this monomer with $BF_3·OEt_2$ in methylene chloride, toluene, and mixtures of these solvents in the range –20 to –78°C. The molecular weight (intrinsic viscosity) of the samples increased by lowering the temperature, for example, in toluene solvent with 10 vol. % monomer $[\eta]$ rose from ~0.6 at –20°C to ~1.8 at –78°C. In CH_2Cl_2 solvent higher molecular weights were obtained over the entire temperature range, $[\eta]$ ~3.0 at –78°C. The precipitation temperature of samples prepared in methylene chloride was about 10°C higher than those synthesized in toluene and having the same intrinsic viscosities. This difference in precipitation temperatures was explained by postulating a higher degree of stereoregularity of the samples obtained in the polar solvent. The mechanism of chain transfer reaction in the polymerization of p-methoxystyrene was examined by Kamath and Haas (547) and later by Imanishi et al. (548) using $BF_3·OEt_2$ in CCl_4, $CHCl_3$, CH_2Cl_2, and $(CH_2Cl)_2$ in the range –20 to +30°C. These authors determined the intrinsic viscosity of samples obtained under various conditions and from these values calculated the molecular weights and the degrees of polymerization. Finally, by assuming a steady state kinetics and plotting $1/\overline{DP} = k_{tm}/k_p + k_t/k_p\ (1/[M])$ they obtained the rate constant ratios k_{tm}/k_p and k_t/k_p. For the polymerization of p-methoxystyrene they found that k_{tm}/k_p decreased with decreasing temperature and with increasing dielectric constant of the solvent, and k_t/k_p decreased with decreasing temperature. These findings suggested that highest molecular weights should be obtained at lowest temperatures and in highest dielectric constant solvent. It was stated that these predictions were borne out by experiment, but unfortunately no additional data have appeared.

The kinetics of polymerization of p-methoxystyrene induced by iodine in $(CH_2Cl)_2$ and CCl_4 solutions has also been studied by Higashimura and coworkers (549). Together with a variety of other para substituted styrene derivatives Heublein and Dawczynski (686) determined the overall rate constant for p-methoxystyrene polymerization initiated by $SnCl_4/H_2O$ in CH_2Cl_2 at $0°C$. The $\overline{M}_n$ of the polymer was 19,000.

Sigwalt and coworkers (303) initated the polymerization of p-methoxystyrene with triphenyl methyl hexachloroantimonate salt at $-70°C$ in CH_2Cl_2

and obtained respectable molecular weight product, $[\eta] = 2.3$, $\overline{DP} = 2720$.

Copolymerization experiments by Japanese authors (513, 548) and by Overberger et al. (520) demonstrated the superior reactivity of this monomer over o-methoxystyrene, styrene, and other styrene derivatives.

In the course of their investigation on the stereoregular polymerization of o-methoxystyrene, Natta et al. (550) also investigated the cationic polymerization of p-methoxystyrene. These authors obtained rather high molecular weight poly(o-methoxystyrene) ($[\eta] \sim 2.0$) with EtAlCl$_2$ and Et$_2$AlCl in toluene solvent at $-78°C$, however, in contrast to the ortho isomer, this polymer was not stereoregular.

$$H_2C=CH$$

26. o-Methoxystyrene

Imanishi et al. (548) investigated the kinetic details of o-methylstyrene polymerization with BF$_3 \cdot$OEt$_2$ in CCl$_4$, CHCl$_3$, CH$_2$Cl$_2$, and (CH$_2$Cl)$_2$ solvents in the range -20 to $+30°C$. Assuming steady state kinetics, from the Mayo equation $1/\overline{DP} = k_{tm}/k_p + k_t/k_p \cdot (1/[M])$ they calculated reaction rate constant ratios. It was found that k_{tm}/k_p was unaffected by both the temperature and dielectric constant of the solvent, however, k_t/k_p decreased with decreasing temperature and dielectric constant.

According to copolymerization studies (548) o-methoxystyrene is much more reactive than styrene. The reactivity ratios obtained with BF$_3 \cdot$OEt$_2$ at $-20°C$ were $r_{oMeS} = 3.6 \pm 0.8$ and $r_S = 0.11 \pm 0.4$. However this monomer was less reactive than p-methoxystyrene: $r_{pMeS} = 3.9 \pm 1$ and $r_{oMeS} = 0.35 \pm 0.09$ (under the same conditions).

Natta et al. (550) investigated the polymerization of o-methoxystyrene. Impetus for this research was provided by the discovery that the alkylaluminum halides (EtAlCl$_2$) yield stereoregular vinyl ethers by a cationic coordinated mechanism in homogeneous solution. Thus it was of interest to examine whether or not a similar mechanism obtains when the coordinating oxygen atom is not in the adjacent position to the vinyl group as in vinyl ethers but is farther, such as in o-methoxystyrene. Experiments showed that EtAlCl$_2$ in toluene solvent at $-78°C$ readily yielded high conversions (up to 90%), whereas the less acidic Et$_2$AlCl gave much lower yields ($\sim 4\%$). The molecular weights were rather low ($[\eta] \sim 0.1$) and the products were amorphous on x-ray analysis. Significantly, however, when the amorphous products were hydrogenated to poly(o-methoxyvinylcyclohexane), x-ray diffraction analysis indicated a significant degree of crystallinity. This experiment established that the starting compound poly(o-methoxystyrene) was of a stereoregular architecture.

This transformation by hydrogenation of poly(o-methoxystyrene) into a crystalline poly(o-methoxyvinylcyclohexane) raised the softening temperature of the polymer from 110–130°C for the former to 190–195°C for the latter. The intrinsic viscosity remained unchanged. The infrared spectrum of the hydrogenated product confirmed the proposed structure. In anology with stereoregular poly(vinyl ethers) the structure of the original poly(o-methoxystyrene) must have been isotactic and therefore the polymerization mechanism must have been a cationic coordinated one.

27. m-Methoxystyrene

$$H_2C=CH$$

Natta et al. (550) state in their paper on the stereoregular polymerization of o-methoxystyrene that the meta isomer does not polymerize under the influence of cationic initiators. This bold statement cannot be true since the copolymerization of this monomer with styrene, p-chlorostyrene, and α-methylstyrene has been established by Overberger et al. (520) using $SnCl_4$ in CCl_4 at 0°C. Interestingly, according to the reactivity ratios:

r_{mMOS} = 1.1 ± 0.15 r_S = 0.90 ± 0.15

r_{mMOS} = 2.6 ± 0.4 r_{pClS} = 0.38 ± 0.05

r_{mMOS} = 0.3 ± 0.1 $r_{\alpha MS}$ = 5 ± 1

m-Methoxystyrene is a quite reactive cationic monomer, the reason for which has caused some speculation by these authors.

28. p-Benzyloxystyrene $CH_2=CH$ — ⬡ — $O-CH_2$ — ⬡

Heublein and Dawczynski (686) mention the polymerization of this monomer by $SnCl_4/H_2O$ in CH_2Cl_2 at 0°C. The overall rate of polymerization of p-benzyloxystyrene was somewhat higher and the molecular weights almost twice as high as that of the p-methoxy derivative.

29. p-Acetoxystyrene $CH_2=CH$ — ⬡ — $OCOCH_3$

Heublein and Dawczynski mention (686) that this monomer cannot be polymerized by $SnCl_4/H_2O$ at 0°C.

30. 2,5-Dimethoxystyrene

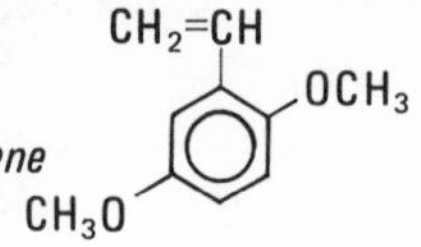

$$CH_2=CH$$

$$OCH_3$$

$$CH_3O$$

Kamogawa and Cassidy (551) studied the polymerization of this monomer using $BF_3 \cdot OEt_2$ in CH_2Cl_2 for 3 hr at $-78°C$. Under these conditions they obtained 58% conversion, $[\eta] = 0.33$. These authors also described a successful copolymerization experiment with styrene ($BF_3 \cdot OEt_2$ in ethylene dichloride at $-10°C$) and implied that this monomer is more reactive than styrene. No physical property data were given.

31. 3,4-Dimethoxystyrene

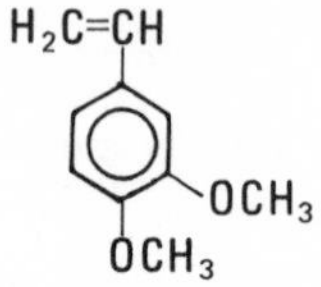

$$H_2C=CH$$

$$OCH_3$$

$$OCH_3$$

Maréchal et al. (552) studied the polymerization and copolymerization of 3,4-dimethoxystyrene with various Lewis acid/Brønsted acid combinations, for example, BF_3/HCl, $AlBr_3/HCl$, and $BF_3 \cdot OEt_2/HF$, in CH_2Cl_2 and C_2H_5Cl at low temperatures. Protogenic compounds (H_2O, HCl) were found to be co-initiators. The rate of polymerization was higher in C_2H_5Cl than in CH_2Cl_2. Respectable molecular weights were obtained ($[\eta]$ range = 0.1–0.6) during these investigations.

32. 2-Methyl-4-methoxy-5-isopropylstyrene (2-Vinyl-5-methoxy-p-cymene)

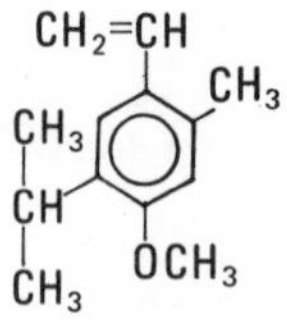

$$CH_2=CH$$

$$CH_3$$

$$CH_3$$

$$CH$$

$$CH_3 \quad OCH_3$$

Lalande et al. (527) mention the polymerization of this monomer by the use of $AlCl_3$ in n-hexane at $20°C$ with a yield of 40% after 10 hr, $\bar{M}_n = 1000–5000$.

33. 2-Isopropyl-4-methoxy-5-methylstyrene (2-Methoxy-5-vinyl-p-cymene)

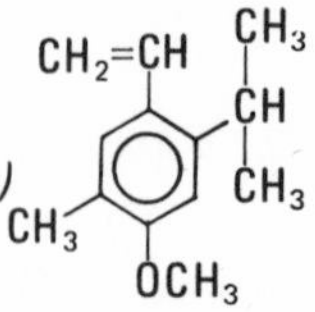

$$CH_2=CH$$

$$CH_3$$

$$CH$$

$$CH_3$$

$$CH_3$$

$$OCH_3$$

Lalande et al. (527) mention the polymerization of this monomer with $AlCl_3$ in n-hexane at $20°C$ with a yield of 38% after 10 hr; $\bar{M}_n = 1000–5000$.

34. m-*Nitrostyrene*

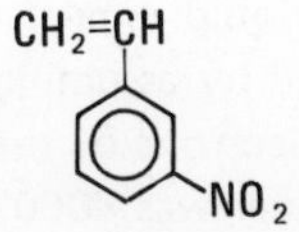

$$CH_2=CH$$

（structure: benzene ring with vinyl group and NO_2 in meta position）

Wiley and Smith (553) described the polymerization of *m*-nitrostyrene with BF_3 and $AlCl_3$. With BF_3 in bulk at $0°C$, just above the melting point of the monomer, 80% yield was obtained. Polymerization also proceeded in ethyl chloride solvent at -20 or $-78°C$ but no product was formed in ether or chloroform. The polymer is soluble in acetone, methyl ethyl ketone, and nitromethane. It can be molded to brittle transparent films. According to viscosity measurements in nitromethane, the molecular weights of these products is quite low, ~ 870. $AlCl_3$ is a less desirable initiator than BF_3 because it yields much lower conversions.

Overberger et al. (520) studied the copolymerization of *m*-nitrostyrene with styrene using $SnCl_4$ in CCl_4 at $0°C$. They determined the reactivity ratios, $r_{mNO_2St} = 0.03 \pm 0.03$ and $r_{St} = 20 \pm 4$, which indicate a very sluggish cationic monomer.

35. p-*Nitrostyrene*

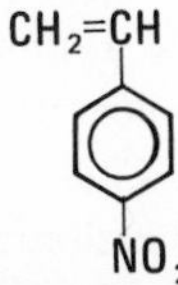

$$CH_2=CH$$

（structure: benzene ring with vinyl group and NO_2 in para position）

According to Overberger et al. (520) only an insignificant amount ($\sim 1.5\%$) of *p*-nitrostyrene was incorporated into a copolymer with styrene when a $50:50$ mixture of these two monomers was reacted with $SnCl_4$ in CCl_4 at $0°C$. German authors (686) could not homopolymerize this monomer with $SnCl_4/H_2O$ at $0°C$.

36. p-*Dimethylaminostyrene*

$$CH_2=CH$$

（structure: benzene ring with vinyl group and $H_3C-N-CH_3$ in para position）

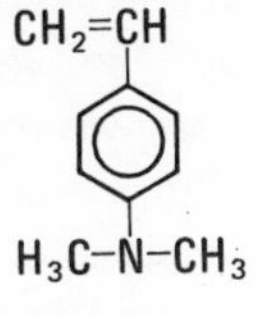

Heublein and Dawczynski (686) examined the homopolymerization of *p*-dimethylaminostyrene together with a variety of other para substituted styrene derivatives, using $SnCl_4/H_2O$ in CH_2Cl_2 at $0°C$. A Hammett plot was constructed (log overall rate constant vs. σ) and while the values for all the other derivatives fell on a straight line, the value for *p*-dimethylaminostyrene was off by a wide margin. Evidently the overall rate constant for this monomer was significantly lower than the value one would expect on the basis of the Hammett plot. This discrepancy was explained by assuming the formation of a donor–acceptor complex between the free electrons of the N of the dimethyamino group and $SnCl_4$. The $\overline{M}_n$ of the polymer was 4000.

The copolymerization of this monomer with α-methylstyrene using $SnCl_4$ in CCl_4 at $0°C$ was investigated by Overberger et al. (520). The reactivity ratios reveal the very high cationic reactivity of this monomer: $r_{pMe_2NSt} = 31 \pm 19$ and $r_{\alpha MSt} + 0.035 \pm 0.015$.

$$CH_3$$
$$H_2C=C$$

37. α-Methylstyrene (Isopropenylbenzene)

α-Methylstyrene is a cationically readily polymerizable monomer. The cationic synthesis of very high molecular weight (up to perhaps 2×10^6) polymers has been reported (554) and seems to be achievable under conventional laboratory conditions without undue difficulties. [In regard to the molecular weights obtained by Hershberger et al. (554) see Bywater's comment (555)].

Cationically polymerized α-methylstyrene has only a limited commercial use in certain resin formulations. A copolymer of vinyltoluene and α-methylstyrene (Picotex R) is available from Pennsylvania Industrial Chemical Company (556).

The scientific literature up to about 1962 on the cationic polymerization of α-methylstyrene has been completely reviewed by Bywater (557). Indeed, since this review was written, except for irradiation polymerization of α-methylstyrene not much scientific information has appeared in this field. An essentially complete list of publications dealing with the cationic polymerization of α-methylstyrene excluding irradiation studies is given in Table 20.

Among the more interesting recent developments is the controversy over the structure of cationically synthesized poly(α-methylstyrene). Brownstein et al. (558) studied the structure of poly(α-methylstyrene) synthesized by $SnCl_4$, $TiCl_4$, BF_3, and $AlCl_3$ in toluene solvent at $-78°C$ and proposed a predominantly syndiotactic configuration. This proposition was mainly based on inspection of Hirschfelder/Taylor atomic models and on correlating the models with NMR spectra. These authors assigned the highest, intermediate, and lowest field signals to the syndiotactic, heterotactic, and isotactic methyl groups, respectively. Although the syndiotactic content of the polymers was estimated to be 80–90%, the products could not be induced to crystallize.

Braun and coworkers (575, 584) also believe that their polymers obtained by cationic techniques ($AlCl_3$ in CS_2 at -40 and $-60°C$, H_2SO_4 and $SbCl_4$ in CH_2Cl_2 from -20 to $-78°C$, and $BF_3 \cdot OEt_2$ in toluene/n-hexane at -5 and

TABLE 20 POLYMERIZATION OF α-METHYLSTYRENE

Initiators	Solvent	Temperature (°C)	Remarks	References
BF$_3$	Toluene	–78	Structure studies	558
BF$_3$·OEt$_2$	Methylcyclohexane/CH$_2$Cl$_2$	–78	Structure analysis	559
BF$_3$·OEt$_2$	Toluene	–78	Structure analysis	560, 561
BF$_3$·OEt$_2$·H$_2$O	(CH$_2$Cl)$_2$	+10 to –20	Kinetics and structure	562
BF$_3$·OEt$_2$	nC$_6$H$_{14}$, toluene, CHCl$_3$	–78	Effect of solvent, molecular weights, etc.	563, 564–566
BF$_3$·OEt$_2$	CHCl$_3$/nC$_6$H$_{14}$ mixture	–70	Structure studies	567
BF$_3$·OEt$_2$	nC$_6$H$_{14}$, toluene, CHCl$_3$, CH$_2$Cl$_2$	–20, –50 –70	Kinetic studies	564, 565 568
BF$_3$·OEt$_2$	Toluene	–78	DP = 5570	569
BF$_3$·OEt$_2$	nC$_6$H$_{14}$	–78	Atactic polymer	570
BF$_3$·OEt$_2$	Toluene, methylcyclohexane and mixtures	–30 and –75	Structure studied	571
BF$_3$·OEt$_2$	Toluene, CH$_2$Cl$_2$	–78	Stereoregular polymer	570
SnCl$_4$	Bulk	Room temp.	Early study, octamers obtained	572
SnCl$_4$	Methylcyclohexane/CH$_2$Cl$_2$	–78	Structure studies	559
SnCl$_4$·CCl$_3$COOH	Toluene, CH$_2$Cl$_2$	0 to –78	Stereoregular polymer, copolymerizes	500, 559, 573
SnCl$_4$·H$_2$O	Methylcyclohexane/CH$_2$Cl$_2$	–78	Structure studies	559
SnCl$_4$·H$_2$O	Methylcyclohexane, toluene and mixtures	–75	Structure studies	572
AlBr$_3$, CCl$_3$COOH	Toluene, CH$_2$Cl$_2$	–78	Stereoregular polymer	570
TiCl$_4$	Toluene, CH$_2$Cl	–78	Stereoregular polymer	570
TiCl$_4$	Toluene	–78	DP = 1500	560, 561, 569
TiCl$_4$	Toluene, methylcyclohexane and mixtures	–75	Structure studies	572
TiCl$_4$ (H$_2$O or HCl)	CH$_2$Cl$_2$	+10 to –78	Coinitiation studies	574

Table 20 (continued)

$AlCl_3$	Toluene	-78	Structure studies	558
$AlCl_3$	EtCl, CS_2	$+23$ to -130	High molecular weight polymer	554
$AlCl_3$	CS_2	-40	Heat degradation studies	575
$AlCl_3$	CCl_4, EtCl	$+25, 0$	Kinetic studies	576
$AlCl_3$	Toluene, methylcyclohexane, and mixtures	-75	Structure studies	572
$SbCl_5$	Methylcyclohexane/CH_2Cl_2	-78	Structure studies	559
$SbCl_4$	Methylcyclohexane/CH_2Cl_2	-78	Structure studies	559
$AlEt_3/TiCl_4$	Toluene	-78	Effect of Al/Ti ratio, mixing, catalyst aging, MWD	569, 577 578,560,561
$SbCl_5$	CH_2Cl_2	-78	Heat degradation studies	575
$AlEtCl_2$	Toluene	0 to -78	Molecular weights, NMR, stereoregularity	561, 569 577–579
$AlEt_2Cl$	Toluene	-78	Molecular weights, NMR, stereoregularity	561, 569 577–579
$SnCl_4$	Toluene, $(CH_2Cl)_2MeNO_2$	-78	Structure studies	558
$SnCl_4(+H_2O)$	EtCl	$0-100$	Kinetic studies	580
$SnCl_4(+H_2O)$	EtCl	-78	Kinetic studies	581
$Ph_3C^{\oplus}AlCl_4^{\ominus}$, $Ph_3C^{\oplus}AlBr^{\ominus}$, $Ph_3C^{\oplus}SnCl_5^{\ominus}$, $Ph_3C^{\oplus}BF_4^{\ominus}$	Methylcyclohexane/CH_2Cl_2, CH_2Cl_2/CH_3CN	-78	NMR, structure studies	559
$Ph_3C^{\oplus}SbCl_6^{\ominus}$	CH_2Cl_2	-70	Mentioned	303
CCl_3COOH, CCl_2HCOOH, $CClH_2COOH$	EtBr, $(CH_2Cl)_2$, CH_3NO_2	$0, 25, 40$	Kinetic studies	4 90, 582
H_2SO_4	CH_2Cl_2	-40	Heat degradation studies	575
I_2	?	25	A brief kinetic study	583
Ph_3CSnCl_5	$(CH_2Cl)_2$, mixtures of C_6H_5 plus $(CH_2Cl)_2$	30	Initiation rate studies	447
Light + tetracyanobenzene	$(CH_2Cl)_2$, CH_2Cl_2	-30 to $+30$	Rate and structure studies	585a

−40°C) are more than 90% syndiotactic. Braun's NMR assignments for the syndiotactic, heterotactic, and isotactic configurations agree with those originally proposed by Brownstein et al. (558). These authors compared the thermal degradation rates of highly syndiotactic products with those of atactic materials obtained in a typical anionic system, Na—K in THF, and found the syndiotactic material more thermally stable than the atactic type (575). Braun et al. attributed this difference to the large internal strain in the chain caused by the frequent isotactic placements in the atactic product.

Ramey et al. (585) studied the structure of poly(α-methylstyrene) by high resolution, 220 MHz, NMR and in their latest paper (585) in which they corrected earlier conclusions (586), they assigned the three methyl peaks to the isotactic, heterotactic, and syndiotactic triads in the order of increasing magnetic fields. Kunitake and coworkers (559, 571) who also investigated the steric course of the cationic polymerization of α-methylstyrene analyzed and adopted the data of Brownstein et al. (558) and Ramey et al. (585) and on the basis of these, assigned a largely syndiotactic structure for their products. Kunitake et al. (559, 571) used a variety of initiators, for example, conventional Lewis acids $AlCl_3$, $BF_3 \cdot OEt_2$, $SnCl_4$ triphenylmethyl salts $(C_6H_5)_3C^{\oplus}X^{\ominus}$ where $X^{\ominus} = AlCl_4^{\ominus}$, $SnCl_5^{\ominus}$, $AlBr_4^{\ominus}$, $BF_4^{\ominus}$, and assumed a predominantly syndiotactic propagation in all these systems. These authors employed methylcyclohexane/methylene chlorine mixtures, pure methyl chloride, and methylene chloride/nitromethane solvents and obtained, in the main, homogeneous polymerization systems and respectable molecular weights. They confirmed the observations of Higashimura and coworkers [see later, (570)] that the nature of the Lewis acid is not important in affecting the steric structure of the polymer.

According to Hayashi (585a) the photoinduced polymerization of α-methylstyrene probably proceeds by cation radicals and yields highly syndiotactic polymer even at 0°C.

In contrast to these investigators, Sakurada et al. (579) and, by adopting this work, Higashimura and coworkers (567, 570) believe that cationically prepared poly(α-methylstyrenes) possess a largely isotactic configuration. Sakurada polymerized α-methylstyrene with a variety of cationic initiators ($BF_3 \cdot OEt_2$, $TiCl_4$, Et_2AlCl), cationic Ziegler-Natta systems ($Et_3Al/TiCl_4$ where Al/Ti ratio was 1.0–1.2), and anionic initiators (K, Na, BuLi) (557, 560, 561, 569, 578, 579). NMR spectroscopy indicated that the most stereoregular (they believe isotactic product (80–90%) was obtained by conventional Lewis acid initiators $EtAlCl_2$, and $BF_3 \cdot OEt_2$ at low temperatures. The cationic Ziegler-Natta system gave a less stereoregular product (60–70%), whereas the anionic initiators yielded the lowest stereoregularity (~60%). According to Sakurada et al. (579) the isotactic, heterotactic, and syndiotactic methyl signals appear with increasing field strength. These workers (569) were able to crystallize by heat stretching under "special" conditions the polymer obtained with the cationic Ziegler-Natta system; attempts to crystallize the other samples failed. The x-ray fiber diagram exhibited a fiber identity period of 6.6 Å and indicated a rhombohedral structure, as is the case with isotactic polystyrene. They concluded that the crystalline component was isotactic.

Higashimura and coworkers (570) investigated the effect of various reaction parameters on the stereoregularity of poly(α-methylstyrene). Interestingly, they found that the nature of the solvent decisively influences the steric course of the polymerization. In n-hexane, a nonsolvent for the polymer, that is, in a heterogeneous system, largerly atactic product formed, whereas in toluene, methylene chloride, or nitroethane, good solvents for the polymer, that is, in homogeneous systems, stereoregular polymer was obtained. The nature of the initiator, for example, $BF_3 \cdot OEt_2$, $SnCl_4 \cdot CCl_3COOH$, $AlBr_3 \cdot CCl_3COOH$, and $TiCl_4$, did not much affect stereoregularity. The polymerization temperature was also an important parameter: highly stereoregular polymer formed only at temperatures below about $-60^\circ C$.

These findings are in contrast to those obtained in the cationic polymerization of isobutylvinyl ether where stereoregularity of the polymer is strongly affected by the nature of the initiator and polarity of the solvent or homogeneity of the medium (587, 588). With alkylvinyl ethers the highest degree of stereoregularity (isotactic polymer) is obtained with $BF_3 \cdot OEt_2$ in nonpolar, poorly solvating media.

Sigwalt and coworkers (574) investigated the initiation mechanism of α-methylstyrene with $TiCl_4$ in CH_2Cl_2 solvent in the range $+10$ to $-72^\circ C$. These authors attempted to stop completely the polymerization of this monomer by rigorous drying of all the chemicals involved and subsequently to effect initiation by introducing cationogenic agents like H_2O or HCl ("stopping" experiments). They were successful in reducing the initial yield of polymer to 1–29% and in obtaining complete conversion (100%) upon the addition of H_2O or HCl. On the basis of these data they concluded that a coinitiator is necessary for polymerization. Kennedy speculated that the absence of a complete stopping of the polymerization may be due to self-initiation (114).

Sigwalt et al. (303) also mention the polymerization initiation of α-methylstyrene with $(C_6H_5)_3C \cdot SbCl_6$ in CH_2Cl_2 at $-70^\circ C$ and report $[\eta] = 0.34$, and $\overline{DP}_n = 470$.

Smets and DeHaes (589) described the $SnCl_4$ induced copolymerization at $-78^\circ C$ of α-methylstyrene and p-chlorostyrene and obtained $r_{\alpha St} = 28 \pm 2$ and $r_{pSt} = 0.12 \pm 0.03$.

α-Methylstyrene and p-methyl-α-methylstyrene were copolymerized by Banas and Juveland (590) using the synthesis technique of Brownstein et al. (558) ($SnCl_4$, toluene, $-78^\circ C$) and chemical shift assignments for the analysis of the NMR spectra. Unfortunately, no experimental details are given (conversion data) so that it is difficult to assess the spectral contribution of possible homopolymers.

$$CH_2=C \begin{matrix} CH_2CH_3 \\ | \\ \end{matrix}$$

38. α-Ethylstyrene

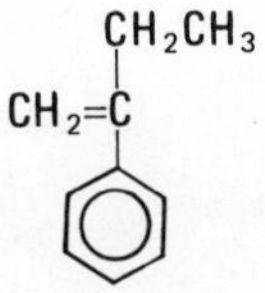

Overberger et al. (520a) report that under those conditions where α-methylstyrene polymerizes readily, α-ethylstyrene does not yield polymer. Rather, when α-ethylstyrene was treated with $SnCl_4$ in a mixture of nitrobenzene and carbon tetrachloride at $0°C$ for 1 hr or 24 hr, a mixture of dimers (~78%) and trimers (~16%) was formed. Evidently the addition of a methylene unit in the α-position (i.e., going from α-methyl to α-ethylstyrene) renders the propagation of the carbenium ion sterically impossible.

The same authors also attempted the copolymerization of α-ethylstyrene with p-chlorostyrene, however, irreproducible results were obtained. Experimental scatter was explained by the presence of uncontrollable amounts of dimers and trimers in the polymerizing charge.

$$CH_2=C \begin{matrix} CH_3 \\ | \\ \end{matrix}$$

39. α-Methyl-p-methylstyrene

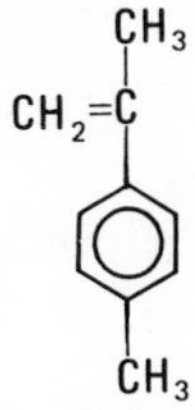

The copolymerization of this compound with α-methylstyrene has been mentioned by Banas and Juveland (590) (see also under α-methylstyrene).

$$CH_2=C \begin{matrix} CH_3 \\ | \\ \end{matrix}$$

40. p-Methoxy-α-methylstyrene (p-Methoxyisopropenylbenzene)

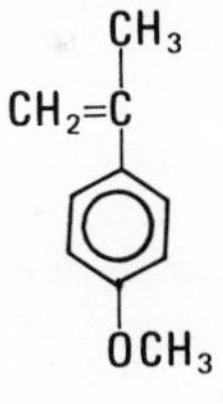

Van der Zanden and Rix (591) described the synthesis and dimerization of this monomer by refluxing with 43% sulfuric acid.

$$CH_2=C-CH_3$$

41. p-Diisopropenylbenzene

$$CH_3-C=CH_2$$

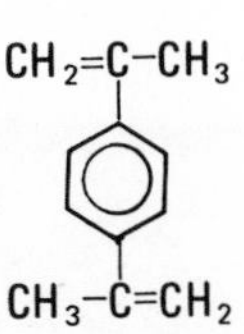

The cationic polymerization of p-diisopropenylbenzene has been studied and patented (592) by Brunner et al. These authors visualize the polymerization as being similar to the dimerization of α-methylstyrene:

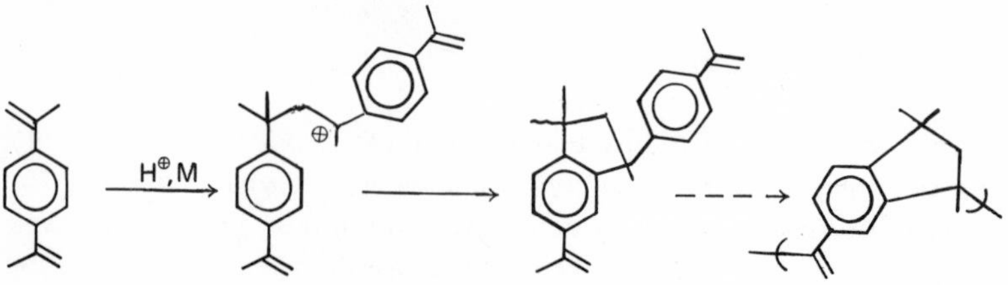

According to a recent discovery *p*-diisopropenylbenzene can be efficiently polymerized to linear polyindanes with a catalyst system consisting of BuLi/TiCl$_4$/HCl (593). The simultaneous presence of all three components is necessary for successful polymerization. The products are almost completely soluble in toluene (up to 93%) and are of respectable molecular weight (reduced viscosity at a concentration of 0.2 g/100 ml was in the range 0.3–0.8).

42. *α,2,6-Trimethyl-4-tert-butylstyrene*

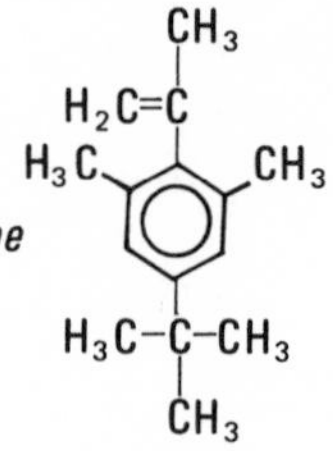

43. *2,4,5-Triisopropyl-α-methylstyrene*

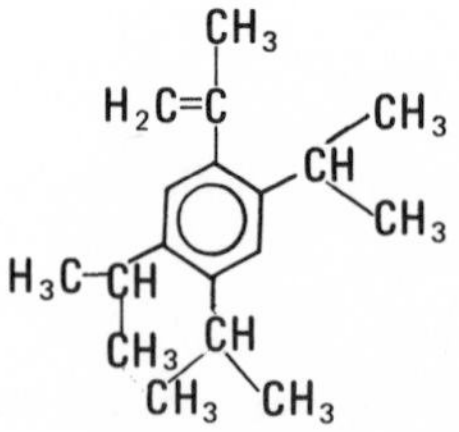

44. *2,6-Dimethoxy-α-methylstyrene*

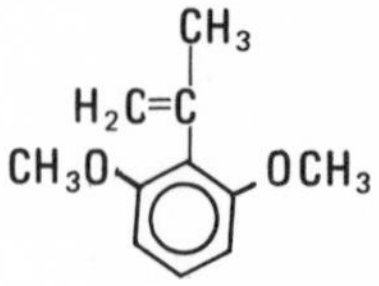

According to Korshak and Matveeva (594, 595) these monomers cannot be polymerized by cationic (AlCl$_3$) initiators; evidently steric hindrance prevents reaction.

45. 1,1-Diphenylethylene

$CH_2=C$ (with two phenyl groups)

This olefin cannot be polymerized to high polymer, obviously because of unfavorable steric compression. The cationic dimerization of this molecule and some of its derivatives, however, has been studied in great detail by Evans and coworkers (596 and previous 10 papers).

46. and 47. cis- and trans-β-Methylstyrene (Propenylbenzene) $CH_3-CH=CH-\!\langle\bigcirc\rangle$

The polymerization of β-methylstyrene (no isomer distribution specified) was examined by Staudinger and Dreher (545). According to these authors this monomer could be polymerized to solid products in the molecular weight range of 1000–3000 only with the most effective initiator BF_3 in toluene at $-70°C$. In contrast, $SnCl_4$ was inactive. On the basis of pyrolysis, ozonolysis experiments, and viscosity measurements, the structure of the polymer was proposed to consist of 1,3 units:

$$-CH_2-CH_2-\underset{\underset{C_6H_5}{|}}{CH}-$$

Later Müller et al. (683) and Kennedy (597) reinvestigated the work of Staudinger et al. and by chemical analysis and by infrared spectroscopy established the presence of CH_3-C- groups in the chain. As a consequence these authors propose for the repeat structure of β-methylstyrene a conventional 1,2 enchainment:

$$-\underset{\underset{C_6H_5}{|}}{CH}-\overset{\overset{CH_3}{|}}{CH}-$$

Available information on the structure of cationically synthesized poly-β-methylstyrene has been assembled and criticized by Kennedy and Langer (41). On the basis of older data and their own infrared studies (41) these authors feel that most of the evidence favors a conventional 1,2 enchainment for this polymer. Similar conclusions have been reached by Higashimura and coworkers (598) on the basis of infrared spectra of poly(β-methylstyrenes) prepared with $BF_3 \cdot OEt_2$ in $(CH_2Cl)_2$ solvent at $30°C$.

The conclusions of a publication by Murahashi et al. (599) on the "rearrangement polymerization" of β-methylstyrene with $BF_3 \cdot OEt_2$ at 0 and $78°C$ in various solvents have been corrected by the same authors in a later paper (600) in which they state that the β-methylstyrene monomer used in their first investigation contained indene impurities leading to erroneous results.

β-Methylstyrene ($\sim$87% *trans* plus 13% *cis* isomer) was polymerized by Shimizu et al. (601) using $AlBr_3$ in CH_2Cl_2 at 10 and $-78°C$ (yields, 01.–0.3% after $\sim$18 days) and $BF_3 \cdot OEt_2$ in bulk at 0–15°C (yields 5–8% after $\sim$10 days). The latter product was a colorless solid of $\bar{M}_n$ $\sim$720 whose structure, according to infrared analysis, was mainly a conventional 1,2 chain $-CH(CH_3)-CH(C_6H_5)-$.

The copolymerization of *cis*- and *trans*- β-methylstyrene with p-chlorostyrene using $SnCl_4$ has been investigated by Overberger et al. (536) and the following reactivity ratios were obtained: $r_{tr\beta} = 0.74 \pm 0.6$, $r_{pCl} = 0.32 \pm 0.4$, $r_{cis\beta} = 0.32 \pm 0.4$, and $r_{pCl} = 1.0 \pm 0.1$. Evidently the trans isomer is more reactive than the cis toward the p-chlorostyrene carbenium ion. The higher reactivity of the trans isomer can be attributed to a more favorable steric interaction of this isomer with the propagating carbenium ion than in the case of cis isomer. Also, the transition state is more resonance stabilized with the trans isomer than with the cis isomer.

Higashimura and coworkers (602) carried out a series of investigations on the cationic polymerization behavior of α,β-disubstituted olefins. Their summary paper (602) reviews progress until about 1968 and the work is still in progress [the last paper is No. 15 in the series (603)]. These authors were interested in elucidating by extensive copolymerization experiments the effects of substituents on the relative reactivity of styrene and styrene derivatives (p-methyl-, p-methoxy-) versus β-methylstyrene and its derivatives (p-methyl-, and p-methoxy-) and combinations of these.

The reactivity ratios of two comprehensive sets of experiments in which $BF_3 OEt_2$ in $(CH_2Cl)_2$ at 30°C was used are shown in Table 21 (519). According

TABLE 21   MONOMER REACTIVITY RATIOS

M_1	M_2	$-R$	r_1	r_2
$CH_2=CH$ (phenyl, R)	CH_3 $CH=CH$ (phenyl, R)	$-H$	1.8 ± 0.2	0.07 ± 0.02
		$-CH_3$	1.3 ± 0.3	0.04 ± 0.04
		$-OCH_3$	1.2 ± 0.2	0.04 ± 0.02
$CH_2=CH$ (phenyl)	CH_3 $CH=CH$ (phenyl, R)	$-H$	1.08 ± 0.2	0.07 ± 0.02
		$-CH_3$	0.75 ± 0.1	0.36 ± 0.05
		$-OCH_3$	0.45 ± 0.2	2.8 ± 0.2

to the first three sets of values $r_1 > 1$ and $1/r_2 > 1$, which indicates that the reactivity of β-methylstyrenes (mixture of cis and trans isomers) is lower than that of the unsubstituted compounds, probably because of steric compression due to the β-methyl substituent. However, it was recognized that the β-methyl group in the cation is sufficiently far away from the propagating center to invoke a purely steric argument for the reduced monomer reactivity. Rather, the very low r_2 values were attributed to reduced monomer reactivity caused by the β-methyl substituent inducing a partial negative charge on the α-carbon of the olefin.

According to the last three sets of values in Table 21, the introduction of an electron releasing substituent in the ring of the β-methylstyrene strongly increases its reactivity. The fact that the product of reactivity ratios is close to unity (random copolymers) for the styrene–anethole pair may indicate that the reactivity increasing effect of the p-methoxy group compensates the reactivity reducing effect of the β-methyl substituent.

In an independent study Higashimura and coworkers (519) established the relative reactivities of cis- and trans-β-methylstyrene by copolymerizing these two species and by copolymerizing them with styrene. They also investigated the relative reactivities of cis- and trans-p-methoxy-β-methylstyrene. The reaction conditions in these experiments were $SnCl_4 \cdot CCl_3COOH$ initiator in $(CH_2Cl)_2$ or toluene at $0°C$. Consistent with the results of Overberger et al. (536) the Japanese workers also found that in copolymerization with styrene, the trans-β-methylstyrene isomer was somewhat more reactive than the cis-β-methylstyrene isomer. Reactivity ratios were $r_{St} = 1.95 \pm 0.1$, $r_{cis\beta} = 0.03 \pm 0.05$, $r_{St} = 1.4 \pm 0.15$, and $r_{tr\beta} = 0.2 \pm 0.1$.

48. and 49. cis- and trans-p-*Methyl-β-methylstyrene* $CH_3-CH=CH-\langle\bigcirc\rangle-CH_3$

Higashimura and coworkers (519) mention the copolymerization of this monomer (see under β-methylstyrene).

50. and 51. cis- and trans-β-*Ethylstyrene* $CH_3CH_2-CH=CH-\langle\bigcirc\rangle$

The cationic copolymerization of mixtures of this monomer with p-chlorostyrene using $SnCl_4$ was described by Overberger et al. (536). According to the reactivity ratios obtained, β-ethylstyrene cannot add to its own active end: $r_{\beta Et} = 0$ and $r_{pCl} = 0.88 \pm 0.3$.

52. and 53. cis and trans-β-n-*Propylstyrene* $CH_3CH_2CH_2-CH=CH-\langle\bigcirc\rangle$

The copolymerization of mixtures of this monomer with p-chlorostyrene using

SnCl$_4$ has been mentioned (536).

54. and 55. cis- *and* trans-α,β-Dimethylstyrene

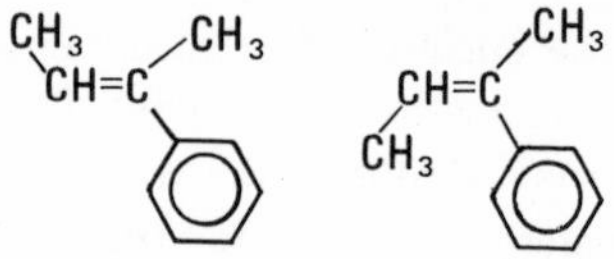

According to Overberger et al. (520a) the cis isomer gave 31% yield of dimers with SnCl$_4$ at 0°C after 24 hr, whereas the trans form did not yield any dimers under the same conditions.

56. and 57. cis *and* trans-p-*Methoxy-β-methylstyrene (Anethole)*

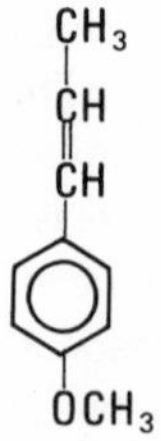

The ready polymerizability of anethole was recognized long ago when Gerhardt described a polymerization induced by SnCl$_4$ as early as 1845 (604); Staudinger and coworkers (545, 605) were also among the early investigators. These authors noted that anethole readily yielded solid polymer even with SnCl$_4$ at −78°C in toluene (β-methylstyrene did not give polymer under the same conditions). Ozonization and pyrolysis experiments as well as peculiarities in the viscosity constant of this product led the early workers to postulate (545) that the main repeat unit of their product was:

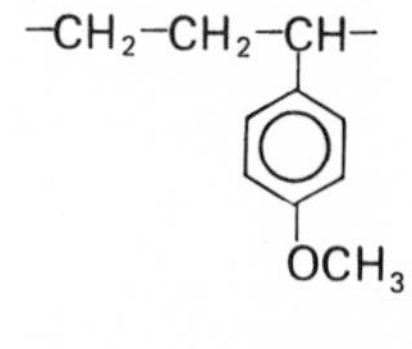

Subsequent investigators Müller et al. (683) found chemical evidence, and Kennedy (597) who tried to repeat the experiments of Staudinger et al. found spectroscopic evidence for the presence of CH$_3$–C groups in the chain and consequently proposed a conventional repeat unit for the polymer:

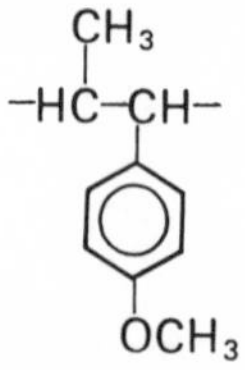

Müller et al. (683) used $TiCl_4$ in benzene at ambient temperatures, Secci and Mameli (606) induced the polymerization by BF_3, and Sigwalt (607) used a variety of Lewis acids (BF_3, BF_3OEt_2, $SnCl_4$, and $TiCl_4$) under a variety of conditions. The last worker obtained low molecular weight, soluble products in toluene. High polymer ($[\eta]$ = 1.3–1.4) formed with $BF_3 \cdot OEt_2$ at $-72^{\circ}C$. Interestingly, Sigwalt claims that some of his polymers exhibited some crystallinity under x-rays that increased when swollen with "crystallizing" solvents. This author has also claimed but has not proven the cationic copolymerization of anethole with styrene, indene, isobutylvinyl ether, and so on.

Information concerning the structure of cationically synthesized polyanethole has been considered by Kennedy and Langer (41) who concluded that the weight of evidence favors a conventional 1,2 enchainment in contrast to an isomerized structure proposed by Staudinger and Dreher (545).

Alfrey et al. (540) compared the copolymerization of anethole with o-chlorostyrene using cationic ($SnCl_4$ in CCl_4 at $0^{\circ}C$) and radical (benzoyl peroxide at $70^{\circ}C$) initiators. The following reactivity ratios were obtained

	Radical	Cationic
r_A	0 ± 0.01	18 ± 3
r_{oClS}	22 ± 8	0.03 ± 0.005

Thus, under cationic conditions anethole is much more reactive with either $(anethole)^{\oplus}$ or $(o\text{-chlorostyrene})^{\oplus}$ than o-chlorostyrene. This may be attributed to the induction effects exerted by the p-methoxy and β-methyl groups.

None of these earlier workers specified the cis/trans isomer distribution in their monomers.

Cerrai et al. (608) studied the kinetics of trans-anethole polymerization initiated by $BF_3 \cdot OEt_2$ in 1,2-dichloroethane in the range 5–$35^{\circ}C$. The steric purity of the monomer was established by its infrared spectrum, refractive index, and specific gravity. The reaction is second order in both monomer and initiator and the overall activation energy was 7.0 ± 0.1 kcal/mole. The polymers were of low molecular weight, $\overline{M}_n$ = 3000, and were found to be amorphous by x-rays. Infrared analysis indicated a conventional 1,2 enchainment.

The kinetic studies of these Italian authors were extended to an investigation of anethole polymerization with $BF_3 \cdot OEt_2$ in $(CH_2Cl)_2$ at $25^{\circ}C$ under high voltage dc electric field (609). Interestingly, under the influence of a 0.5 kV/cm field, the rate of polymerization increased very strongly (by a factor of 3–4) and the reaction was reduced to first order in the Lewis acid, that is, lower than without the electric field. At the same time the molecular weights increased only slightly (~5%). According to the authors the increased polymerization rate under the electric field is not caused by an increase in free ion concentration but to an increase in the propagation rate constant which in turn is caused by an increased stabilization of the propagating ion pairs.

Higashimura and coworkers (519) investigated the relative reactivity of anethole isomers by copolymerization experiments using $SnCl_4 \cdot CCl_3COOH_3$ in $(CH_2Cl)_2$ at $0°C$ and found the *cis*-anethole 1.5–2.0 times as reactive as the trans isomer. This monomer pair was reinvestigated by Higashimura et al. (610) using $BF_3 \cdot OEt_2$ in $EtNO_2$ solvent at $0°C$ and the reactivity ratios obtained were: $r_{cisA} = 1.02 \pm 0.08$ $r_{trA} = 0.29 \pm 0.04$. In contrast to these findings, which indicate that the cis isomer is more reactive than the trans olefin when the two are copolymerized, seemingly contradictory results were obtained by copolymerizing either the cis or the trans isomer with styrene or *p*-methoxystyrene. The reactivities are illustrative: $r_S = 0.42 \pm 0.12$, $r_{cisA} = 8.24 \pm 0.86$, $r_S = 0.13 \pm 0.04$, and $r_{trA} = 1.56 \pm 0.11$. According to these data *cis*-anethole is more reactive than the *trans* monomer to $(anethole)^{\oplus}$, however, the trans isomer is more reactive than the cis to $(styrene)^{\oplus}$ or to $(p\text{-methoxystyrene})^{\oplus}$. Considering a variety of complex facts the authors conclude that *trans*-anethole is intrinsically more reactive than the cis isomer and that this situation obtains when the growing electrophilic species is not bulky; however, with the more sterically hindered $(anethole)^{\oplus}$ the reduction of reactivity is greater with the trans isomer than with the cis, so that the apparent reactivities seem to reverse. These conclusions were also substantiated by C^{13} NMR studies.

58. and 59. cis- *and* trans-p-*Ethoxy-β-methylstyrene* $CH_3-CH=CH-\!\!\langle\bigcirc\rangle\!\!-OCH_2CH_3$
(p-Ethoxypropenylbenzene)

Staudinger and Dreher (545) polymerized mixtures of this monomer (no isomer distribution given) with $SnCl_4$ in benzene at $30°C$. The softening range of the methanol insoluble precipitated product was $172-178°C$.

*60. 1-Hydroxy-2-methoxy-4-propenyl(*cis*)-benzene (Isoeugenol)*

According to Whitby and Katz (611) isoeugenol, because of its conjugated double bond system, is polymerizable by $SbCl_5$.

B. INDENE AND DERIVATIVES

1. Indene

The ready cationic polymerizability of indene has been well documented.

Among the earliest references are those by Whitby et al. (335) and Staudinger et al. (612) which can be used to trace back developments prior to 1928. Whitby, for example, used $SbCl_5$ and $SnCl_4$ in chloroform and found that the calculated $\overline{DP}$s were in the range 15–25; Staudinger carried out extensive study with $SnCl_4$ but also examined $TiCl_4$, $SbCl_5$, $SbCl_3$, $FeCl_3$, and BF_3 initiators. Brown and Mathieson (613a) used CCl_3COOH.

Sigwalt's investigations with indene in the early sixties (614a) have since then been extended by himself and his students, particularly Maréchal, to a detailed study of the polymerization kinetics of this monomer and many of its derivatives. In his first study (614a) Sigwalt, building upon the work of the earlier researchers, worked out suitable conditions and was able to accomplish the synthesis of very high molecular weight polyindenes, indeed, products whose molecular weights were higher than 10^6. He used BF_3, $TiCl_4$, $SnCl_4$, and so on in solvents such as CH_2Cl_2, and $CHCl_3$, in the range 0 to $-90°C$. The intrinsic viscosity of the polymer increased with decreasing temperatures. While Sigwalt claimed the synthesis of various copolymers (i.e., with anethole, vinyl isobutyl ether, etc.), he did not prove their existence. Bkouche-Waksman and Sigwalt (615a) also studied high molecular weight polyindenes by light scattering. These authors prepared a polymer with $TiCl_4$ in CH_2Cl_2 at $-72°C$, $[\eta] \sim 1.50$, fractionated a sample, determined the molecular weight distribution by light scattering, and derived the second virial coefficient, the radius of gyration, and suitable equations to calculate molecular weights from intrinsic viscosities, that is, $[\eta] = 4.52 \times 10^{-4} M^{0.58}$. In a parallel publication Sigwalt's coworkers (298) reported that the high molecular weight polyindenes are branchy. The softening point of a polymer with $[\eta] = 0.2$ is $\sim 240°C$ and that with $[\eta] = 0.7$ is $250–260°C$. Above these temperatures the polymer has a tendency to yellow and oxidize and can be molded into brittle films. In a review Maréchal (616) summarized syntheses of indene and many of its derivatives.

In the late sixties Sigwalt and his coworkers elucidated some of the details of indene polymerization with $TiCl_4$ (616–619), $SnCl_4$ (620–623), and $(C_6H_5)_3C^\oplus SbCl_6^\ominus$ (303) in methylene chloride solvent. One major objective of these studies was to clarify the initiation mechanism by "stopping" experiments (i.e., experiments carried out under the highest possible degree of purity), to see the effect of cationogenic agents. On the basis of carefully conducted experiments (618, 619) these workers could state that H_2O and HCl are initiators of indene polymerization with $TiCl_4$ in CH_2Cl_2, however, polymerization also proceeds, albeit slower, in the absence of these cationogenic agents (114). Similar results have been obtained with the indene/$SnCl_4$/H_2O or HCl systems (620–623). These workers also determined the enthalpy of polymerization initiated by $TiCl_3OBu$ in CH_2Cl_2 in the range $+10$ to $-70°C$ to be 13.9 kcal/mole (301).

Interesting findings in the course of the work with $SnCl_4$ were that polymerization occurred in spite of greatest efforts in purification and that the polymerization was nonterminating, that is, when fresh aliquots of monomer were added to a charge that reached complete conversion, polymerization resumed and the rate of polymerization of the second batch was the same as that of the first charge (622, 623). The first observation may be an indication for direct initiation by hydride abstraction (114). The second finding indicates the presence and survival of active stable cations for long periods of time. In this regard the authors report (623) that the molecular weights of nonterminated charges slowly increase, probably due to alkylating side reactions.

The copolymerization of indene with its 1-methyl-, 2-methyl-, and 3-methyl-derivatives and of styrene has been studied by Sigwalt and Maréchal (624) using $TiCl_4$ in CH_2Cl_2 at $-72°C$. The following reactivity ratios have been obtained:

r_I = 3.18 ± 0.05 r_{1Mel} = 0.14 ± 0.02

r_I = 0.99 ± 0.03 r_{2Mel} = 0.052 ± 0.004

r_I = 0.25 ± 0.005 r_{3Mel} = 0.72 ± 0.15

r_I = 2.16 ± 0.15 r_{St} = 0.97 ± 0.07

The first two systems are easy to understand: The parent compound is more reactive than either the 1- or 2-methyl derivative because of steric compression during propagation and/or unfavorable inductive effects. The relatively higher reactivity of 3-methylindene as compared to indene is due to the favorable polarization of the double bond by the methyl group in the 3-position. Propagation, of course, is severely impeded by the same substituent. The relatively higher reactivity of indene over that of styrene is discussed below.

Maréchal and Evrard further studied the copolymerization of indene with 4,6-, 4,7-, and 5,6-dimethylindene and 4,5,6,7-tetramethylindene (625). The authors again used $TiCl_4$ in CH_2Cl_2 at $-30°C$, and the most reactive monomer was found to be the tetramethyl derivative (625).

Higashimura and his coworkers (626) also carried out significant fundamental studies with indene. They were interested in elucidating the relation between monomer structure and reactivity. They found that the rate of polymerization of indene is greater than that of styrene ($BF_3 \cdot OEt_2$ in ethylene dichloride at $30°C$), that is, indene is the more reactive monomer. Similarly, the average $\overline{DP}$ of polyindene was much larger than that of polystyrene under identical conditions: $\overline{DP}_I \sim 250$ and $\overline{DP}_{St} = 40$. Indene and styrene were copolymerized and the monomer reactivity ratios also reflected the reactivity difference between these monomers: r_I = 3.7 ± 0.2 and r_{St} = 0.6 ± 0.1, $BF_3 \cdot OEt_2$, $(CH_2Cl)_2$ at $30°C$. The increased reactivity of indene above that of styrene was explained by a measure of ring strain in the five-membered ring of the former member.

Schmitt and Schuerch (219) attempted the synthesis of optically active polyindene by the use of BF_3 in conjunction with optically active alcohols, that is, 1-α-methylbenzyl alcohol, in n-hexane or chloroform solvents. While polyindene did form, it did not exhibit optical activity, birefringence, or crystallinity by x-rays, even after stretching a cast film.

The polymerization of indene with stable carbenium ion salt initiators, for example, tropylium hexachlorantimonate and xanthylium perchlorate in methylene chloride solvent, has been studied by Ledwith and coworkers (627). The molecular weights obtained were around 10,000. The tropylium salt was a far superior initiator to the xanthylium compound. Spectrophotometric analysis indicated the incorporation of the troplyium moiety into the polymer:

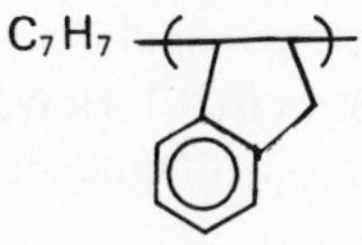

Recently the polymerization of indene by methylethyl ketone peroxide in liquid SO_2 has been described by Filho and Gomes (628, 629). Initially polymerizations were carried out by mixing indene and SO_2 and bubbling O_2 (or N_2) through the yellow solution at 25 or $-10°C$ and subsequently adding a few drops of MEK peroxide. [In later experiments the O_2 bubbling was omitted (629)]. Polymer precipitation ensued immediately. In the absence of SO_2, or in the presence of DMF, no polymer formed. No sulfur was detected in the polymers.

The molecular weights obtained in experiments carried out at $-10°C$ ranged from 55,000 to 60,000. Interestingly, the molecular weight distributions were consistently less than most probable ($\bar{M}_w/\bar{M}_n$ = 1.6–1.8) and they seemed to increase with conversions from 17 to 47%. This might indicate branching at higher conversions. Molecular weight distributions obtained with conventional Lewis acids $AlCl_3$, and $TiCl_4$ in various solvents at 0 to $-76°C$ fall in the range $\bar{M}_n/\bar{M}_n$ = 3.2–4.6.

The authors visualize the formation of a charge transfer complex, (indene)$^{\oplus}$ $SO_2^{\ominus}$, which somehow initiates the polymerization with the MEK peroxide.

2. 1-Methylindene

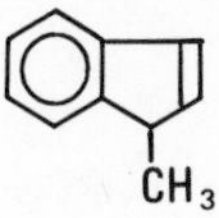

1-Methylindene was synthesized and polymerized with $SnCl_4$, $TiCl_4$, and BF_3 in CH_2Cl_2 at various temperatures by Sigwalt and coworkers (630). Slow polymerization proceeded with $SnCl_4$; the rate was faster with $TiCl_4$. The molecular weights were low and the highest viscosity product was obtained with BF_3 at $-72°C$; $[\eta]$ = 0.17 which might correspond to a molecular weight of 30,000.

3. 2-Methylindene

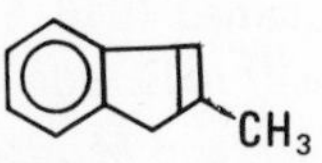

According to Sigwalt and coworkers (630) much lower rates and molecular weight products are obtained with 2-methylindene than with 1-methylindene.

Thus a polymerization with BF_3 in CH_2Cl_2 at $-72°C$ gave 68% yield and $[\eta] =$ 0.042. With $TiCl_4$ methanol insoluble product could be obtained only in the presence of relatively large amounts of water.

4. 3-Methylindene

Neither $TiCl_4$ nor BF_3 produce methanol insoluble polymer and only dimers are obtained of this indene derivative (630). However, it is possible to copolymerize 3-methylindene with indene or 1-methylindene.

5. 5-Methylindene

Maréchal and coworkers (616, 631) studied the polymerization behavior of this monomer with $TiCl_4$, $SnCl_4$, and $BF_3 \cdot OEt_2$ in CH_2Cl_2 for 5 min at $-72°C$. With $0.224M$ monomer and $0.01M$ concentrations of the three Lewis acids examined, the yields were 100%, but the intrinsic viscosities were rather low, 0.5, 0.8, and 0.8, respectively. Practically no yield was obtained with H_2SO_4 and $AlCl_3$. The decomposition range of all the polymers was $250-260°C$.

Copolymerization of 5-methylindene with indene using $TiCl_4$ in CH_2Cl_2 solution at $-70°C$ gave $r_{5MI} = 0.85 \pm 0.05$ and $r_I = 1.15 \pm 0.05$. Thus this monomer is somewhat less reactive than indene. This is in agreement with Maréchal's quantum chemical calculations (632, 633).

Copolymerization of 5-methylindene with 6-methylindene has also been studied under various conditions. High molecular weight product, $[\eta]$ up to 2.1, was obtained with $BF_3 \cdot OEt_2$ at $-70°C$ at 100% conversions.

6. 6-Methylindene

The polymerization of 6-methylindene has been investigated by Maréchal and coworkers (631) who used $TiCl_4$ at very low temperatures. The liquid medium was ClF_2CH alone (down to $-140°C$) or mixtures of C_2H_5Cl/ClF_2CH (down to $-170°C$) in which the monomer is soluble down to $-180°C$. Polymerization readily occurred and 100% yield could be obtained of polymers with $[\eta] = 0.4-0.6$. At $-100°C$, in pure C_2H_5Cl, $[\eta] = 1.0$. At a higher temperature, with $TiCl_4$, $BF_3 \cdot OEt_2$, and H_2SO_4 in CH_2Cl_2 solvent, intrinsic viscosities of 2.2, 2.7, 4.5, and 2.2, respectively, were obtained (616, 631). These values are among the highest ever obtained with indene or indene derivatives. For example, the $[\eta] = 4.5$ corresponds with $\bar{M}_n = 2.5 \times 10^6$ and $\bar{M}_w \sim 7 \times 10^6$. The fusion range of these materials was $250-270°C$.

The effect of temperature on $[\eta]$ was studied using $TiCl_4$ and $BF_3 \cdot OEt_2$ initiators. The molecular weights increased significantly in both instances. Thus, with $TiCl_4$ in the range +15 to $-80°C$, $[\eta]$ increased from 0.31 to 2.35. With $BF_3 \cdot OEt_2$ in the range -40 to $-70°C$ the $[\eta]$ values were 2.2 and 3.5, respecitvely, all at essentially complete conversions. The enthalpy of polymerization was found to be ~ 13.7 kcal/mole (616).

Copolymerization of this monomer with 5-methylindene with various Lewis acids also gave high molecular weight product. Copolymerization with indene by the use of $TiCl_4$ at $-70°C$ gave $r_{6MI} = 2.5 \pm 0.25$ and $r_I = 0.45 \pm 0.05$, indicating that the 6-methyl derivative is a much more reactive monomer than the parent indene. These findings are in agreement with Maréchal's quantum chemical calculations (632, 633).

7. 7-Methylindene

Maréchal (616, 631) reported the polymerization of 7-methylindene with $TiCl_4$, $BF_3 \cdot OMe_2$, and $BF_3 \cdot OEt_2$ in CH_2Cl_2 at $-72°C$. With $0.224 M$ monomer and $0.01 M$ Lewis acid the yields were close to 100%, however, the intrinsic viscosities were mediocre with values of 0.82, 0.62, and 0.48, respectively. The decomposition range of these polymers was $252-258°C$.

Copolymerization of this monomer with indene was investigated by the use of $TiCl_4$ in CH_2Cl_2 at $-70°C$ (631). Reactivity ratios of $r_{7MI} = 1.15 \pm 0.05$ and $r_I = 0.8 \pm 0.005$ were obtained indicating that 7-methylindene is somewhat more reactive than indene.

8. 1,1-Dimethylindene

Maréchal et al. (635) attempted to polymerize this monomer with a variety of Lewis acids, such as $TiCl_4$, $SbCl_5$, $SnCl_4$, $BF_3 \cdot OEt_2$, and VCl_4 in CH_2Cl_2 at 20 or $-65°C$. Qualitative results indicate that low yield ($\sim 15\%$) of low molecular weight ($\bar{M}_n = 770 \pm 70$) oligomer formed only with VCl_4, $TiCl_4$, and $SbCl_5$ at $20°C$. All the products were white powders exhibiting strong fluorescence and yielding colored solutions. NMR spectroscopy indicates that under polymerization conditions the monomer has a tendency to isomerize partially to 2,3-dimethylindene.

9. 2,3-Dimethylindene

Heilbrunn and Maréchal (635) report the polymerization of this monomer with a variety of Lewis acids, such as TiCl$_4$, BF$_3 \cdot$OEt$_2$, and SbCl$_5$, in CH$_2$Cl$_2$ for 24 hr at 20 and -65°C. These authors obtained $\sim$30% yield and $\bar{M}_n$ = 750 $\pm$ 80. On the basis of some NMR evidence the suggestion was made that under the influence of TiCl$_4$ the monomer is partially transformed into 2,3-dimethylindane, however, no further explanation is available.

10. 4,7-Dimethylindene

Maréchal and coworkers (616, 636) examined the effect of various reaction parameters on the yield and intrinsic molecular weight of poly(4,7-dimethylindene). For example, among the various initiators examined (TiCl$_4$, SnCl$_4$, BF$_3 \cdot$OR$_2$, and H$_2$SO$_4$), the highest intrinsic viscosity was produced with BF$_3 \cdot$OMe$_2$ and BF$_3 \cdot$OEt$_2$, [η] = 3.6 and 3.5, respectively, at -30°C. In contrast, at -72°C a much lower [η] was found with BF$_3 \cdot$OMe$_2$ ([η] = 1.5); the lower temperature did not much affect the value obtained with BF$_3 \cdot$OEt$_2$ ([η] = 3.4 at -72°C). According to independent measurements these intrinsic viscosity values are well above the 10^6 mark on the number average molecular weights scale, for example, [η] = 3.5 is equivalent to $\bar{M}_n$ = 1,200,000 (616).

Anton and Maréchal (526) determined the influence of temperature on the intrinsic viscosity of poly(4,7-dimethylindene) synthesized with SnCl$_4$ in CH$_2$Cl$_2$. In the range of +30 to -70°C at essentially complete conversions, [η] decreased from 2.36 to 1.90, showing a minimum around -30°C. The effect of solvents on [η] was also examined and consistently higher values were obtained for polymerization in C$_2$H$_5$Cl than in CH$_2$Cl$_2$.

11. 4,6-Dimethylindene

Maréchal and coworkers (616, 636) examined the effect of various Lewis acids (TiCl$_4$, BF$_3 \cdot$OR$_2$, SnCl$_4$, AlBr$_3$, and SbCl$_5$) and H$_2$SO$_4$ in CH$_2$Cl$_2$ at -30 and -72°C on the yield and intrinsic viscosity of poly(4,6-dimethylindene). The highest intrinsic viscosities were obtained with BF$_3 \cdot$OMe$_2$, [η] $\sim$2.5. The temperature did not greatly affect the values obtained with BF$_3 \cdot$OR$_2$. Relatively high molecular weight polymer was obtained with TiCl$_4$ as well. With this Lewis acid the highest values were [η] = 1.5, $\bar{M}_n$ = 20,000 at -30°C and [η] $\sim$ 1.0, $\bar{M}_n \sim$ 120,000 at -72°C.

The enthalpy of polymerization was reported to be 18.5 $\pm$ 0.9 kcal/mole (616, 636).

12. 5,6-Dimethylindene

Maréchal (610, 636) studied and summarized the effect of various experimental parameters (concentrations, nature of initiator, etc) on the yields and intrinsic viscosities of poly(5,6-dimethylindene). He found that among the initiators tried, for example, $TiCl_4$, $SnCl_4$, $BF_3 \cdot OR_2$, and H_2SO_4, the highest intrinsic viscosities were obtained with $BF_3 \cdot OMe_2$ and $BF_3 \cdot OEt_2$. At $-30°C$ $[\eta] = 0.46$ and 0.45, $\bar{M}_n$ = 64,000 and 65,000, fusion temperature range = 286–288°C, and at $-72°C$ $[\eta]$ = 3.3 and 3.4, $\bar{M}_n$ = 1,000,000, and fusion temperature range = 288–290°C (606). All these values were obtained at 100% yield. Similarly, with $TiCl_4$, the intrinsic viscosities and number average molecular weights increased monotonically from $[\eta] \sim 0.2$, $\bar{M}_n \sim 70,000$ at 0°C to $[\eta] \sim 2.0$, $\bar{M}_n$ = 400,000 at $-72°C$. Under suitable conditions (0.3M monomer, $-72°C$, in CH_2Cl_2) the latter Lewis acid also produces $[\eta]$ = 2.5, $\bar{M}_n \sim 10^6$ (616). The measured polymerization enthalpy was found to be 18.9 ± 0.9 kcal/mole (636).

13. 5,7-Dimethylindene

Maréchal and coworkers (637) were the first to synthesize this monomer and to examine its cationic polymerization behavior. In their polymerization experiments these authors used a variety of Lewis acids, for example, $TiCl_4$, $SnCl_4$, $AlBr_3$, and $SbCl_5$, and 1,2-dichloroethane or methylene chloride solvents at -30 and $-72°C$. While high yields of low molecular weight ($[\eta]$ = 0.07–0.17) products were obtained with these initiators, no polymerization occurred with $BF_3 \cdot OMe_2$ and $BF_3 \cdot OEt_2$. Copolymerization experiments with indene using $TiCl_4$ at $-30°C$ gave the reactivity ratios $r_{5,7DMI}$ = 3.80 ± 0.04 and r_I = 0.10 ± 0.05. Evidently, 5,7-dimethylindene is much more reactive toward an indene cation than indene itself. This is an unexpected result and is difficult to explain.

In a subsequent publication, Maréchal (616) examined and summarized the effect of various parameters on the yield, intrinsic viscosities, and number average molecular weights of poly(5,7-dimethylindene). Among the various active initiators used, for example, $TiCl_4$, $SnCl_4$, $AlBr_3$, $SbCl_5$, and H_2SO_4, the relatively highest intrinsic viscosities at 100% yield were obtained with $AlBr_3$ in CH_2Cl_2 where $[\eta]$ = 0.12 at $-30°C$ and 0.17 at $-72°C$.

14. 6,7-Dimethylindene

The synthesis and polymerization of this monomer have been studied by

Maréchal and coworkers (526, 637, 638). Polymerization with TiCl$_4$ at $-70°$C in CH$_2$Cl$_2$ and C$_2$H$_5$Cl gave $[\eta]$ = 0.37, softening range = 288–291$°$C, and $[\eta]$ = 0.53, softening range = 288–289$°$C, respectively. Yields were 100% (638). By decreasing the temperature, the intrinsic viscosity rose from 0.29 at +25$°$C to 0.37 at $-70°$C. Higher molecular weights were obtained with BF$_3$·OEt$_2$ and SnCl$_4$. Indeed, the polymerization with SnCl$_4$ in CH$_2$Cl$_2$ for 5 min showed that at essentially complete conversion levels the intrinsic viscosity increases from 0.29 at +25 to 1.23 at $-70°$C (526). Highest molecular weights were obtained with SnCl$_4$ and BF$_3$·OEt$_2$ at $-70°$C, $\bar{M}_n$ = 138,000, softening range = 287–306$°$C. The copolymerization of this monomer with indene has been studied (638) and the reactivity ratios determined: $r_{6,7\text{DMI}}$ = 5.1 ± 0.3 and r_I = 0.18 ± 0.05.

15. *4,6,7-Trimethylindene*

Maréchal and coworkers (637) were the first to synthesize this monomer and to investigate its polymerization behavior. The polymerizations were induced by TiCl$_4$, SnCl$_4$, BF$_3$·OEt$_2$, BF$_3$·OMe$_2$, and H$_2$SO$_4$ in ethyl chloride and methylene chloride at $-72°$C. The highest polymer $[\eta]$ = 1.73 was obtained with BF$_3$·OEt$_2$, but respectable intrinsic viscosity values were also produced by the other initiators as well, $[\eta]$ = 0.6–0.9. The effect of temperature in the range +25 to $-70°$C on $[\eta]$ clearly indicated a maximum at ${\sim}-15°$C ($[\eta] \sim 0.6$) with TiCl$_4$ and at ${\sim}0°$C ($[\eta] \sim 2.0$) with SnCl$_4$ in CH$_2$Cl$_2$. Copolymerization of 4,6,7-trimethylindene with indene using TiCl$_4$ at 0$°$C gave $r_{5,6,7\text{TMI}}$ = 3.4 ± 0.1 and r_I = 0.15 ± 0.05. According to these data the 4,6,7-trimethylindene is much more reactive toward an indene cation than the parent compound itself. Quantum chemical calculations predict a very high reactivity for the 4,6,7-trimethyl derivative (632, 633).

In a subsequent publication Maréchal (616) once again summarized the effect of a variety of reaction variables (nature of initiator, temperature, etc). on the yields and intrinsic viscosities of poly(4,6,7-trimethylindene).

16. *4,5,6,7-Tetramethylindene*

Maréchal et al. (671) worked out a new synthesis for this monomer and also investigated its polymerization behavior (616, 671). The influence of reaction parameters, that is, temperature, initiator, and monomer concentration, on the intrinsic viscosity, number average molecular weight, and softening point of poly(4,5,6,7)tetramethylindene were determined and compared to those of polyindene. The tetraalkylated monomer is only sparingly soluble at low temperatures in methylene chloride and ethyl chloride. With 0.03M monomer with TiCl$_4$ in CH$_2$Cl$_2$ solvent for 5 min at $-75°$C they obtained 100% yield and $[\eta] = 0.22$. The temperature variation in the range -45 to $0°$C did not much affect the intrinsic viscosities, $[\eta] \sim 0.25$, and softening range which was 317–320°C. The products were amorphous. $[\eta] = 0.25$ corresponds to $\overline{M}_n = 43,600$. Interestingly, by using H$_2$SO$_4$ as initiator at $-30°$C the authors report (580, 581) $[\eta] = 0.38$ and $\overline{M}_n = 94,500$, softening range = 317–320°C. In contrast, with indene much lower molecular weight polymers and lower yields are obtained by using H$_2$SO$_4$ than with conventional Lewis acids TiCl$_4$ or SnCl$_4$. The enthalpy of polymerization was found to be 18.2 kcal/mole (636).

17. *3,4,5,6,7-Pentamethylindene*

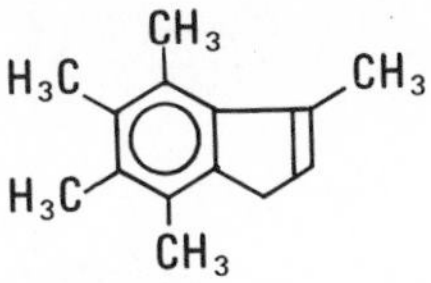

Quere and Maréchal (639) first synthesized this compound to study its polymerization behavior, however, no information has yet become available.

18. *5-Vinylindene*

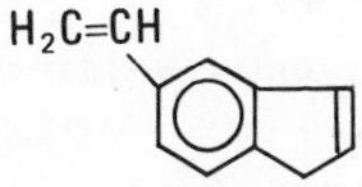

According to Maréchal and Quere (639, 672) the polymerization of this monomer with TiCl$_4$ and BF$_3$·OEt$_2$ in CH$_2$Cl$_2$ at $-70°$C gives high yields of totally insoluble and infusible polymer. Copolymerization with styrene also results in insoluble product. Theoretical calculations showed that the two double bonds in this monomer are of about equal activity toward cations. Thus, under cationic conditions, two independent polymerizations proceed which lead to crosslinked products.

19. *1-Phenylindene*

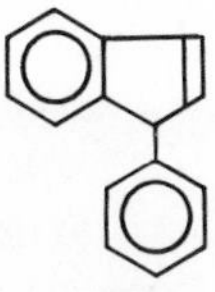

Maréchal and Hamy (640) briefly examined the polymerization behavior of this monomer under the influence of BF_3 and $TiCl_4$ at $-72°C$ and $AlCl_3$ at 15°C. Yields were 90–95% and all the products were insoluble and infusible exhibiting good thermal stability. Gel formation was attributed to the growing cation being able to add to the olefinic double bond or to the 4-position of the phenyl substituent.

20. 1,3-Diphenylindene

All efforts to polymerize this olefin by Lewis acids have failed, according to Maréchal and Hamy (640). Dimerization with $TiCl_4$ at ambient temperature is possible with a yield of ~80%.

21. 1,1′-Biindenyl (1-Indenylindene)

Sigwalt and coworkers (630) first reported the cationic polymerization of biindenyl with $TiCl_4$ in CH_2Cl_2 at $-72°C$ to high molecular weight soluble material, $[\eta] = 1.2$. Other Lewis acids employed were BF_3 and $SnCl_4$. Of these, $SnCl_4$ gave very low yields and BF_3 consistently yielded badly crosslinked product. This brief initial report was followed up by Maréchal and coworkers (641–644) with a series of papers on the polymerization behavior of biindenyl and its derivatives. First, the synthesis of soluble high molecular weight biindenyl with $TiCl_4$ was confirmed. The thermal stability of poly(biindenyl) is not outstanding: a polymer sample with $[\eta] = 0.67$ had a softening range of 325–330°C, however, the product discolored (yellow) at ~280°C. Copolymerization of biindenyl with indene was effected; the reactivity ratios indicated that indene was more reactive than biindenyl: $r_{BI} = 0.045$ and $r_I = 0.49$. Copolymerizations of biindenyl with styrene and α-methylstyrene using $TiCl_4$ in CH_2Cl_2 at $-72°C$ were studied (643). Again biindenyl was found to be the less reactive monomer:

r_{BI} = 0.17 ± 0.05 r_{St} = 1.95 ± 0.1°C

r_{BI} = 0.15 ± 0.15 $r_{\alpha MeSt}$ = 3.60 ± 0.2°C

The softening points of these copolymers were found to depend on the composition of the copolymers rising monotonically from ~120 and ~205°C [for polystyrene and poly(α-methylstyrene), respectively] to ~320°C, the softening point of polybiindenyls. However, the intensity of the yellow discoloration in air was also a function of the polybiindenyl content of the copolymer.

22. *3,3'-Biindenyl (3-Indenylindene)*

This monomer was synthesized and polymerized by Maréchal and Sigwalt (642). The polymerization experiments were somewhat hampered by the insolubility of this compound at low temperatures. Only low molecular weight products have been obtained: using $TiCl_4$ in CH_2Cl_2 for 5 min at $-40°C$, yield = 30%, $[\eta]$ = 0.08. The polymer exhibited a nice blue fluorescence. The structure of the polymer has not been investigated in detail. To explain the presence of fluorescence, three possibilities were considered: *trans*-stilbene like units, or highly conjugated end-groups:

23. *1,1'-Biindenyl-1,2-ethane*

The synthesis of this monomer was undertaken by Maréchal and Lepert (644). Their product was a mixture of two isomers: Isomer I, m.p. = $100°C$, constituting 63% of the mixture and Isomer II, m.p. = $66°C$, constituting 37% of the mixture. The authors propose that Isomer I is the optically inactive form, whereas Isomer II is the racemic mixture.

These two isomers were polymerized separately each by $TiCl_4$ in CH_2Cl_2 for 5 min at $-70°C$. Isomer I gave a partially (70%) soluble product that crosslinked on standing in benzene solution. The intrinsic viscosity range was 0.1–0.34, $\overline{M}_n$ = 3600–14,000, softening range = 235–255°C. Isomer II gave lower intrinsic viscosity product, $[\eta]$ = 0.07, $\overline{M}_n$ = 4600, softening range = 220–242°C. The residual double bond content of this polymer was much higher than that obtained from Isomer I. The products crosslink when heated.

Isomers I and II were copolymerized to give $[\eta]$ = 0.075.

24. *3,3'-Biindenyl-1,2-ethane*

Maréchal and Lepert (644) synthesized this monomer and polymerized it with

$TiCl_4$ in CH_2Cl_2 for 5 min at $-30°C$, yield = 70%, $[\eta]$ = 0.044, $\bar{M}_n$ ~1800. The product is crosslinked when heated, softening range = 250–260°C.

25. 1,1'-Biindenyl-1,4-butane

$$CH_2-CH_2-CH_2-CH_2$$

Maréchal and Lepert (644) synthesized this monomer and studied its polymerization behavior. They found that under the influence of $TiCl_4$ in CH_2Cl_2 for 5 min at $-72°C$ polymerization was rapid and a crosslinked product formed. Yields up to 100% were reported and $[\eta]$ = 0.13–0.21, $\bar{M}_n$ = 5000–14,500. Softening starts at 240°C. The polymer crosslinks when heated.

26. 1,1'-Biindenyl-1,4-trans-butene-2

$$CH_2-\overset{H}{\underset{H}{C=C}}-CH_2$$

Maréchal and Lepert (644) reported the synthesis and polymerization of this compound. The polymerization was carried out by $TiCl_4$ in CH_2Cl_2 for 5 min at $-72°C$ and conversions up to 97% were readily obtained. Number average molecular weights of the soluble products were in the range 6700–14,500, softening range = 137–315°C depending on $\bar{M}_n$. Infrared spectra indicated that the aliphatic double bond remained unchanged. Crosslinking occurs upon heating.

27. 3,3'-Biindenyl-1,4-butane

Maréchal and Lepert (644) described the synthesis and polymerization of this monomer by $TiCl_4$ in CH_2Cl_2 for 5 min at $-37°C$, yield = 5%, completely benzene soluble polymer, $[\eta]$ = 0.09, $\bar{M}_n$ = 1700, softening range = 175–225°C. Unsaturation determination indicated that only one of the two available double bonds reacted. The polymer is crosslinked upon heating.

28. 6-Bromoindene

Quere and Maréchal (639) synthesized and polymerized this monomer using BF_3, $TiCl_4$, $SnCl_4$, and H_2SO_4 initiators in CH_2Cl_2 and C_2H_5Cl for 5 min in the range 0 to $-117°C$. The effect of various reaction conditions (nature of the initiator and solvent, temperature, concentration of monomer, etc.) on the yield and product intrinsic viscosities was studied. With TiCl$_4$ the highest $[\eta]$ obtained was about 0.4, however, with BF$_3$ intrinsic viscosities close to 3.0 were reported under suitable conditions (-80°C, 0.2M monomer, in C$_2$H$_5$Cl). The intrinsic viscosities obtained with SnCl$_4$ and H$_2$SO$_4$ were consistently low. The dimethyl or diethyl etherates of BF$_3$ did not give any methanol-precipitable polymer. Carbon disulfide was found to be the only solvent that was able to dissolve the polymer. The softening range was 275–285°C.

Interestingly, the authors state that the 5-isomer, 5-bromoindene, cannot be polymerized by their techniques. It is proposed that the Lewis acids strongly complex with the bromine in the 5-position. The use of larger than 1:1 amounts of Lewis acids (i.e., after saturation of the bromine substituent) has not been tried, and experiments with H$_2$SO$_4$ have also not been mentioned.

The copolymerization of 6-bromoindene and styrene has been described (639) and reactivity ratios were obtained: $r_{5BrI} = 0.21 \pm 0.02$ and $r_{St} = 4.2 \pm 0.04$.

29. *3-Bromopropyl-1-indene*

CH$_2$CH$_2$CH$_2$Br

This monomer, a pale yellow liquid, was synthesized and polymerized by Maréchal and Lepert (644) who used TiCl$_4$ in CH$_2$Cl$_2$ for 5 min at -72°C and obtained 80% yield, $[\eta] = 0.045$, $\bar{M}_n \sim 5860$, softening range = 195–210°C. No bromine loss occurred during polymerization. Crosslinking occurs upon heating.

30. *4-Bromobutyl-1-indene*

CH$_2$CH$_2$CH$_2$CH$_2$Br

This monomer, a yellow oil, was prepared and polymerized by Maréchal and Lepert (644). These authors used TiCl$_4$ in CH$_2$Cl$_2$ for 5 min at -72°C. The monomer contained about 1% of the 3-indene isomer. They obtained 20% yield, $[\eta] = 0.015$, $\bar{M}_n = 5400$, softening range = 95–130°C. Some bromine loss occurred during polymerization. Crosslinking takes place upon heating.

OCH$_3$

31. *4-Methoxyindene*

Maréchal and coworkers (552, 645) first synthesized this monomer and examined its cationic polymerization behavior. Due to the scarcity of the compound only a limited number of studies have been made. Polymerizations have been carried out at $-78°C$ in CH_2Cl_2 solution for 5 min with various Lewis acids, for example $TiCl_4$, $AlBr_3$, $TaCl_5$, $BF_3 \cdot OEt_2$, and H_2SO_4. Largely insoluble polymer was obtained. In one case, using $TiCl_4$ in C_2H_5Cl, soluble product was produced, $[\eta] = 1.05$. In all cases the softening points were $\sim 260°C$.

32. 5-Methoxyindene

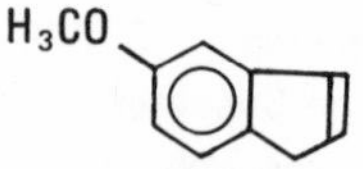

Maréchal and coworkers (552, 645) studied the polymerization behavior of 5-methoxyindene with various acid initiators. Polymer was obtained with BF_3, $TiCl_4$, and H_2SO_4 ($[\eta] \sim 0.1$), but not with $SnCl_4$ or $BF_3 \cdot OR_2$ in CH_2Cl_2 in the range -30 to $-50°C$. The softening range of the polymers was $216-220°C$.

Copolymerization experiments ($TiCl_4$ at $-50°C$) with styrene yielded $r_{5MI} = 6.3 \pm 0.5$ and $r_{St} = 0.35 \pm 0.5$.

33. 6-Methoxyindene

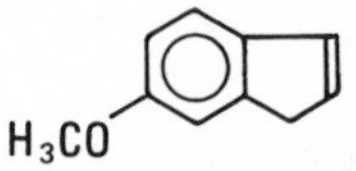

Maréchal and coworkers (552, 645) investigated the polymerization of 6-methoxyindene and used $TiCl_4$, gaseous BF_3, and $BF_3 \cdot OEt_2$ initiators in CH_2Cl_2 for 5 min at $-78°C$. Polymer formed in high yield. Insoluble product was obtained with $TiCl_4$, however, soluble polymer was obtained ($[\eta] \sim 0.35-0.5$) with BF_3 and $BF_3 \cdot OEt_2$. The softening range of the products was $230-240°C$. Copolymerization experiments ($TiCl_4$, $-78°C$) with styrene gave $r_{6MI} = 4.1 \pm 0.5$ and $r_{St} = 0.1 \pm 0.05$.

Among the methoxyindenes examined (4-, 5-, and 6-methoxyindenes) the 6-methoxyindene isomer is the most reactive. The sequence of comparative reactivities is as follows:

6-MeOI (10) > 4-MeOI (6.7) > 5-MeOI (2.8) > indene (1)

where the numbers in parentheses indicate the relative reactivities ($1/r$) *vis-a-vis* the styrene cation (645).

C. MISCELLANEOUS AROMATIC OLEFINS

1. Allylbenzene

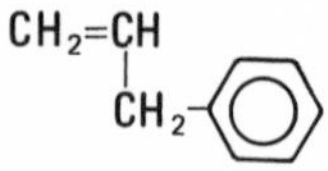

Staudinger and Dreher (545) mention that allylbenzene and, similarly, its

p-methoxy and p-ethoxy derivatives do not polymerize under conditions yielding polypropenylbenzenes. This difference in polymerizability, they suggest, may be useful in separating allyl and propenylbenzene derivatives.

Earlier successful attempts to oligomerize allylbenzene with Lewis acids were made by Schmidt and Schöller (646) who used solid $AlCl_3$ at room temperature and obtained low molecular weight product ($\overline{DP}$ = 6.2). D'Alelio et al. (647) used a variety of conventional Lewis acids such as $TiCl_4$, $AlCl_3$, and BF_3, and obtained low molecular viscous oils. With $Et_3Al/TiCl_4$ (1:1) systems solid products were obtained.

In the course of studies concerning isomerization polymerizations, Kennedy (597) investigated the cationic polymerization of allylbenzene ($AlCl_3$ in methyl chloride at $-60°C$). Solid product was obtained (61% yield) and the structure was analyzed by infrared spectroscopy. The presence of methyl groups in the spectra suggested a partial rearrangement to β-methylstyrene and subsequent 1,2 polymerization of the latter. In contrast, Murahashi et al. (614), who oligomerized ($\overline{DP}$ = 4.7) allylbenzene by BF_3 in CH_2Cl_2 at $-78°C$, proposed on the basis of NMR spectra and other analytical data a polyalkylation process:

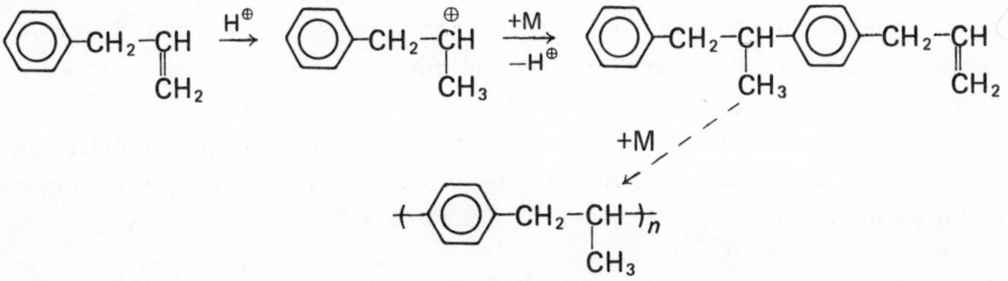

Davidson (648), who used $AlBr_3$ in CS_2 at $-30°C$ to induce the polymerization of allylbenzene, came to similar conclusions. The solid product had $\overline{M}_n$ = 4200 ($\overline{DP}$ ~35). Careful NMR work suggested a polyalkylation mechanism resulting in the repeat unit:

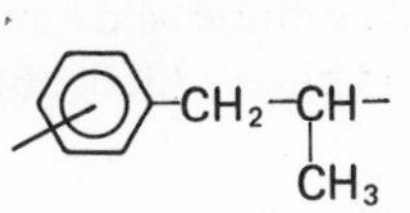

Shimizu et al. (601) also investigated the polymerization behavior of allylbenzene and reported that neither polymerization nor isomerization (to propenylbenzene) occurs upon treatment with $AlBr_3$ in CH_2Cl_2 at 10 or $-78°C$, or with $BF_3 \cdot OEt_2$ at $10°C$. These findings are very difficult to reconcile with Davidson's observations that allylbenzene readily polymerizes with $AlBr_3$ in CS_2 $-30°C$. Shimizu obtained solid product with an $Et_3Al/TiCl_3$ (1:1) mixture and noted that $TiCl_3$ alone isomerized allylbenzene to propenylbenzene, which yielded a viscous oligomer

2. *p-Allyltoluene*

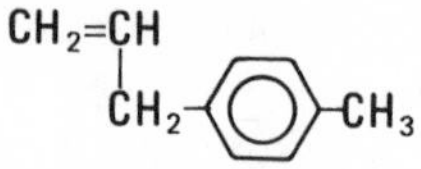

$$CH_2=CH$$
$$CH_2-\langle\bigcirc\rangle-CH_3$$

According to Davidson (648) *p*-allyltoluene and allylbenzene behave in a similar manner toward $AlBr_3$ in CS_2 and both give polyalkylated products of rather ill defined structures.

3. *3-Phenyl-1-butene*

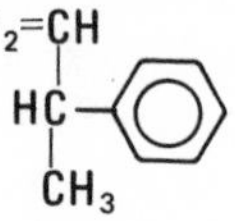

$$CH_2=CH$$
$$HC-\langle\bigcirc\rangle$$
$$CH_3$$

This monomer was oligomerized by Kennedy et al. (649) using $AlCl_3$ in ethyl chloride at $-78°C$, $\bar{M}_n = 570$, $\overline{DP} \sim 4$. The product starts to soften at $69°C$, softening range $= 73-76°C$, and is readily soluble in benzene, *n*-pentane and so on. At $-100°C$ no polymerization occurs and at temperatures higher than $-78°C$ oily products are obtained. Spectroscopic analysis suggests a number of possible contributing structures due to facile isomerization and alkylation reactions during oligomerization. The published spectral data were reanalyzed by Davidson (648) who concluded that the tetramer probably forms by polyalkylation and not by an electrophilic attack on the double bond.

4. *p-Methoxyallylbenzene (Methylchavicol)*

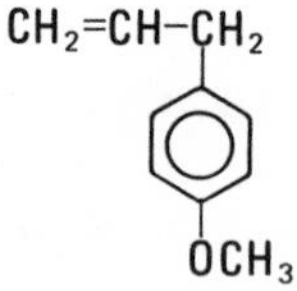

$$CH_2=CH-CH_2$$
$$\langle\bigcirc\rangle$$
$$OCH_3$$

Under the influence of 43% sulfuric acid *p*-methoxyallylbenzene gives a mixture of various isomers and liquid dimers (650, 651).

5. *o-Methoxyallylbenzene*

$$CH_2=CH-CH_2$$
$$OCH_3$$
$$\langle\bigcirc\rangle$$

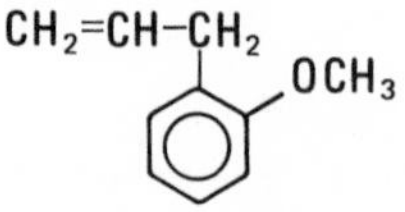

When *o*-methoxyallylbenzene is treated with 43% sulfuric acid at reflux, a viscous liquid product is obtained which on chemical analysis was found to be a hydrated dimer (652).

6. 1,2-Dihydronaphthalene

Higashimura and coworkers (626) polymerized this monomer (containing ~17% decalin) with $SnCl_4 \cdot CCl_3COOH$ in $(CH_2Cl)_2$ for 30 min at $30°C$ and obtained 10% yield, of which 60% was methanol soluble. Under the same conditions $BF_3 \cdot OEt_2$ was inactive. According to comparative rate studies, the rate of polymerization of 1,2-dihydronaphthalene is much lower than that of indene and styrene. The molecular weight of the methanol insoluble solid indicates the presence of a tetramer. Copolymerization with styrene [$BF_3 \cdot OEt_2$ in $(CH_2Cl)_2$ at $30°C$] gave $r_{DHN} = 0.4 \pm 0.2$ and $r_{St} = 1.0 \pm 0.3$.

Although the rate of homopolymerization of 1,2-dihydronaphthalene is much slower than that of styrene, according to the reactivity ratios the relative reactivities of these two monomers are not very different. This may be due to the removal of the steric compression of 1,2-dihydronaphthalene propagation in the copolymerization system. That steric hindrance is important in the homopolymerization is also indicated by the fact that the product of the reactivity ratios is less than unity, that is, 0.4.

These authors were interested in the cationic reactivity of 1,2-dihydronaphthalene in systems where steric hindrance cannot occur. Thus they investigated the protonation of this chemical in competition with styrene and indene. Interestingly the data show that the rates of protonation (HI in acetic acid at $30°C$) are 1,2-dihydronaphthalene > indene > styrene. This is interpreted to mean that the releasing of the ring strain in 1,2-dihydronaphthalene increases the rate of protonation and that a stable cation is formed. In the case of indene, protonation also releases ring strain but, presumably, the indene (five-membered ring) cation that is formed is somewhat more strained than the 1,2-dihydronaphthalene (six-membered ring) cation.

7. Acenaphthylene

The first ionic polymerization of acenaphthylene to relatively low molecular weight products was carried out by Dziewonski and Stolyhow (653). Higher polymer was obtained by Flowers and Miller (654) by reacting this monomer with BF_3 gas in chlorobenzene solvent at $0°C$. Low molecular weight product was obtained by Jones (655) who reacted acenaphthylene in CS_2 with $AlCl_3$ at $-50°C$. This early paper also describes copolymerization experiments with isobutylene. The mechanism of acenaphthylene polymerization initiated by $BF_3 \cdot OEt_2$ in benzene solvent in the range 20–50°C was investigated by Imoto and Takemoto (656). The overall rate was found to be first order in monomer and initiator and the overall activation energy was calcualted to be 1.5 kcal/mole.

The number average molecular weights of reprecipitated (from benzene into methanol) polymers were in the range 125,000–200,000 indicating a DP range of 840–1300. The polymers appeared to be heat stable up to 280°C. Imoto and Soematsu (657) also polymerized acenaphthylene with $Et_3Al/TiOnBu_4$ mixtures, which at low Al/Ti ratios were probably cationic in character.

Story and Canty (658) polymerized acenaphthylene with BF_3 in chlorobenzene solvent at −5 and −23°C. Initiation was attributed to the presence of either adventitious water or purposely added initiator (H_2O or CH_3OH). The molecular weights obtained by these authors were rather low ([η] ~0.03 and $\bar{M}_n$ ~ 3000, i.e., $\overline{DP}$ ~ 20). The difference in molecular weight levels obtained by these authors and those obtained by earlier investigators was recognized but could not be explained. The polymers were found to be crystalline by x-rays and on the basis of this and other spectroscopic evidence a stereoregular enchainment was proposed. From an inspection of Fisher-Hirschfelder models and other circumstantial evidence the authors further propose that the polymer produced by solid hydroxyfluoroboric acid ($H^\oplus BF_3OH^\ominus$) is predominantly of a trans syndiotactic structure, whereas with methoxyfluoroboric acid ($H^\oplus BF_3OCH_3{}^\ominus$) trans isotactic structure is formed. These conclusions should be confirmed.

An Italian group of authors investigated the polymerization of acenaphthylene induced by gaseous BF_3 (464a) and, by the use of dilatometry, the kinetics of iodine initiated polymerization in $(CH_2Cl)_2$ in the range 0–30°C (464). The rate law was $= k$ [M] [I]^2 and the overall activation energy was 7.0 kcal/mole. Low reaction rates and low conductivity found during the polymerization led the authors to suggest a pseudocationic mechanism. For the chain carrier they visualize a 1-iodoacenaphthylene derivative in which the iodine is activated by coordination with one or two molecules of iodine and is stabilized by the monomer. These studies were followed up by kinetic investigations conducted in the presence of dc fields (electric field strengths from 0.33 to 1.5 kV/cm). Results of these studies confirmed earlier conclusions in regard to a pseudo-cationic polymerization (681). In contrast, the kinetics of the system acenaphthylene/CH_2Cl_2, or $(CH_2Cl)_2/BF_3 \cdot OEt_2$ which has also been studied by the same techniques, for example, dilatometry in the presence of dc fields, presented a more complex picture and no firm conclusion as to the nature of the propagating entities could be drawn (682).

The polymerization of acenaphthylene by tropylium salts $C_7H_7{}^\oplus SbCl_6{}^\ominus$ and $C_7H_7{}^\oplus BF_4{}^\ominus$ has been mentioned by Bawn and coworkers (675). Barrales-Rienda and Pepper (676, 677) polymerized this monomer by $BF_3 \cdot OEt_2$ in benzene at 80°C and obtained 100% yield. The material was dissolved in benzene and precipitated into methanol. The dilute solution properties of the highest molecular weight fraction ($\bar{M}_n$ = 23,000, $\bar{M}_w$ = 33,000, $\bar{M}_v$ = 31,000, $\bar{M}_w/\bar{M}_n$ = 1.44) were investigated. On the basis of their experiments these authors came to the unexpected conclusion that the unperturbed coil dimension of poly-acenaphthylene is smaller than that for polystyrene.

Very recently Belliard and Maréchal studied in some depth the cationic polymerization of acenaphthylene (and its derivatives)(680). First they developed a new synthesis of this olefin (dehydration of acenaphthenol with anhydrous MgSO$_4$) giving rise to chromatographically pure material. This was a necessary precondition for meaningful polymerization studies. According to these authors the presence of acenaphthene, a common impurity, traces of which accompany even carefully purified acenaphthylene, is detrimental to the polymerization yield and molecular weight of polyacenaphthylene. A variety of Lewis acids (TiCl$_4$, SnCl$_4$, BF$_3$·OEt$_2$, BF$_3$, AlCl$_3$, and SbCl$_3$) have been examined under various conditions, nonetheless, the highest intrinsic viscosity they were able to obtain was only ~0.14 (TiCl$_4$ in CH$_2$Cl$_2$ at -70°C; yield = 94%).

The cationic copolymerization of acenaphthylene and styrene has been investigated by Saotome and Imoto (659, 660) and Belliard and Maréchal (680). The monomer reactivity ratios determined by the Japanese authors for a series of copolymerizations with BF$_3$·OEt$_2$ in benzene at 30°C were r_{St} = 0.3 ± 0.1 and r_A = 4.4 ± 0.3. The significantly reduced rate (as compared to the rate of the homopolymerization) and molecular weight (as compared to the molecular weight of the homopolymer) were explained by a "cross termination" step. The same authors also explored the copolymerization with n-butyl vinyl ether using BF$_3$·OEt$_2$, which is of interest since it involves a copolymerization between an olefin and a vinyl ether (678). It is also of interest that in this system the reactivity ratios were strongly affected by the polymerization temperature. This unusual phenomenon was discussed at some length by this writer (679).

8. 1-Methylacenaphthylene

Belliard and Maréchal (680) synthesized, characterized by NMR, and examined the polymerization behavior of 1-methylacenaphthylene. Among the four Lewis acid systems tried (TiCl$_4$, BF$_3$·OEt$_2$, BF$_3$, and SnCl$_4$) in CH$_2$Cl$_2$ at -72°C only SnCl$_4$ gave low yields (2–3%) of a precipitate in methanol. The other acids gave, after solvent evaporation, oily products which could be purified by recrystallization from pentane. All these products were found to be crystalline trimers. The detailed structure of these oligomers has not been determined. Evidently considerable steric hindrance prevents propagation beyond the trimer stage in this system.

To overcome the propagation inhibiting effect of the 1-methyl substituent, the copolymerization of this olefin with styrene, α-methylstyrene, indene, and acenaphthylene has also been explored. While the exact structure of the polymer has not been determined, the authors imply that copolymerization has taken place with styrene. On the basis of infrared spectroscopy they determined the overall composition of the product and calculated the reactivity ratios

r_{1MeAc} = 0.4 ± 0.5 and r_{St} = 0.11 ± 0.05, but they did not comment on these unusual values.

9. *3-Methylacenaphthalene*

Belliard and Maréchal (680) synthesized, characterized by NMR, and examined the polymerization behavior of this olefin. Best results in terms of highest yields (>90%) and highest molecular weights ([η] = 0.36–0.47, $\bar{M}_n$ = 56,000–90,000) were obtained with $SnCl_4$ in CH_2Cl_2 at −72°C. To obtain highest molecular weights large amounts of $SnCl_4$, indeed practically equimolar monomer/$SnCl_4$ systems, had to be employed. The effect of temperature on molecular weight was investigated and the usual relationship was found: the molecular weights increased with decreasing temperatures.

The copolymerization with styrene has also been examined and while very little information is available, reactivity ratio calculations substantiate the authors belief of copolymer formation, r_{3MeAc} = 9.6 ±0.8 and r_{St} = 0.2 ± 0.05.

10. *5-Methylacenaphthylene*

The polymerization behavior of this olefin has been briefly examined by Belliard and Maréchal (680). By the use of $BF_3 \cdot OEt_2$ in CH_2Cl_2 at −72°C they harvested 57% polymer of [η] = 0.55. Copolymerization with styrene gave r_{5MeAc} = 11.7 ± 1 and r_{St} = 0.15 ± 0.05.

11. *9-Vinylanthracene*

According to Bergman and Katz (661) the cationic polymerization of 9-vinylanthracene yields a conventional 1,2 vinyl repeat unit. These authors used $SnCl_4$ in benzene solvent at 20°C and obtained dark-brown insoluble solid with 1.0–2.5% $SnCl_5$, whereas a soluble polymer of pure blue fluorescence was obtained with 0.15–0.25% $SnCl_5$. Michel (662) investigated the polymerization of this monomer in some detail using $TiCl_4$, BF_3, and $SnCl_4$ for 2 hr at −70 and −130°C. Methylene chloride and carbon disulfide were the solvents at −70°C and dichlorofluoromethane was used at −130°C. Under most conditions respectable conversions (20–100%) of soluble polymer were obtained and the number average molecular weights were in the range 1000–10,000 or perhaps even higher as can be judged from available viscosity data. According to infrared, NMR, and model compound evidence, the predominent repeat structure of these polymers is a dihydroanthracene unit:

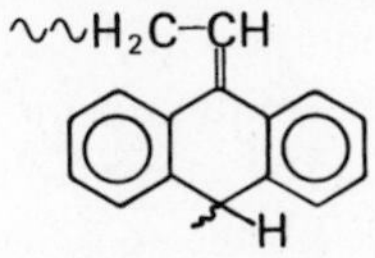

According to this evidence, 9-vinylanthracene polymerizes like a conjugated triene, 1,3,5-hexatriene, giving a 1,6 enchainment. The vinyl group and the center ring provide the three double bonds. The product softened at 210–230°C with some discoloration.

This polymer could be converted by treatment with CF_3COOH in CH_2Cl_2 solution to a product with essentially 9,10-dimethyleneanthracene repeat units:

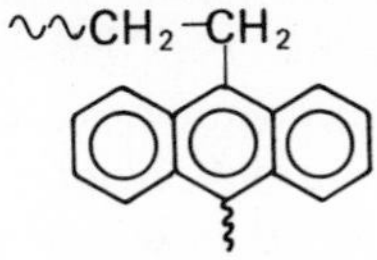

The latter product was insoluble and had a fusion range of 290–320°C.

12. *1-Vinylpyrene*

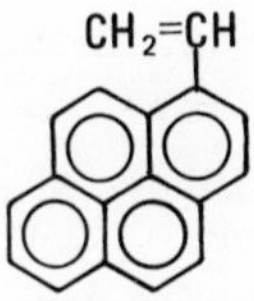

Flowers and Nichols (663) described the synthesis and cationic polymerization of this monomer by the use of $BF_3 \cdot OEt_2$ in benzene solvent probably at room temperature. The polymer was a white powder with a softening point of ~220°C. It could be molded into a disc at 160°C. Its dielectric constant was 3.2 and the power factor was 0.07% at 1 megacycle.

13. *9-[2-Vinylfluoroenylidenemethyl]fluorene*

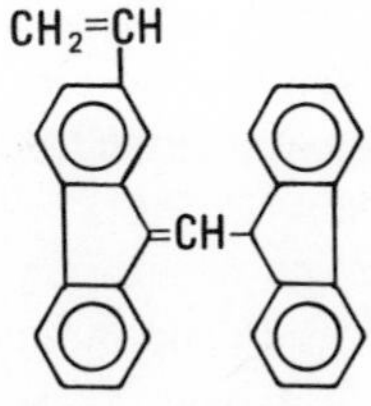

Förster and Maneke (664) synthesized and polymerized this monomer with

$BF_3 \cdot OEt_2$ in toluene solution at $-70°C$. They obtained an 80% yield of a soluble polymer whose decomposition point was $240°C$. According to its number average molecular weight, 3350, the average DP of the product was 9–10.

The most remarkable property of these polymers is their ability to give highly colored polymer anions upon treatment with bases. For example, the addition of NaOH to a DMSO or DMF solution of the polymer produces a deep red–violet color. The polymer anion can be converted to a yellowish-green polymer radical by treatment with $K_3 [Fe(CN)_6]$.

14. o-Divinylbenzene

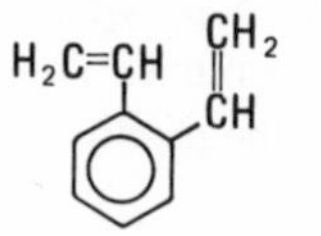

According to Aso and Kita (665) the $BF_3 \cdot OEt_2$ induced polymerization of o-divinylbenzene in the range -78 to $+20°C$ proceeds readily (10–70% yield) to give $\bar{M}_n \sim 3000$–14,000 ($[\eta]$ = 0.06–0.23) and the soluble product contains less than one double bond (pendant vinyl) per repeat unit. Infrared spectroscopy is the basis for the suggestion that, besides a conventional 1,2 polymerization, an intramolecular cyclization polymerization also proceeds:

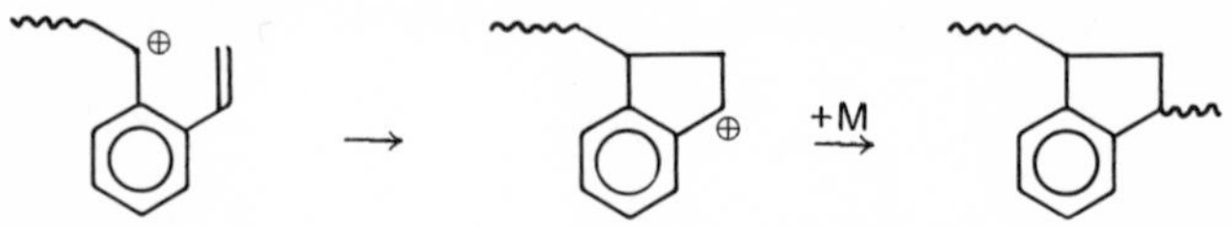

The ease of cyclization by different initiators followed the sequence: $SnCl_4 \cdot CCl_3COOH > TiCl_4 \cdot CCl_3COOH > BF_3 \cdot OEt_2$. Cyclization tendency was expressed by the cyclization constant $r_c = k_c/k_p$ where k_c and k_p are the rate constants for intramolecular cyclization and intermolecular propagation, respectively. Numerical values for r_c were obtained by infrared spectroscopy by determining the residual pendant double bond content in the polymer.

In later publications (666, 667) the effects of various solvents were investigated. Experiments showed that cyclization was favored in solvents with low dielectric constants, for example, CCl_4 and toluene. Very little cyclization occurred in highly polar solvents such as $C_6H_5NO_2$ and CH_3CN (667). It is remarkable that polymerization proceeded at all in these media. The number of initiator systems was also extended; the tendency for cyclization followed the sequence: $AlCl_3 > AlBr_3 > SnCl_4 > TiCl_4 > FeCl_3 > BF_3 \cdot OEt_2 > ZnCl_2$.

The driving force for cyclization (difference in activation energy between intramolecular cyclization and monomer addition) was found to be 5.3 and 6.2 kcal/mole with $BF_3 \cdot OEt_2$ and $(C_6H_5)_3CBF_4$, respectively (665, 667).

The effect of solvent polarity, nature of the initiator, and temperature on the cyclization rate was discussed with reference to a very speculative model (667).

$$CH_2=\underset{\underset{CH_2\text{———}CH_2}{|}}{\overset{\overset{C_6H_5}{|}}{C}} \quad \underset{}{\overset{\overset{C_6H_5}{|}}{C}}=CH_2$$

15. 2,5-Diphenylhexadiene-1,5

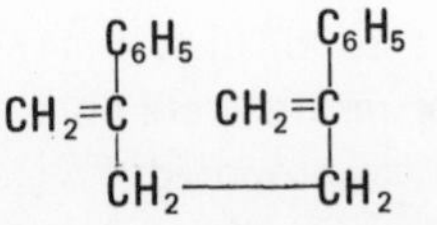

According to Marvel and Gall (668) this monomer can be polymerized by cationic initiators (BF$_3$) and the polymer (83% yield) exhibits only 5–10% residual unsaturation, m.p. = 185–210°C, η_{inh} = 0.083. A cyclized repeat structure was proposed for the polymer:

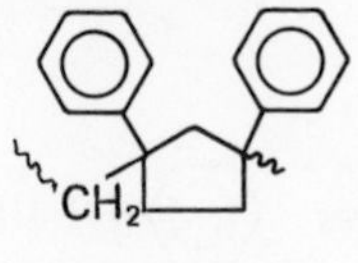

$$CH_2=\underset{\underset{CH_2}{|}}{\overset{\overset{C_6H_5}{|}}{C}} \qquad \underset{\underset{CH_2}{|}}{\overset{\overset{C_6H_5}{|}}{C}}=CH_2$$

16. 2,6-Diphenylheptadiene-1,6

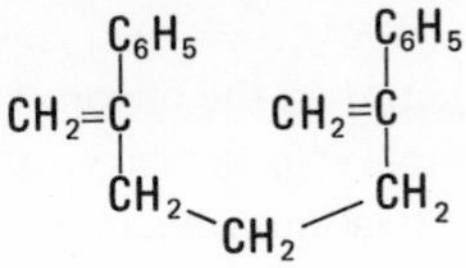

This monomer was polymerized with BF$_3$ in CH$_2$Cl$_2$ solution at −78°C by Field (669). The polymer (94% yield) had a melting point of 225°C and η_{inh} = 0.19. No residual unsaturation was detected by infrared in the high polymer. On the basis of this finding an inter-intramolecular polymerization mechanism was postulated:

The polymer could be cast or hot pressed to clear but brittle films that exhibited good thermal stability. It is of interest that free radical, anionic, and Ziegler-Natta polymerizations all gave the same repeat units. Field's findings were corroborated by Marvel and Gall (668). These authors obtained with BF$_3$ in CH$_2$Cl$_2$, 67% conversion, a white amorphous solid, m.p. = 240–270°C, η_{inh} = 0.145. With TiCl$_4$ in n-heptane they report 52% yield, m.p. = 115–130, η_{inh} = 0.051.

$$CH_2=\underset{\underset{CH_2}{|}}{\overset{\overset{C_6H_5}{|}}{C}} \qquad \underset{\underset{CH_2}{|}}{\overset{\overset{C_6H_5}{|}}{C}}=CH_2$$

17. 2,7-Diphenyloctadiene-1,7

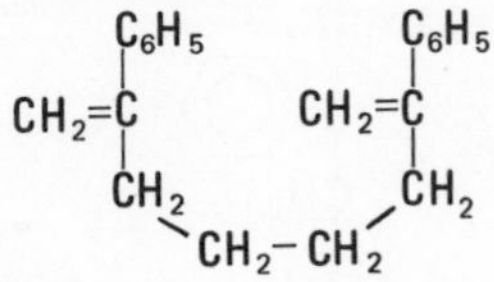

Marvel and Gall (668) polymerized this monomer with BF$_3$ in CH$_2$Cl$_2$ at −78°C and obtained 6% yield of a solid, m.p. = 115–150°C, η_{inh} = 0.021, with no residual unsaturation. Similar results were obtained with $TiCl_4$. A fully cyclized structure for the repeat unit was proposed:

18. and 19. cis- and trans-Stilbenes

Brachman and Plesch (634) studied the oligomerization of *cis*- and *trans*-stilbenes using $TiCl_4 \cdot CCl_3COOH$ in toluene, benzene, or *n*-hexane. In *n*-hexane at 0°C, *cis*-stilbene gives a mixture of various oligomers up to a pentamer. *Trans*-stilbene, in a mixture of benzene/hexane yields a mixture of dimers and trimers; in toluene dimers are formed and toluene is alkylated to yield 1,2-diphenyl-1-*p*-tolylethane.

Heublein and Agatha (684) studied the kinetics of the stilbene/$SbSl_5$/benzene or ethylene chloride system at 25°C. They observed that to obtain complete conversion, they had to charge stoichiometric amounts of $SbCl_5$ in regard to monomer when benzene was used, and two equivalents of $SbCl_5$ per mole of *trans*-stilbene when ethylene chloride was employed. The need for such high $SbCl_5$ concentrations and HCl liberation during polymerization indicated a polyalkylation of the aromatic ring. On the basis of elemental analysis, NMR and infrared spectroscopy, as well as thermogravimetry and molecular weight determinations ($\bar{M}_n$ ~ 1000 or $\overline{DP}$ ~ 5), the authors propose the following possibilities for their products:

These polymers are powdery materials soluble in benzene, chloroform, ethylene chloride, and so on.

20. p-*Methoxy*-trans-*Stilbene*

Heublein and Agatha (684) mention the polyreaction of this monomer by the use of $SbCl_5$ in ethylene chloride at 25°C (see under Stilbene).

21. p-*Dimethoxy*-trans-*Stilbene*

Heublein and Agatha (684) mention the polyreaction of this monomer by the use of $SbCl_5$ in ethylene chloride at 25°C (see under Stilbene).

22. p-*Nitro*-trans-*Stilbene*

Heublein and Agatha (684) mention the polyreaction of this monomer by the use of $SbCl_5$ in ethylene chloride at 25°C (see under Stilbene).

23. α-*Cyclopropyl*-p-*isopropylstyrene*

Ketley et al. (210) have polymerized this monomer with $AlBr_3$ in C_2H_5Cl at −78°C. According to (infrared and) NMR analysis the polymer ($\overline{M}_n$ = 4000-16000) consists exclusively of the repeat unit:

that is, the cyclopropyl group does not get involved in the polymerization.

24. *α-Cyclopropyl-2,4-dimethylstyrene*

Ketley et al. (210) polymerized this monomer with $AlBr_3$ in C_2H_5 Cl at $-78°C$. High conversion of polymer was obtained in 15 min. On the basis of NMR spectroscopy the structure of the polymer was proposed to consist of the repeat units:

(60%)   (30%)

No NMR evidence for a conventional vinyl enchainment was found.

25. *α-Cyclopropyl-p-fluorostyrene*

This monomer has been polymerized by Ketley et al. (210) with $AlBr_3$ in C_2H_5Cl at $-78°C$. Yield was > 90% in 15 min. On the basis of infrared and NMR spectroscopy the polymer structure was proposed to be comparable to the following units:

(20%)   (50%)   (30%)

CH$_2$=CH

O-CH$_2$-CH=CH$_2$

26. o-*Allyloxystyrene [Ally(o-vinylphenyl)ether]*

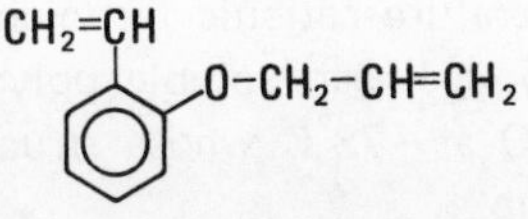

Kato and Kamogawa (670) polymerized this monomer in CH$_2$Cl$_2$ with BF$_3$·OEt$_2$ and SnCl$_4$ at $-78°$C and obtained good yields (20–97%) of soluble product. The intrinsic viscosities were in the 0.2–0.38 range. Infrared spectroscopy and the solubility behavior of these materials indicated that polymerization proceeded only by the vinyl unsaturation and that the allyl group was not involved in the reaction. Evidently, at low temperatures, the reactivity difference between these two functions is sufficiently large to produce a selective polymerization by the more reactive vinyl group.

Postpolymerization crosslinking of the unreacted allyl function by free radical sources (benzoyl peroxide and azobis-isobutyronitrile) was effective and crosslinked insoluble films could be obtained. Similarly, crosslinking of the soluble polymers by reaction with maleic anhydride plus benzoyl peroxide has been demonstrated.

Upon treatment of the soluble linear polymer with diethylaniline, the unreacted allyl groups undergo a Claisen rearrangement to give poly(2-hydroxy-3-allylstyrene):

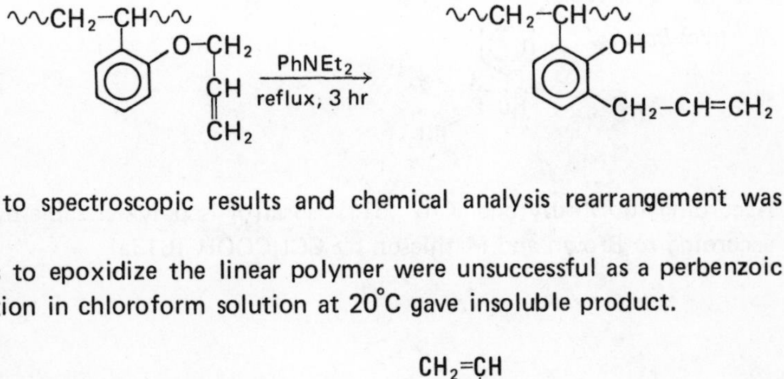

∿∿CH$_2$-CH∿∿

O-CH$_2$
CH
‖
CH$_2$

→ PhNEt$_2$
reflux, 3 hr

∿∿CH$_2$-CH∿∿

OH
CH$_2$-CH=CH$_2$

According to spectroscopic results and chemical analysis rearrangement was complete.

Attempts to epoxidize the linear polymer were unsuccessful as a perbenzoic acid oxidation in chloroform solution at 20°C gave insoluble product.

CH$_2$=CH

27. p-*Allyloxystyrene [Ally-(p-vinylphenyl)ether]*

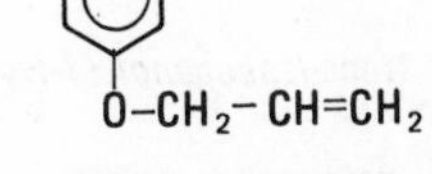

O-CH$_2$-CH=CH$_2$

Kato and Kamogawa (670) described the polymerization of this monomer. These authors employed BF$_3$·OEt$_2$ initiator in CH$_2$Cl$_2$ at $-78°$C. The purpose of these studies was to find conditions under which the vinyl group could be polymerized selectively without interference by the allyl ether function. While the authors failed to reach their aim by a free radical method, they were successful with low temperature cationic technique. According to solubility data and infrared spectroscopy, linear, soluble polymer is obtained readily in high yield with $BF_3 \cdot OEt_2O$ at $-78°C$ whose structure indicates polymerization only by the vinyl group.

Upon treatment with refluxing diethylaniline an essentially quantitative Claisen rearrangement occurs and poly(3-allyl-4-hydroxystyrene) is obtained:

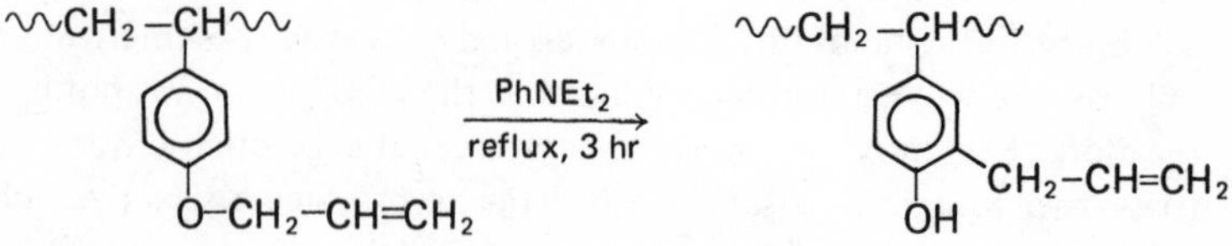

The linear polymer containing pendant allyl groups can be crosslinked by free radical sources (e.g., benzoyl peroxide) to give largely insoluble films. Insolubilization can also be achieved by the addition of maleic anhydride plus benzoyl peroxide.

Attempts to produce epoxides by oxidation with perbenzoid acid gave insoluble products.

28. trans-*Isosafrol*

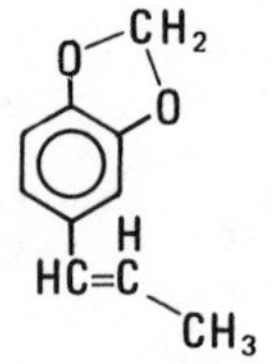

According to Whitby and Katz (611) isosafrol is polymerizable by $SbCl_5$ and according to Brown and Mathieson by CCl_3COOH (613a).

29. cis- *and* trans-*Isoeugenol (1-Hydroxy-2-methoxy-4-propenylbenzene)*

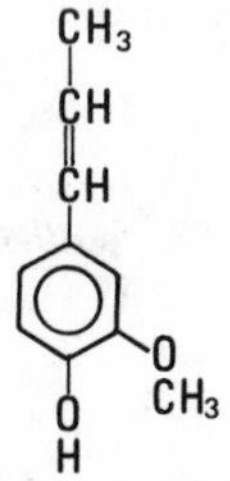

Brown and Mathieson (613a) mention the polymerization of isoeugenol by CCl_3COOH.

REFERENCES

1. V. N. Ipatieff and A. V. Grosse, *J. Am. Chem. Soc.*, **58**, 915 (1936).

2. J. P. Kennedy and R. M. Thomas, *J. Polymer Sci.*, **46**, 481 (1960), see also German Patent 704,038 to I. C. Farben. (1938).

3. H. L. Waterman and A. J. Tulleners, *Chim. Ind.*, Special No., 496-505 (June 1933).

4. V. N. Ipatieff and C. B. Linn, U.S. Patent 2,421,946 (1947).

5. V. N. Ipatieff, R. E. Schaad, and W. B. Shanley, *Science of Petroleum*, Vol. V, Part II, Oxford University Press, 1953, pp. 14-23.

6. W. A. Horne, *Ind. Eng. Chem.*, **42**, 2428 (1950).

7. T. Gramstad and R. N. Haszeldine, *J. Chem. Soc.*, 1957, 4069.

8. G. J. Karabatsos, Reprints, Div. Petroleum Chem., A.C.S. **16**, C22 (1971).

9. L. Schmerling, *J. Am. Chem. Soc.*, **67**, 1152 (1947).

10. L. Schmerling and E. E. Meisinger, *J. Am. Chem. Soc.*, **71**, 753 (1949).

11. L. Schmerling, in *Friedel Crafts and Related Reactions* Vol. II, G. A. Olah, Ed., Interscience, New York, 1964, p. 1133.

12. A. E. Evans, *The Reactions of Organic Halides in Solution*, Manchester University Press, Manchester, 1946, Table on p. 15.

13. J. L. Fry and G. J. Karabatsos in *Carbonium Ions*, Vol. II, G. A. Olah and P. R. Schleyer, Eds., Interscience, New York, 1970, Chap. 14.

14. G. J. Karabatsos and F. M. Vane, *J. Am. Chem. Soc.*, **85**, 729 (1963).

15. G. M. Kramer, *J. Am. Chem. Soc.*, **91**, 4819 (1969).

16. A. W. Nash, *J. Soc. Chem. Ind.*, **49**, 349 (1930).

17. A. W. Nash, H. M. Stanley, and A. R. Bowen, *J. Inst. Petrol. Technol.*, **16**, 830 (1930).

18. F. C. Hall and A. W. Nash, *J. Inst. Petrol. Technol.*, **23**, 679 (1937).

19. F. C. Hall and A. W. Nash, *J. Inst. Petrol. Technol.*, **24**, 471 (1930).

20. E. A. M. Törnquist in *Polymer Chemistry of Synthetic Elastomers*, Vol. I, J. P. Kennedy and E. A. M. Törnquist, Eds., Interscience, New York, 1968, Chap 2.

21. M. Fischer, German Patent 847,215, to BASF (filed 1943, published 1952).

22. A. W. Langer, *J. Macromol. Sci., Chem.*, **A4**, 775 (1970).

23. H. Bestian and K. Clauss, *Angew. Chem.*, **75**, 1068 (1963).

24. A. M. Butlerov and V. Gorianov, *J. Russ. Chem. Soc.*, **5**, 302 (1873); *ibid.*, **9**, 38 (1877).

25. A. V. Topchiev and Y. M. Pauskin, *Compounds of Boron Fluoride as Catalysts in Alkylation, Polymerization and Condensation Reaction*, Gostoptekhizdat, Moskow/-Leningrad, 1949.

26. C. M. Fontana in *The Chemistry of Cationic Polymerization*, P. H. Plesch, Ed., Macmillan, New York, 1963, Chap. 5.

27. A. B. Herschberger and R. G. Heiligmann, U.S. Patent 2,474,670 (1949).

28. C. M. Fontana, A. G. Oblad, and G. A. Kidder, U.S. Patent 2,525,787 (1950).

29. P. H. Emmett, Ed., *Catalysis*, Vols. I-VI. Reinhold, New York, 1958.

30. C. M. Fontana and G. A. Kidder, *J. Am. Chem. Soc.*, **70**, 3745 (1948).

31. C. M. Fontana, G. A. Kidder, and R. J. Herold, *Ind. Eng. Chem.*, **44**, 1688 (1952).

32. C. M. Fontana, R. J. Herold, E. J. Kinney, and R. C. Miller, *Ind. Eng. Chem.*, **44**, 2955 (1952).

33. C. M. Fontana in *Cationic Polymerisation and Related Complexes*, P. H. Plesch, Ed., Heffer and Sons, Cambridge, 1953, Chapt. II, p. 121.

34. C. M. Fontana, *J. Phys. Chem.*, **63**, 1167 (1959).

35. H. Kölbel, D. Klamann, and M. Boldt, *Brennstoff-Chem.*, **42**, 273 (1961).

36. A. D. Ketley and M. C. Harvey, *J. Org. Chem.*, **26**, 4649 (1961).

37. E. H. Immergut, G. Kollman, and A. Malatesta, *J. Polymer Sci.*, S57 (1961).

38. T. Szell and A. M. Eastham, *J. Chem. Soc. B*, **1966**, 30.

39. J. M. Roberts, Z. Katovic, and A. M. Eastham, *J. Polymer Sci.*, A-1, **8**, 3503 (1970).

40. E. I. Tinyakova, T. G. Zhuravleva, T. N. Kurengina, N. S. Kirikova, and B. A. Dolgoplosk, *Dokl. Akad. Nauk SSSR* **144**, 592 (1962).

41. J. P. Kennedy and A. W. Langer, *Advan. Polym. Sci.*, **3**, 508 (1964).

42. G. Henrici-Olive and S. Olive, *Polymer Lett.* **8**, 205 (1970).

43. R. A. Rhein, paper presented at Meeting American Chemical Society, Division of Rubber Chemistry, Washington, 1970, paper 10.

44. C. J. Norton, *Ind. Eng. Chem.*, **3**, 230 (1964).

45. C. J. Norton, *Chem. Ind. (London)*, 1962, 258.

46. E. H. Immergut, G. Kollmann, and A. Malatesta, *Makromol. Chem.*, **41**, 9 (1961).

47. G. F. Leyonmark and P. E. Hardy, U.S. Patent 2,569,383 (1951).

48. H. Hoog, J. Smittenberg, and G. H. Visser, *2nd World Petrol. Congr.*, **2**, 489, 1937.

49. A. V. Topchiev and Y. M. Paushkin, *Zh. Obshch. Khim., Mosk.*, **18**, 1537 (1948).

50. G. A. Olah, H. W. Quinn, and S. J. Kuhn, *J. Am. Chem. Soc.*, **82**, 426 (1960).

51. R. L. Meier, *J. Chem. Soc.* **1950**, 3656.

52. N. Yamazaki, T. Suminoe, and S. Kambara, *Makromol. Chem.*, **65**, 157 (1963).

53. H. R. Allcock, and A. M. Eastham, *Can. J. Chem.*, **41**, 932 (1963).

54. A. V. Topchiev, S. V. Zavogorodni, and Y. M. Paushkin *Boron Fluoride and its Compounds as Catalysts in Organic Chemistry*, Pergammon Press, London, 1959, p. 177.

55. A. Turner, Ph.D. thesis, University of Michigan, 1958.

56. R. G. W. Norris and J. P. Joubert, *J. Am. Chem. Soc.*, **49**, 873 (1927).

57. J. J. Leedertse, A. J. Tulleners, and H. I. Waterman, *Rec. Trav. Chim. Pays-Bas*, **53**, 715 (1934).

58. R. M. Thomas and H. C. Reynolds, U.S. Patent 2,387,784 (1945).

59. J. P. Kennedy, U.S. Patent 3,299,028 (1967).

60. J. P. Kennedy and R. M. Thomas, *Makromol. Chem.*, **53**, 28 (1962).

61. J. P. Kennedy, L. S. Minckler, Jr., G. G. Wanless, and R. M. Thomas, *J. Polymer Sci.*, A-2, 1441 (1964).

62. W. R. Edwards and N. F. Chamberlain, *J. Polymer Sci., A*, **1**, 2299 (1963).

63. B. E. Hudson, Jr., *Makromol. Chem.*, **94**, 1972 (1966).

64. J. P. Kennedy, W. W. Schulz, R. G. Squires, and R. M. Thomas, *Polymer*, **6**, 287 (1965).

65. J. P. Kennedy and R. M. Thomas, *Makromol. Chem.*, **64**, 1, (1963).

66. J. P. Kennedy, J. J. Elliot, and B. Groten, *Makromol. Chem.*, **77**, 26, (1964).

67. J. P. Kennedy and J. J. Elliott, unpublished data.

68. G. J. Quigley, unpublished work.
69. J. Martin and J. K. Gillham, *Polymer Preprints*, 12, 554 (1971).
70. R. B. Kelsey, British Patent 1,147,912 (1969).
71. J. P. Kennedy, L. S. Minckler, Jr., and R. M. Thomas, *J. Polymer Sci.*, *A*, 2, 367 (1964).
72. J. P. Kennedy, *J. Polymer Sci.*, *A*, 2, 381 (1964).
73. R. Bacskai, *J. Polymer Sci.*, *A-1*, 5, 619 (1967).
74. J. P. Kennedy, *Encyclop. Polymer Sci. Technol.*, 7, 754 (1967).
75. J. P. Kennedy, J. J. Elliot, and B. E. Hudson, Jr., *Makromol. Chem.*, 79, 109 (1964).
76. V. V. Maltsev, N. A. Plate, T. Azimov, and V. A. Kargin, *Polymer Sci. USSR*, 11, 248 (1969).
77. J. E. Goodrich and R. S. Porter, *Polymer Letters*, 2, 353 (1964).
78. G. G. Wanless and J. P. Kennedy, *Polymer*, 6, 111 (1965).
79. A. D. Ketley, *Polymer Letters*, 2, 827 (1964).
80. J. P. Kennedy and J. E. Johnston, *Advan. Polym. Sci.*, in press, 1975.
81. R. F. Killey, D. H. Welch, and C. A. McKenzie, U.S. Patent 2,815,334 (1957).
82. J. P. Kennedy, W. Naegele, J. J. Elliot, *Polymer Letters*, 3, 729 (165).
83. G. Sartori, H. Lammens, J. Siffert, and A. Bernard, *Polymer Lett.*, 9, 599 (1971).
84. A. D. Ketley and R. J. Ehrig, *J. Polymer Sci.*, *A*, 2, 4461 (1964).
85. H. Güterbock, *Polyisobutylen*, Springer, Berlin, 1959.
86. J. P. Kennedy and I. Kirshenbaum in *Vinyl and Diene Monomers*, E. C. Leonard, Ed., Wiley, New York, 1971, Part 2, chap. 3, p. 691.
87. P. H. Plesch in *The Chemistry of Cationic Polymerization*, P. H. Plesch, Ed., Macmillan, New York, 1963, chap. 4, p. 139.
88. R. H. Biddulph, W. R. Longworth, J. Penfold, P. H. Plesch, and P. P. Rutherford, *Polymer*, 1, 521 (1960).
89. A. M. Butlerov, *Liebigs Ann. Chem.*, 189, 47 (1877).
90. Y. T. Eidus and B. K. Nefedov, *Usp. Khim.*, 29, 833 (1960).
91. F. C. Whitmore and J. M. Church, *J. Am. Chem. Soc.*, 54, 3710 (1932).
92. E. D. Hughes and C. K. Ingold, *Trans. Faraday Soc.*, 37, 657 (1941).
93. C. K. Ingold, *Structure and Mechanism in Organic Chemistry*, Cornell University Press, Ithaca, New York, 1953.
94. H. C. Brown, *J. Chem. Soc.*, 1956, 1248.
95. J. F. Bunnett in *Progress in Chemistry*, E. Russell, Ed., Wiley, New York, 1969.
96. H. C. Brown and H. L. Berneis, *J. Am. Chem. Soc.*, 75, 10 (1953).
97. F. C. Whitmore, C. D. Wilson, J. V. Capinjola, C. O. Tonberg, G. H. Fleming, R. V. McGrew, and J. N. Cosby, *J. Am. Chem. Soc.*, 63, 2035 (1941).
98. O. Wichterle, Z. Laita and M. Pazlar, *Chem. Listy*, 49, 1612 (1955); and the same paper in *Collection Czech. Chem. Commun.*, 21, 614 (1956).
99. O. Wichterle, M. Kolinski and M. Marek, *Chem. Listy*, 52, 1049 (1958); and the same paper in *Collection Czech. Chem. Commun.*, 36, 2473 (1959).
100. E. T. Child and W. P. Doyle, paper presented at Symposium on Polymer Kinetics and Catalyst Systems: Part II, 44th Annual Meeting American Institute of Chemical Engineering, New York, 1961, Preprint 31.

101. K. Iimura, R. Endo, M. Takeda, *Bull. Chem. Soc. Japan*, 37, 874 (1964).

102. F. S. Dainton and G. B. B. M. Sutherland, *J. Polymer Sci.*, 4, 37 (1949).

103. M. St. C. Flett and P. H. Plesch, *J. Chem. Soc.*, 1952, 3353.

104. N. Tokura, M. Matsuda, and I. Shirai, *Bull. Chem. Soc. Japan*, 35, 371 (1962).

105. Y. Imanishi, Ph. D. Thesis, Kyoto University, 1964, p. 94.

106. P. H. Biddulph, P. H. Plesch, and P. P. Rutherford, *J. Chem. Soc.*, 1965, 275.

107. H. Cheradame and P. Sigwalt, *Compt. Rend.*, 259, 4273 (1964).

108. M. Szwarc, *Carbanions, Living Polymers and Electron Transfer Processes,* Interscience, New York, 1968.

109. T. Shimomura, K. J. Tölle, J. Smid, and M. Szwarc, *J. Am. Chem. Soc.,* 89, 796 (1967).

110. J. P. Kennedy and R. M. Thomas, *J. Polymer Sci.*, 55, 331 (1961).

111. J. P. Kennedy and R. M. Thomas, *Advan. in Chem. Ser.*, 34, 111 (1962).

112. C. J. Patton, P. H. Plesch, and P. P. Rutherford, *J. Chem. Soc.*, 1964, 2856.

113. J. H. Beard, P. H. Plesch, and P. P. Rutherford, *J. Chem. Soc.*, 1964, 2566.

114. J. P. Kennedy, *J. Macromol. Sci., Chem.*, A6, 329 (1972).

115. H. Cheradame and P. Sigwalt, *Bull. Soc. Chim. France*, 3, 843 (1970).

116. H. Cheradame and P. Sigwalt, *Compt. Rend.*, 259, 4273 (1964).

117. N. A. Ghanem and M. Marek, *European Polymer J.*, 8, 999 (1972).

118. M. Chmelir, M. Marek and O. Wichterle, *J. Polymer Sci., C*, 16, 833 (1967).

119. J. M. Solich, M. Chmelir, and M. Marek, *Collection Czechoslovak Chem. Commun.*, 34, 2611 (1969).

120. M. Marek and M. Chmelir, *J. Polymer Sci., C*, 23, 223 (1968).

121. M. Marek and M. Chmelir, *J. Polymer Sci., C*, 22, 177 (1968).

122. P. Lopour and M. Marek, *Makromol. Chem.*, 134, 23 (1970).

123. M. Chmelir and M. Marek, *Collection Czech. Chem. Commun.*, 32, 3047 (1967).

124. P. Lopour, J. Pecka, and M. Marek, *Makromol. Chem.*, 151, 139 (1972).

125. J. P. Kennedy, S. Sivaram, and N. Desai, *J. Org. Chem.*, 38, 2262 (1973).

126. G. A. Olah and de Member *J. Am. Chem. Soc.*, 92, 2562 (1970) and references therein.

127. P. Trivedi and J. P. Kennedy, unpublished data, 1972.

128. M. Marek and L. Toman, paper presented at International Symposium on Macromolecules, Helsinki, 1972; *Polymer Preprints*, 2, 601 (1972).

129. Y. Imanishi, T. Higashimura, and S. Okamura, *Kobunshi Kagaku*, 18, 333 (1961).

130. J. Feeney, A. K. Holliday, and F. J. Marsden, *J. Chem. Soc.*, 1961, 356.

131. R. G. W. Norris and K. E. Russell, *Trans. Faraday Soc.*, 48, 91 (1952).

132. P. H. Plesch in *The Chemistry of Cationic Polymerization,* P. H. Plesch, Ed., Macmillan, New York, 1963, p. 174.

133. R. H. Biddulph and P. H. Plesch, *J. Chem. Soc.*, 1960, 3913.

134. R. F. Bauer, R. T. LaFlair, and K. E. Russell, *Can. J. Chem.*, 48, 1251 (1970).

135. R. F. Bauer and K. E. Russell, *J. Polymer Sci., A*, 9, 1451 (1971).

136. J. P. Kennedy, A. Shinkawa, and F. Williams, *J. Polymer Sci., A-1*, 9, 1551 (1971).

137. R. B. Taylor and F. Williams, *J. Am. Chem. Soc.*, 91, 3278 (1969).

137.a. J. J. Sparapany, *J. Am. Chem. Soc.*, **88**, 1357 (1966).

138. F. L. Dalton, *Polymer*, **6**, 1 (1965).

139. I. A. Bartlett and F. L. Dalton, *Polymer*, **7**, 107 (1966).

140. Z. Zlamal, L. Ambroz, and K. Vesely, *J. Polymer Sci.*, **24**, 285 (1957).

141. M. Otto and M. Müller-Cunradi, German Patent 641,281 (1931).

142. R. M. Thomas, W. J. Sparks, P. K. Frölich, M. Otto, and M. Müller-Cunradi, *J. Am. Chem. Soc.*, **62**, 276 (1940).

143. P. H. Plesch, *J. Chem. Soc.*, **1964**, 104.

144. Z. Zlamal and A. Kazda, *J. Polymer Sci., A-1*, **4**, 1783 (1966).

145. J. P. Kennedy and R. M. Thomas, *J. Polymer Sci.*, **45**, 227 (1960).

146. J. P. Kennedy and R. M. Thomas, *J. Polymer Sci.*, **45**, 229 (1960).

147. J. P. Kennedy and R. M. Thomas, *J. Polymer Sci.*, **46**, 233 (1960).

148. J. P. Kennedy and R. M. Thomas, *J. Polymer Sci.*, **46**, 481 (1960).

149. J. P. Kennedy and R. M. Thomas, *J. Polymer Sci.*, **49**, 189 (1961).

150. J. P. Kennedy and R. M. Thomas, *J. Polymer Sci.*, **55**, 311 (1961).

151. J. P. Kennedy, I. Kirshenbaum, R. M. Thomas, and D. C. Murray, *J. Polymer Sci., A*, **1**, 313 (1963).

152. I. Kirshenbaum, J. P. Kennedy, and R. M. Thomas, *J. Polymer, Sci., A*, **1**, 789 (1963).

153. J. P. Kennedy and R. M. Thomas, *Advan. Chem. Ser.*, **34**, 111 (1962).

154. J. P. Kennedy and F. P. Baldwin, *Polymer*, **6**, 237 (1965).

155. J. P. Kennedy and R. G. Squires, *Polymer*, **6**, 579 (1965).

156. J. P. Kennedy and R. G. Squires, *J. Macromol. Sci.*, **A1**, 805 (1967).

157. J. P. Kennedy and R. G. Squires, *J. Macromol. Sci.*, **A1**, 831 (1967).

158. J. P. Kennedy and R. G. Squires, *J. Macromol. Sci.*, **A1**, 847 (1967).

159. J. P. Kennedy and R. G. Squires, *J. Macromol. Sci.*, **A1**, 861 (1967).

160. J. P. Kennedy, S. Bank, and R. G. Squires, *J. Macromol. Sci.*, **A1**, 961 (1967).

161. J. P. Kennedy, S. Bank, and R. G. Squires, *J. Macromol. Sci.*, **A1**, 977 (1967).

162. J. P. Kennedy and R. G. Squires, *J. Macromol. Sci.*, **A1**, 995 (1967).

163. K. Vesely, *J. Polymer Sci.*, **52**, 277 (1961).

164. Z. Zlamal, J. Ambroz, and L. Ambroz, *Collection Czech. Chem. Commun.*, **21**, 586 (1956).

165. K. Vesely, *J. Polymer Sci.*, **30**, 375 (1958).

166. L. Ambroz and Z. Zlamal, *J. Polymer Sci.*, **30**, 381 (1958).

167. Z. Zlamal and L. Ambroz, *J. Polymer Sci.*, **29**, 595 (1958).

168. Z. Zlamal and A. Kazda, *J. Polymer Sci.*, **53**, 203 (1961).

169. Z. Zlamal and A. Kazda, *J. Polymer Sci., A1*, 3199 (1963).

170. Z. Zlamal, A. Kazda, and L. Ambroz, *J. Polymer Sci., A-1*, **4**, 367 (1966).

171. P. J. Flory in *Principles of Polymer Chemistry,* Cornell University Press, Ithaca, New York, 1953, p. 218.

172. H. C. Brown and R. S. Fletcher, *J. Am. Soc.*, **71**, 1845 (1949).

173. J. Hoffman, *J. Org. Chem.*, **29**, 3039 (1964).

174. C. Horrex and F. T. Perkins, *Nature*, **163**, 486 (1949).

175. J. P. Kennedy in *Polymer Chemistry of Synthetic Elastomers,* J. P. Kennedy and E. G. M. Törnquist, Eds., Interscience, New York, 1968, p. 291.

176. S. Cesca, A. Priola, and G. Ferraris, *Makromol. Chem.,* **156**, 325 (1972).

177. J. P. Kennedy, Belgan Patent 663,319 (April 30, 1965).

178. J. P. Kennedy and F. P. Baldwin, Belgan Patent (April 30, 1965).

179. J. P. Kennedy and G. E. Milliman, *Advan. Chem. Ser.,* **91**, 287 (1969).

179.a. J. P. Kennedy, U.S. Patent (1968).

179.b. J. P. Kennedy and S. Sivaram, *J. Macromol. Sci., Chem.,* **A7**, 969 (1973).

179.c. J. P. Kennedy, Intl. Union Pure Appl. Chem., *Macromol. Chem.,* **8**, 179 (1972).

179.d. M. Baccaredda, M. Bruzzone, S. Cesca, M. DiMaina, G. Ferraris, P. Giusti, P. L. Magagnini, and A. Priola, *Chim. Ind.,* **55**, 109 (1973).

180. Z. Zlamal in *Kinetics and Mechanism of Polymerization Series,* Vol. I, G. E. Ham, Ed., Dekker, New York, 1969, p. 231 chap. 6, Part. II.

181. P. H. Plesch, Appendix on p. 197 of ref. 63., also ref. 129.

182. V. M. Sobolev, Y. N. Prokofev, and G. N. Spiridinova, paper presented at International Symposium on Macromolecular Chemistry, Preprint P564, Prague, 1965.

183. V. M. Sobolev, Y. N. Prokofev, and G. N. Spiridinova, paper presented at International Symposium on Macromolecular Chemistry, Preprint P565, Prague, 1965.

184. O. Wichterle, M. Marek, and J. Trekoval, paper presented at IUPAC Symposium on Macromolecules, Wiesbaden, Sect. III A13, 1959.

185. O. Wichterle, M. Marek, and J. Trekoval, *J. Polymer Sci.,* **53**, 281 (1961).

186. M. Marek, Czechoslovak Patent 88,879 (1959).

187. Y. Imanishi, R. Yamamoto, T. Higashimura, J. P. Kennedy, and S. Okamura, *J. Macromol. Sci., (Chem.),* **A1**, 877 (1967).

188. R. Yamamoto, Y. Imanishi, and T. Higashimura, *Kobunshi Kagaku,* **24**, 397 (1967).

189. R. Yamamoto, Y. Imanishi, and T. Higashimura, *Kobunshi Kagaku,* **24**, 405 (1967).

190. R. Yamamoto, Y. Imanishi, and T. Higashimura, *Kobunshi Kagaku,* **24**, 412 (1967).

191. R. Yamamoto, Y. Imanishi, and T. Higashimura, *Kobunshi Kagaku,* **24**, 479 (1967).

192. R. Yamamoto, Y. Imanishi, and T. Higashimura, *Kobunshi Kagaku,* **24**, 486 (1967).

193. M. Miyoshi, S. Uemura, S. Tsuchiya, and O. Kato, U.S. Patent 3,402,146 (1968).

194. British Patent 1,056,730 (1967).

195. British Patent 1,183,118 (1970).

196. British Patent 1,233,557 (1971).

196.a. S. Matsushima, German Patent 2,163,956 to Sumitomo Chemical Co. (1927).

196.b. S. Matsushima and K. Ueno, German Patent 2,163,957 to Sumitomo Chemical Co. (1972).

197. K. S. B. Addecott, L. Mayor, and C. N. Turton, *European Polymer J.,* **3**, 601 (1957).

198. N. Yamada, K. Shimada, and T. Hayashi, *Polymer Letters,* **4**, 477 (1966).

199. N. Yamada, K. Shimada, and T. Takemura, U.S. Patent 3,326,879 (1967).

199.a. M. M. Hamada and J. H. Gary, *Polymer Preprints,* **9**, 413 (1968).

199.b. A. V. Topchiev, B. A. Krentsel, N. F. Bogolomova, and Y. Y. Goldfarb, *Dokl. Akad. Nauk SSSR,* **111**, 121 (1956).

199.c. B. A. Krentsel, L. G. Siderova, and A. V. Topchiev, *J. Polymer Sci. C,* **3**, 3 (1964).

199.d. R. Bacskai and S. J. Lapporte, *J. Polymer Sci. A,* **1**, 2225 (1966).

200. French Patent 838,064 to Standard Oil Development Co., 1939.

201. O. E. Van Lohuizen and K. S. De Vries, *J. Polymer Sci.*, C, 16, 3943 (1968).

202. Y. Imanishi, H. Imamura, and T. Higashimura, *Kobunshi Kagaku*, 27, 251 (1970).

202.a. J. P. Kennedy and J. K. Gillham, *Polymer Preprints*, 12, 463 (1971).

203. M. Marek, M. Roosova, and D. Doskocilova, *J. Polymer Sci., C*, 16, 971 (1967).

204. M. Marek and J. Peka, *J. Macromol. Sci., Chem.*, A5, 211 (1971).

205. H. Yamaoka, J. P. Kennedy, Y. Yamane, and K. Hayashi, unpublished data, Kyoto, 1965.

206. T. Takahashi and I. Yamashita, *Kogyo Kagaku Zasshi*, 68, 869 (1965).

207. T. Takahashi, *J. Polymer Sci.*, A-1, 6, 403 (1968).

208. T. Takahashi, *J. Polymer Sci.*, A-1, 6, 3327 (1968).

209. T. Takahashi and I. Yamashita, *Polymer Letters*, 3, 251 (1965).

210. A. D. Ketley, A. J. Berlin, and L. P. Fisher, *J. Polymer Sci.*, A-1, 5, 227 (1967).

211. V. M. Zhulin, A. R. Volchek, M. G. Gonikberg, A. S. Shashkov, and S. V. Zotova, *Vysokomol. Soedin., Ser. A*, 14, 1484 (1972).

212. A. D. Ketley, A. J. Berlin, and L. P. Fisher, *J. Polymer Sci.*, A-1, 5, 227 (1967).

213. J. P. Kennedy, J. J. Elliott, and P. E. Butler, *J. Macromol. Chem.*, A2, 1415 (1968).

214. C. Pinazzi and J. Brossas, *Compt. Rend. Acad. Sci. Paris*, 261, 3410 (1965).

215. C. P. Pinazzi and J. Brossas, *Makromol. Chem.*, 122, 105 (1969).

216. C. P. Pinazzi, J. C. Brosse, A. Pleurdeau, and J. Brossas, *Polymer Preprints*, 13, 445 (1972).

216.a. C. P. Pinazzi and J. Brossas, *Makromol. Chem.*, 147, 15 (1971).

217. F. Hoffman, *Chem. Ztg.*, 57, 5 (1933).

218. J. Boor, E. A. Youngman, and M. Dimbat, *Makromol. Chem.*, 90, 26 (1966).

219. G. J. Schmitt and C. Schuerch, *J. Polymer Sci.*, 49, 287 (1961).

220. O. Ohara, *Kobunshi Kagaku*, 28, 285 (1971).

221. W. J. Roberts and A. R. Day, *J. Am. Chem. Soc.*, 72, 1266 (1950).

222. J. P. Kennedy, unpublished data.

223. J. P. Kennedy, J. J. Elliot, and W. Naegele, *J. Polymer Sci., A*, 2, 5029 (1964).

224. W. Marconi, S. Cesa, G. D. Fortuna, *Polymer Letters*, 2, 301 (1964).

225. G. B. Butler and M. L. Miles, *J. Polymer Sci., A*, 3, 1609 (1965).

226. L. E. Ball and H. J. Harwood, *Polymer Preprints*, 2, 59 (1961).

227. G. B. Butler, M. L. Miles, and W. S. Brey, *J. Polymer Sci., A*, 3, 723 (1965).

228. C. Aso, T. Kunitake, and H. Uchio, German Patent 1,955,519 (1970).

229. G. B. Butler and M. L. Miles, *Polymer Eng. Sci.*, 6, 1 (1966).

230. C. Pinazzi and J. C. Brosse, *Compt. Rend. Acad. Sci., Paris*, 266, 1136 (1968).

230.a. R. L. Lipnick and E. W. Garbisch, Jr., *J. Am. Chem. Soc.*, 95, 6370 (1973).

231. W. J. Bailey in *Vinyl and Diene Monomers* Part Two, E. C. Leonard, Eds., Interscience, New York, 1971, p. 757.

232. W. M. Saltman, *Encycl. Polymer Sci. Technol.*, 2, 678 (1965).

233. W. Cooper in *The Chemistry of Cationic Polymerization*, P. H. Plesch, Ed., Macmillan, New York, 1963, p. 249.

234. L. A. Eldib, *Hydrocarbon Process, Petrol. Refiner*, 42, 165 (1963).

318 THE ORGANIZATION OF CATIONICALLY POLYMERIZABLE OLEFINS

235. C. S. Marvel, R. Gilkey, C. R. Morgan, J. F. Noth, R. D. Rands, Jr., and C. H. Young, *J. Polymer Sci.*, 6, 483 (1951).

236. T. E. Ferrington and A. V. Tobolsky, *J. Polymer Sci.*, 31, 25 (1958).

237. W. S. Richardson, *J. Polymer Sci.*, 13, 325 (1954).

238. H. Weber, *Angew. Chem.*, 72, 274 (1960).

238.a.T. Ashara and H. Kise, *Bull. Chem. Soc. Japan*, 40, 2739 (1966).

239. T. Asahara and H. Kise, *Bull. Chem. Soc. .Japan*, 40, 1941 (1967).

240. T. Asahara and H. Kise, *Bull. Chem. Soc. Japan*, 40, 2664 (1967).

241. F. Runge and R. Aust, *Chem. Technol. (Berlin)*, 9, 701 (1957).

242. The last of a series of twenty publications: N. G. Gaylord and M. Svestka, *J. Polymer Sci., B*, 7, 55 (1969).

243. N. G. Gaylord, I. Kössler, M. Stolka, and J. Vodehnal, *J. Polymer Sci., A*, 2, 3969 (1964).

244. N. G. Gaylord, I. Kössler, M. Stolka, and J. Vodehnal, *J. Am. Chem. Soc.*, 85, 641 (1963).

245. M. Stolka, J. Vodehnal, and I. Kössler, *J. Polymer Sci., A*, 2, 3987 (1964).

246. N. G. Gaylord, I. Kössler, and M. Stolka, *J. Macromol. Sci., Chem.*, A2, 421 (1968).

247. H. Krauserova, K. Mach, B. Matyska, and I. Kössler, *J. Polymer Sci., C*, 16, 469 (1967).

248. B. Matyska, K. Mach, N. Vodehnal, and I. Kössler, *Collection Czech. Chem. Commun.*, 30, 2569 (1965).

249. N. G. Gaylord, B. Matyska, K. Mach, and J. Vodehnal, *J. Polymer Sci., A-1*, 4, 2493 (1966).

250. I. Kössler, M. Stolka, and K. Mach, *J. Polymer Sci., C*, 4, 977 (1964).

251. H. Sinn, H. Winter, and W. Tirpitz, *Makromol. Chem.*, 48, 59 (1961).

252. A. Takahashi, K. Takahashi, T. Hirose, and S. Kambara, *Polymer Letters*, 5, 415 (1967).

253. M. Gippin, *Ind. Eng. Prod. Res. Develop.*, 1, 32 (1962).

254. M. Gippin, *Ind. Eng. Prod. Res. Develop.*, 4, 160 (1965).

255. J. P. Kennedy, unpublished research, 1971.

256. H. Z. Friedlander, *J. Polymer Sci., C*, 4, 1291 (1963).

257. S. H. Pinner, in *The Chemistry of Cationic Polymerization*, P. H. Plesch, Ed., Macmillan, New York, 1963, p. 611.

258. V. Stepan, J. Vodehnal, I. Kössler, and N. G. Gaylord, *J. Polymer Sci., A-1*, 5, 503 (1967).

259. W. J. Bailey in *Vinyl and Diene Monomers*, Part Two, E. C. Leonard, Ed., Interscience, New York, 1971, p. 997.

260. G. Holden and R. H. Mann, *Encycl. Chem. Technol.*, 12, 64 (1967).

261. J. Vohlidal, V. Bohackova, and B. Matyska, *J. Chim. Phys.*, 1972, 556.

262. V. Bohckova, J. Polacek, and H. Benoit, *J. Chem. Phys.*, 66, 197 (1969).

263. E. Drahoradova, D. Doskocilova, and B. Matyska, *Collection Czech. Chem. Commun.*, 36, 1301 (1971).

264. J. Vohlidal, B. Matyska, and V. Bohackova, *J. Chim. Phys.*, 1972, 1348.

265. N. G. Gaylord and M. Svestka, *Polymer Letters*, 7, 55 (1969).

266. N. G. Gaylord and M. Svestka, *Polymer Letters*, 7, 455 (1969).

267. J. P. Varion and P. Sigwalt, *Bull. Soc. Chim. France*, 1971, 569.

268. B. Matyska, M. Svestka and K. Mach, *Coll. Czech. Chem. Commun.*, 31, 659 (1966).

269. G. S. Whitby and W. Gallay, *Can. J. Res.*, 6, 280 (1932).

270. N. G. Gaylord and I. Kössler, *J. Polymer Sci.*, *C*, 16, 3097 (1968).

271. D. W. Young and H. D. Hineline, U.S. Patent 2,521,437 to S. O. Dev. Co. (1950).

272. R. M. Thomas and E. J. Dahlke, U.S. Patent 2,291,510 to Jasco, Inc. (1942).

273. R. M. Thomas and W. J. Sparks, U.S. Patent, 2,356,130 to Jasco, Inc. (1944).

274. W. J. Sparks and D. W. Young, U.S. Patent 2,609,359 to S. O. Dev. Co. (1952).

275. W. J. Sparks and R. M. Thomas, U.S. Patent 2,585,867 (1952).

276. M. Kamachi, K. Matsumura, and S. Murahashi, *Polymer J.*, 1, 499 (1970).

277. D. Cuzin, Y. Chauvin, and G. Lefebre, *European Polymer J.*, 3, 367 (1967).

278. D. Cuzin, Y. Chauvin, and G. Lefebre, U.S. Patent 3,476,731 (Nov. 4, 1969).

279. Y. Chavin, D. Cuzin, and G. Lefebre, paper presented at Symposium on Kinetics and Mechanisms of Polyreactions, International Symposium on Macromolecular Chemistry, Preprints, 2, 321 (1969).

280. K. Hara, Y. Imanishi, T. Higashimura, and M. Kamachi, *J. Polymer Sci.*, *A-1*, 9, 2933 (1971).

281. F. B. Moody, *Polymer Preprints*, 2, 285 (1961).

282. J. F. Jones, *J. Polymer Sci.*, 33, 513 (1958).

283. C. S. Marvel and P. E. Kiener, *J. Polymer Sci.*, 61, 311 (1962).

284. C. S. Marvel and C. C. L. Hwa, *J. Polymer Sci.*, 45, 25 (1960).

285. P. de Radzitski and G. Smets, *Bull. Soc. Chim. Belg.*, 62, 320 (1953).

286. P. de Radzitski and M. C. de Wilde, and G. Smets, *J. Polymer Sci.*, 13, 477 (1954).

287. A. D. Aliev, B. A. Krentsel, and M. V. Shishkina, *Azerb. Khim. Zh.*, 1, 4 (1967).

288. A. D. Aliev, B. A. Krentsel, G. M. Mamediarov, I. P. Solomatina, and E. P. Tiurina, *European Polymer J.*, 7, *1721* (1971).

289. G. M. Mamedyarov, A. D. Aliev, and B. A. Krentsel, *Fiziol. Opt. Aktiv. Polim. Veshchestra*, 1971, 24, *Chem. Abstr.*, 77, (11) 6 (1972) 140607t.

290. F. C. Foster, *J. Polymer Sci.*, 5, 369 (1950).

291. C. G. Overberger and V. G. Kamath, *J. Am. Chem. Soc.*, 85, 446 (1963).

291.a. V. L. Bell, *J. Polymer Sci.*, *A*, 2, 5291 (1964).

292. D. E. Applequist and J. D. Roberts, *J. Am. Chem. Soc.*, 78, 4012 (1956).

293. C. C. Wu and R. W. Lenz, *Polymer Preprints*, 12, 209 (1971).

294. H. Staudinger and H. A. Burson, *Ann.*, 447, 110 (1926).

295. H. A. Burson and H. Staudinger, *Ind. Eng. Chem.*, 18, 381 (1926).

296. B. Eisler, A. Wassermann, S. D. Farnsworth, D. Kedrick, and R. Schnurmann, *Nature*, 168, 459 (1951).

297. B. Eisler, A. Wassermann, S. D. Farnsworth, D. Kedrick, and R. Schnurmann, *J. Polymer Sci.*, 3, 157 (1952).

298. H. Cheradame and J. P. Vairon, *Peintures, Pigments, Vernis*, 42, 353 (1966).

299. J. P. Vairon and P. Sigwalt, *Bull. Soc. Chim. France*, 1964, 482.

300. J. P. Vairon and P. Sigwalt, *Bull. Soc. Chim. France*, 1971, 559.

301. H. Cheradame, J. P. Vairon, and P. Sigwalt, *European Polymer J.*, **4**, 13 (1968).

302. G. Sauvet, J. P. Vairon, and P. Sigwalt, *J. Polymer Sci.*, *A-1*, **7**, 983 (1969).

303. G. Sauvet, J. P. Vairon, and P. Sigwalt, *Compt. Rend.*, **265**, 1090 (1967).

304. C. Aso, T. Kunitake, and S. Ushio, *Chem. High Polymers*, **19**, 734 (1962).

305. T. Kunitake, C. Aso, and K. Ito, *Makromol. Chem.*, **97**, 40 (1966).

306. C. Aso, T. Kunitake, K. Ito, and Y. Ishimoto, *Polymer Letters*, **4**, 701 (1966).

307. C. Aso, T. Kunitake, and Y. Ishimoto, *J. Polymer Sci.*, *A-1*, **6**, 1163 (1968).

308. C. Aso, T. Kunitake, and Y. Ishimoto, *J. Polymer Sci.*, *A-1*, **6**, 1175 (1968).

309. M. A. Bonin, W. R. Busler, and F. Williams, *J. Am. Chem. Soc.*, **87**, 199 (1965).

310. A. Schmidt and G. Kolb, *Makromol. Chem.*, **130**, 90 (1969).

311. Z. Momiyama, Y. Imanishi, and T. Higashimura, *Kobunshi Kagaku*, **23**, 56 (1966).

312. Y. Imanishi, S. Kohjiya, Z. Momiyama, and T. Higashimura, *Kobunshi Kagaku*, **23**, 119 (1966).

313. Y. Imanishi, S. Kohjiya, and S. Okamura, *J. Macromol. Sci., Chem.*, **A2**, 471 (1968).

314. M. J. Hayes and D. C. Pepper, *Proc. Roy. Soc. (London), Ser. A.*, **263**, 63 (1961).

315. C. Aso, T. Kunitake, Y. Ishimoto, *Bull. Chem. Soc. Japan*, **40**, 2894 (1967).

316. C. Aso and O. Ohara, *Makromol. Chem.*, **109**, 161 (1967).

317. C. Aso and O. Ohara, *Makromol. Chem.*, **127**, 78 (1969).

318. S. Kohjiya, Y. Imanishi, and S. Okamura, *J. Polymer Sci.*, *A-1*, **6**, 809 (1968).

319. C. Aso and O. Ohara, *Kobunshi Kagaku*, **23**, 895 (1966).

320. R. S. Mitchell, S. McLean, and J. E. Guillet, *Macromol.*, **1**, 417 (1968).

321. H. Maines and J. H. Day, *Polymer Letters*, **1**, 347 (1963).

322. D. A. Frey, M. Hasegawa, and C. S. Marvel, *J. Polymer Sci.*, *A*, **1**, 2057 (1963).

323. G. LeFebvre and F. Dawans, *J. Polymer Sci.*, A, **2**, 3277 (1964).

324. F. Dawans, *J. Polymer Sci.*, A, **2**, 3297 (1964).

325. Y. Imanishi, K. Matsuzaka, T. Yamane, S. Kohjiya, and S. Okamura, *J. Macromol. Sci., Chem.*, **A3**, 249 (1969).

326. Y. Imanishi, T. Yamane, S. Kohjiya, and S. Okamura, *J. Macromol. Sci., Chem.*, **A3**, 223 (1969).

327. W. J. Bailey and J. C. Grossens, paper presented at Meeting *American Chemical Society*, Division of Polymer Chemistry, Cincinnati, March 29-April 7, 1955, Paper 2R.

328. K. Mabuchi, T. Saegusa, and J. Furukawa, *Makromol. Chem.*, **81**, 112 (1965).

329. Y. Imanishi, K. Matsuzaki, S. Kohjiya, and S. Okamura, *J. Macromol. Sci., Chem.*, **A3**, 237 (1969).

330. M. A. S. Mondal and R. N. Young, *European Polymer J.*, **7**, 523 (1971).

331. T. H. Bates, J. V. F. Best, and T. F. Williams, *J. Chem. Soc.*, **1962**, 1531.

332. D. A. Frey, M. Hasegawa, and C. S. Marvel, *J. Polymer Sci.*, *A*, **1**, 2057 (1963).

333. L. A. Errede, J. M. Hoyt, and R. S. Gregorian, *J. Am. Chem. Soc.*, **82**, 5224 (1960).

334. E. Maréchal, *Compt. Rend.*, **260**, 6898 (1965).

335. G. S. Whitby and M. Katz, *J. Am. Chem. Soc.*, **50**, 1160 (1928).

336. S. Cesa, A. Roggero, N. Palladino, and A. DeChirico, *Makromol. Chem.*, **136**, 23 (1970).

337. S. Cesca, A. Priola, A. De Chirico, and G. Santi, *Makromol. Chem.*, **143**, 211 (1971).

338. C. F. H. Tipper and D. A. Walker, *J. Chem. Soc.*, **1959**, 1352.

339. V. N. Ipatieff, H. Pines, and L. Schmerling, *J. Org. Chem.*, **5**, 253 (1940).

340. A. V. Grosse and V. N. Ipatieff, *J. Org. Chem.*, **2**, 447 (1937).

341. C. P. Pinazzi, J. Brossas, J. C. Brosse, and A. Pleurdeau, *Makromol. Chem.*, **144**, 155 (1971).

342. H. Pines, W. D. Huntsman, and V. N. Ipatieff, *J. Am. Chem., Soc.*, **75**, 2315 (1953).

343. W. Naegele and H. Haubenstock, *Tetrahedron Letters*, **48**, 4283 (1965).

344. A. D. Ketley, *Polymer Eng. Sci.*, **6**, 1 (1966) *Polymer Letters*, **1**, 313 (1963).

345. S. Aoki, Y. Harita, T. Otsu, and M. Imoto, *Bull Chem. Soc. Japan*, **39**, 889 (1966).

346. C. P. Pinazzi, J. Brossas, J. C. Brosse, and F. Clouet, *Compt. Rend. Acad. Sci., Paris*, **272**, 2131 (1971).

347. C. P. Pinazzi, J. C. Brosse, J. Brossas, and A. Pleurdeau, *Compt. Rend.*, **274**, 140 (1972).

348. C. P. Pinazzi, J. C. Brosse, and A. Pleurdeau, *Makromol. Chem.*, **142**, 273 (1971).

349. C. Pinazzi, A. Pleurdeau, and J. C. Brosse, paper presented at Symposium on Kinetics and Mechanisms of Polyreactions, International Symposium on Macromolecular Chemistry, Preprints, **2**, 273 (1969).

350. S. Murahashi, S. Nozakura, and Y. Kotake, *Kobunshi Kagaku*, **22**, 652 (1965).

351. C. P. Pinazzi, A. Pleurdeau, and J. C. Brosse, *Makromol. Chem.*, **142**, 259 (1971).

352. C. P. Pinazzi, A. Pleurdeau, J. C. Brosse, and J. Brossas, *Makromol. Chem.*, **156**, 173 (1972).

353. C. Pinazzi, A. Pleurdeau, and J. P. Villette, *Compt. Rend.*, **274**, 350 (1972).

354. T. H. Bates and T. F. Williams, *Nature*, **187**, 665 (1960).

355. A. Sivola and H. Harva, *Sum. Kemistilehti B*, **43m**, 475 (1970).

356. C. S. Marvel, J. R. Hanley, Jr., and D. T. Longone, *J. Polymer Sci.*, **40**, 551 (1959).

356.a. M. Modena, R. B. Bates, and C. S. Marvel, *J. Polymer Sci., A*, **3**, 949 (1965).

357. J. M. Huet and E. Maréchal, *Compt. Rend. Acad. Sci. Paris*, **271**, 1058 (1970).

358. H. Pietila, A. Sivola, and H. Sheffer, *J. Polymer Sci., A-1*, **8**, 727 (1970).

359. T. Chou and J. P. Kennedy, unpublished data, 1972.

360. W. R. Longworth and P. H. Plesch, *J. Chem. Soc.*, **1959**, 1618.

361. P. H. Plesch, *Advan. Polymer Sci.*, **8**, 137 (1971).

362. S. Rengachary, Ph.D. thesis, The University of Akron, 1973.

362.a. T. H. Bates, J. V. F. Best, and T. F. Williams, *Nature*, **188**, 469 (1960).

363. S. H. Pinner in *The Chemistry of Cationic Polymerization*, P. H. Plesch, Ed., Macmillan, New York, 1963, chap. 4, p. 139.

364. G. Sartori, F. Ciampbelli, and N. Cameli, *Chim. Ind. (Milano)*, **45**, 1478 (1963).

365. T. Tsunio, T. Saegusa, S. Kobayashi, and T. Furukawa, *Kogyo Kagaku Zasshi*, **67**, 1961 (1964).

366. T. Oshika, S. Murai, and U. Koga, *Kobunshi Kagaku*, **22**, 633 (1965).

367. J. P. Kennedy and H. S. Makowski, *J. Macromol. Sci. (Chem.)*, **A1**, 345 (1967).

368. M. Saunders, P. R. Schleyer, and G. A. Olah, *J. Am. Chem. Soc.*, **86**, 5679 (1964) and references therein.

369. A. Takada, T. Otsu, and M. Imoto, *J. Chem. Soc. Japan, Ind. Chem. Sect.*, **69**, 715 (1966).

370. F. Landolph, *Ber. Deut. Chem. Ger.*, **10**, 1312 (1877).

371. M. Imoto, T. Otsu, A. Takata, *J. Soc. Chem. Ind. Japan*, **68**, 369 (1965).

372. J. P. Kennedy and H. S. Makowski, *J. Polymer Sci., C*, **22**, 247 (1968).

373. G. Sartori, A. Valvassori, V. Turba, and M..P. Lachi, *Chim. Ind. (Milano)*, **45**, 1529 (1963).

374. J. P. Kennedy and J. A. Hinlicky, *Polymer*, **6**, 133 (1965).

375. M. B. Roller, J. K. Gillham, and J. P. Kennedy, *J. Appl. Polymer Sci.*, **17**, 2223 (1973).

376. S. Cesca, A. Priola, and G. Santi, *Polymer Letters*, **8**, 573 (1970).

377. S. Kobayashi, T. Saegusa, and J. Furukawa, *Kogyo Kagaku Zasshi*, **70**, 372 (1967).

378. T. Corner, R. G. Foster, and P. Hepworth, *Polymer*, **10**, 393 (1969).

379. A. Takada, T. Otsu, and M. Imoto, *J. Chem. Soc. Japan, Ind. Chem. Sect.*, **69**, 711 (1966).

380. R. G. Foster and P. Hepworth, paper presented at Symposium on Kinetics and Mechanisms of Polyreactions, International Symposium on Macromolecules, Preprints, **1**, 339 (1969).

381. G. Williams, *J. Chem. Soc.*, **1938**, 246.

382. G. Williams, *J. Chem..Soc.*, **1938**, 1046.

383. G. Williams, *J. Chem. Soc.*, **1940**, 775.

384. D. C. Pepper, paper presented at International Symposium on Macromolecular Chemistry, Wiesbaden, 1959 paper III, A9.

385. D. C. Pepper and A. E. Sommerfeld in *Cationic Polymerisation and Related Complexes*, P. H. Plesch, Ed., Heffer, Cambridge, 1953, p. 75.

386. M. J. Hayes and D. C. Pepper, *Proc. Chem. Soc.*, **1958**, 228.

387. J. J. Throssell, S. P. Sood, M. Szwarc, and V. Stannett, *J. Am. Chem. Soc.*, **78**, 1122 (1956).

388. D. C. Pepper and R. E. Burton, *Proc. Roy. Soc. (London), Ser. A.*, **263**, 58 (1961).

389. M. J. Hayes and D. C. Pepper, *Proc. Roy. Soc. (London), Ser. A.*, **263**, 63 (1961).

390. A. Albert and D. C. Pepper, *Proc. Roy, Soc. (London), Ser. A*, **263**, 75 (1961).

391. K. Ikeda, T. Higashimura, and S. Okamura, *Kobunshi Kagaku*, **26**, 364 (1969).

392. K. Ikeda, T. Higashimura, and S. Okamura, *Kobunshi Kagaku*, **26**, 369 (1969).

393. S. Bywater and D. J. Worsfold, *Can. J. Chem.*, **44** (1966).

394. Z. Lisicki, J. Polaczek, and D. Gornig, *Roczniki Chem.*, **46**, 99 (1972).

395. A. Gandini and P. H. Plesch, *J. Chem. Soc.*, **1965**, 4765..

396. A. Gandini and P. H. Plesch, *Proc. Chem. Soc.*, **1964**, 240.

397. A. Gandini and P. H. Plesch, *Polymer Letters*, **3**, 127 (1965).

398. D. C. Pepper and P. J. Reilly, *Proc. Chem. Soc.*, **1961**, 200.

399. D. C. Pepper, paper presented at International Symposium on Macromolecular Chemistry, Boston, 1971 Preprints, **1**, 98 (1971).

400. L. E. Darcy, W. P. Millrine, and D. C. Pepper, *Chem. Commun.*, **1968**, 1441.

401. L. E. Darcy and D. C. Pepper, Abstracts, International Symposium on Macromolecular Chemistry, Tokyo, 1966, Vol. I, p. 42.

402. B. McCarthy, W. P. Millrine, and D. C. Pepper, *Chem. Commun.*, **1968**, 1442.

403. D. C. Pepper and P. J. Reilly, *Proc. Royal Soc. (London), Ser. A*, **291**, 41 (1966).

404. D. C. Pepper and P. J. Reilly, *J. Polymer Sci.*, **58**, 639 (1962).

405. A. Gandini and P. H. Plesch, *J. Chem. Soc.*, **1965**, 4824.

406. W. R. Longworth, P. H. Plesch, and P. P. Rutherford, *Proc. Chem. Soc.*, **1960**, 68.

407. P. H. Plesch, *J. Chem. Soc.*, **1953**, 1653.

408. P. H. Plesch in *Cationic Polymerisation and Related Complexes*, P. H. Plesch, Ed., Heffer and Sons, Cambridge, 1953, p. 85.

409. P. H. Plesch, *J. Chem. Soc.*, **1953**, 1659.

410. P. H. Plesch, *J. Chem. Soc.*, **1953**, 1662.

411. Y. Sakurada, T. Higashimura, and S. Okamura, *J. Polymer Sci.*, **33**, 496 (1958).

412. W. R. Longworth, C. J. Panton, and P. H. Plesch, *J. Chem. Soc.*, **1965**, 5579.

413. C. G. Overberger, G. F. Endres, *J. Am. Chem. Soc.*, **75**, 6349 (1953).

414. G. F. Endres and C. G. Overberger, *J. Am. Chem. Soc.*, **77**, 2201 (1955).

415. C. G. Overberger and G. F. endres, *J. Polymer Sci.*, **16**, 283 (1955).

416. C. G. Overberger, G. F. Endres, and A. Monaci, *J. Am. Chem. Soc.*, **78**, 1969 (1956).

417. C. G. Overberger, R. J. Ehring, and R. A. Marcus, *J. Am. Chem. Soc.*, **80**, 2456 (1958).

418. C. G. Overberger, and M. G. Newton, *J. Am. Chem. Soc.*, **82**, 3622 (1960).

419. G. F. Endres, V. G. Kamath, and C. G. Overberger, *J. Am. Chem. Soc.*, **84**, 4813 (1962).

420. Y. Sakurada, T. Higashimura, and S. Okamura, *J. Polymer Sci.*, **33**, 496 (1958).

421. J. T. Atkins and F. W. Billmeyer, *J. Phys. Chem.*, **63**, 1966 (1959).

422. J. George and H. Wechsler, *J. Polymer Sci.*, **6**, 725 (1951).

423. J. George, H. Wechsler, and H. Mark, *J. Am. Chem. Soc.*, **72**, 3891 (1950).

424. J. George, H. Mark, and H. Wechsler, *J. Am. Chem. Soc.*, **72**, 3896 (1950).

425. S. Okamura and T. Higashimura, *J. Polymer Sci.*, **21**, 289 (1956).

426. R. O. Colclough and F. S. Dainton, *Trans. Faraday Soc.*, **54**, 886 (1958).

427. R. O. Colclough and F. S. Dainton, *Trans. Faraday Soc.*, **54**, 901 (1958).

428. T. Higashimura and S. Okamura, *Kobunshi Kagaku*, **13**, 338 (1956).

429. D. C. Pepper, p. 70 in ref. 408.

430. T. Higashimura and S. Okamura, *Kobunshi Kagaku*, **13**, 342 (1956).

431. T. Higashimura and S. Okamura, *Kobunshi Kagaku*, **17**, 57 (1960).

432. T. Higashimura and S. Okamura, *Kobunshi Kagaku*, **13**, 397 (1956).

433. D. C. Pepper, *Trans. Faraday Soc.*, **45**, 397 (1949).

434. D. C. Pepper, *Trans. Faraday Soc.*, **45**, 404 (1949).

435. N. Kanoh, T. Higashimura, and S. Okamura, *Kobunshi Kagaku*, **19**, 181 (1962).

436. R. O. Colclough and F. S. Dainton, *Trans. Faraday Soc.*, **54**, 894 (1958).

437. T. Masuda and T. Higashimura, *J. Polymer Sci., Part A-1*, **9**, 1563 (1971).

438. S. S. Medvedev and A. R. Gantmakher, *Zh., Fiz. Khim.*, **23**, 516 (1949).

439. S. S. Medvedev and A. R. Gantmakher, *Zh., Fiz. Khim.*, **25**, 1328 (1951).

440. R. O. Colclough and F. S. Dainton, *Trnas. Faraday Soc.*, **54**, 898 (1958).

441. K. Heald and G. Williams, p. 78 in ref. 408.

442. H. Imai, T. Saegusa, and J. Furukawa, *Makromol. Chem.*, **81**, 92 (1965).

443. J. P. Kennedy, *J. Macromol. Sci., Chem.*, **A3**, 861 (1969).

444. J. P. Kennedy, *J. Macromol. Sci., Chem.*, **A3**, 885 (1969).

445. T. Saegusa, H. Imai, and J. Furukawa, *Makromol. Chem.*, **79**, 207 (1964).

446. T. Masuda and T. Higashimura, *Polymer Letters*, **9**, 783 (9171).

447. T. Higashimura, F. Fukushima, and S. Okamura, *J. Macromol. Sci., Chem.*, **A1**, 683 (1967).

448. T. Kagiya, M. Izu, H. Maruyama, and K. Fukui, *J. Polymer Sci., A-1*, **7**, 917 (1969).

449. D. S. Trifan and P. D. Bartlett, *J. Am. Chem. Soc.*, **81**, 5573 (1959).

450. S. Okamura, N. Kanoh, and T. Higashimura, *Makromol. Chem.*, **47**, 19 (1961).

451. N. Kanoh, T. Higashimura, and S. Okamura, *Makromol. Chem.*, **56**, 65 (1962).

452. N. Kanoh, K. Ikeda, A. Gotoh, T. Higashimura, and S. Okamura, *Makromol. Chem.*, **86**, 200 (1965).

453. S. Aoki, T. Otsu, and M. Imoto, *Chem. Ind.*, **1965**, 1761.

454. J. A. Bittles, A. K. Chaudhur, and S. W. Benson, *J. Polymer Sci., A*, **2**, 1221 (1964).

455. A. Asami and N. Tokura, *J. Polymer Sci.*, **42**, 545 (1960).

456. A. Asami and N. Tokura, *J. Polymer Sci.*, **42**, 553 (1960).

457. N. Sakota, H. Nakamura, and K. Nishihara, *Makromol. Chem.*, **129**, 56 (1969).

458. Y. Minoura and H. Toshima, *J. Polymer Sci., A-1*, **11**, 1109 (1972).

459. A. R. Mathieson in *The Chemistry of Cationic Polymerization*, P. H. Plesch, Ed., Macmillan, New York, p. 235.

460. K. Matsuzaki, T. Uryu, K. Osada, and T. Kawamura, *Macromol.*, **5**, 816 (1972).

461. D. O. Jordan and F. E. Treolar, *J. Chem. Soc.*, **1961**, 737.

462. J. P. Kennedy, *J. Org. Chem.*, **35**, 532 (1970).

463. P. H. Plesch, *Advan. Polymer Sci.*, **8**, 139 (1971).

463.a. D. C. Pepper, *European Polymer J.*, **1**, 11 (1965).

464. P. Giusti, G. Puce, and F. Andruzzi, *Makromol. Chem.*, **98**, 170 (1966).

464.a. P. Giusti and F. Andruzzi, *Gazz. Chim. Ital.*, **96**, 1563 (1966).

465. S. D. Hamann, A. J. Murphy, D. H. Solomon, and R. I. Willing, *J. Macromol. Sci., Chem.*, **A6**, 771 (1972).

466. T. Higashimura in *Structure and Mechanism in Vinyl Polymerization*, T. Tsuruta and K. F. O'Driscoll, Eds., Dekker, New York, 1969, chap. 10.

467. P. Giusti, F. Andruzzi, P. Cerrai, and G. L. Possanzini, *Makromol. Chem.*, **136**, 97 (1970).

468. T. Higashimura, H. Kusano, T. Masuda, and S. Okamura, *Polymer Letters*, **9**, 463 (1971).

469. W. R. Longworth, C. J. Panton, and P. H. Plesch, *J. Chem. Soc.*, **1965**, 5579.

470. I. M. Panayotov, I. K. Dimitrov, and I. E. Bakerdjiev, *J. Polymer Sci., A-1*, **7**, 2421 (1969).

471. N. Ise, *Advan. Polymer Sci.*, **6**, 347 (1969) and references therein.

472. I. Sakurada, N. Ise, and Y. Hayashi, *J. Macromol. Sci., Chem.*, **A1**, 1039 (1967).

473. N. Ise, Y. Hayashi, S. X. Chen, and I. Sakurada, *Makromol. Chem.*, **117**, 180 (1968).

474. P. Giusti, P. Cerrai, M. Tricoli, P. L. Magagnini, and F. Andruzzi, *Makromol. Chem.*, **113**, 299 (1970).

475. K. Takaya, H. Hirohara, and N. Ise, *Makromol. Chem.*, **139**, 277 (1970).

476. J. W. Breitenbach and C. Srna, *Pure Appl. Chem.*, **4**, 245 (1962).

477. S. H. Pinner in *The Chemistry of Cationic Polymerization*, P. H. Plesch, Ed., Macmillan, New York, chap. 17, p. 611.

477.a. J. F. Westlake and R. Y. Huang, *J. Polymer Sci., Polymer. Chem. Ed.*, **10**, 3053 (1972) and six previous papers in this series.

478. F. Williams in *Fundamental Processes in Radiation Chemistry*, P. Ausloos, Ed., Interscience, 1968, p. 515.

479. R. C. Potter, C. Schneider, M. Ryska, and D. O. Hummel, *Angew. Chem. Intern. Ed.*, **7**, 845, 1968.

480. F. Williams, *Quart. Rev.*, **17**, 101 (1963).

481. D. J. Metz, *Advan. Chem. Ser.*, **91**, 202 (1969).

482. A. Charlesby and J. Morris, *Proc. Royal Soc., (London) Ser. A*, **281**, 392 (1964).

483. K. Ueno, H. Yamaoka, K. Hayashi, and S. Okamura, *Intern. Appl. Radiation Isotopes*, **17**, 513 (1966).

484. A. Chapiro and V. Stannett, *J. Chem. Phys.*, **57**, 35 (1960).

485. M. Magat, *J. Polymer Sci.*, **48**, 379 (1960).

486. A. Gandini and P. H. Plesch, *European Polymer J.*, **4**, 55 (1968).

487. C. P. Brown and A. R. Mathieson, *Ric. Sci., Suppl. Simp. Intern. Chim. Macromol.*, **25**, 154 (1955).

488. C. P. Brown and A. R. Mathieson, *J. Chem. Soc.*, **1957**, 3612.

489. C. P. Brown and A. R. Mathieson, *J. Chem. Soc.*, **1957**, 3631.

490. C. P. Brown and A. R. Mathieson, *J. Chem. Soc.*, **1958**, 3445.

491. C. P. Brown and A. R. Mathieson, *J. Chem. Soc.*, **1957**, 3620.

492. C. P. Brown and A. R. Mathieson, *Trans. Faraday Soc.*, **53**, 1033 (1957).

493. D. O. Jordan and F. E. Treolar, *J. Chem. Soc.*, **1961**, 729.

494. D. O. Jordan and A. R. Mathieson, *J. Chem. Soc.*, **1952**, 611.

495. D. O. Jordan and A. R. Mathieson, *Nature*, **161**, 523 (1951).

496. D. O. Jordan and A. R. Mathieson, *J. Chem. Soc.*, **1952** (621).

497. D. Clark, p. 99 in ref. 408.

498. S. Okamura, T. Higashimura, and Y. Ogawa, *Kobunshi Kagaku*, **16**, 239 (1959).

499. M. Imoto and S. Aoki, *Makromol. Chem.*, **48**, 72 (1961).

500. M. Imoto and S. Aoki, *Makromol. Chem.*, **63**, 141 (1963).

501. S. Aoki and M. Imoto, *Makromol. Chem.*, **65**, 243 (1963).

502. S. Okamura and T. Higashimura, *Kobunshi Kagaku*, **13**, 262 (1956).

503. S. Okamura and T. Higashimura, *J. Polymer Sci.*, **20**, 581 (1956).

504. S. Okamura and T. Higashimura, *Kobunshi Kagaku*, **15**, 708 (1958).

505. G. Williams and H. Thomas, *J. Chem. Soc.*, **1948**, 1867.

506. G. Williams and H. Bardsley, *J. Chem. Soc.*, **1952**, 1707.

507. K. Heald and G. Williams, *J. Chem. Soc.*, **1954**, 357.

508. Y. Landler, *Proc. Intern. Coll. Macromol., Amsterdam*, **1949**, 70.

509. R. O. Colclough, *J. Polymer Sci.*, **8**, 467 (1952).

510. M. Kamachi and H. Miyama, *J. Polymer Sci., A*, **3**, 1337 (1965).

511. M. Kamachi and H. Miyama, *J. Polymer Sci., A-1*, **6**, 1537 (1968).

512. M. Kamachi and H. Miyama, *J. Polymer Sci., A-1*, **5**, 3207 (1967).

513. N. Kanoh, A. Gotoh, T. Higashimura, and S. Okamura, *Makromol. Chem.*, **63**, 106 (1963).

514. N. Kanoh, A. Gotoh, T. Higashimura, and S. Okamura, *Makromol. Chem.*, **63**, 115 (1963).

515. N. Kanoh, K. Ikeda, A. Gotoh, T. Higashimura, and S. Okamura, *Makromol. Chem.*, **86**, 200 (1965).

516. N. D. Prishchepa, Y. Y. Goldfarb, and B. A. Krentsel, *Vysomol. Soedin.*, **8**, 1658 (1966).

517. N. D. Prishchepa, Y. Y. Goldfarb, B. A. Krentsel, and M. V. Shishkina, *Vysomol. Soedin.*, **9**, 2426 (1967); and *Polymer Sci. USSR*, **9**, 2743 (1967).

518. J. P. Kennedy, P. L. Magagnini, and P. H. Plesch, *J. Polymer Sci., A-1*, **9**, 1635 (1971).

519. A. Mizote, T. Higashimura, and S. Okamura, *J. Polymer Sci., A-1*, **6**, 1825 (1968).

520. C. G. Overberger, L. H. Arond, D. Tanner, J. J. Taylor, and T. Alfrey, Jr., *J. Am. Chem. Soc.*, **74**, 4848 (1952).

520.a. C. G. Overberger, E. M. Pierce, and D. Tanner, *J. Am. Chem. Soc.*, **80**, 1761 (1955).

521. R. Disselhoff, Diplomarbeit, Technical University, Darmstadt, Germany, 1964.

522. P. L. Magagnini, P. H. Plesch, and J. P. Kennedy, *European Polymer J.*, **7**, 1161 (1971).

523. C. Aso, T. Kunitake and S. Shinkai, *Kobunshi Kagaku*, **26**, 280 (1969).

524. J. P. Kennedy, P. L. Magagnini, and P. H. Plesch, *J. Polymer Sci., A-1*, **9**, 1647 (1971).

525. A. Anton and E. Maréchal, *Bull. Soc. Chim. France*, **1971**, 3753.

526. A. Anton and E. Maréchal, *Bull. Soc. Chim. France*, **1971**, 2669.

527. R. Lalande, J. P. Pillion, F. Flies, and J. Roux, *Compt. Rend.*, **274**, 2060.

528. F. R. Buck, G. T. Kennedy, F. Morton, and E. M. Tanner, *Nature*, **162**, 103 (1948).

529. V. V. Korshak and N. G. Matveeva, *J. Gen. Chem. USSR*, **22**, 1173 (1952).

530. F. M. Aliev, *Azerb. Khim. Zh.*, **1971**, 109.

531. G. R. Brown and D. C. Pepper, *Polymer*, **6**, 497 (1965).

532. T. Alfrey, Jr., and H. Wechsler, *J. Am. Chem. Soc.*, **70**, 4266 (1948).

533. C. G. Overberger and V. G. Kamath, *J. Am. Chem. Soc.*, **81**, 2910 (1959).

534. K. Saotome and M. Imoto, *J. Chem. Soc. Japan*, **62**, 1130 (1959).

535. C. G. Overberger, R. J. Ehrig, and D. Tanner, *J. Am. Chem. Soc.*, **76**, 772 (1954).

536. C. G. Overberger, D. Tanner, and E. M. Pearce, *J. Am. Chem. Soc.*, **80**, 4566 (1958).

537. G. Smets and L. de Haes, *Bull. Soc. Chim. Belges*, **59**, 29 (1950).

538. C. G. Overberger, L. H. Arond, and J. J. Taylor, *J. Am. Chem. Soc.*, **73**, 5541 (1951).

539. M. A. S. Mondal and R. N. Young, *European Polymer J.*, **7**, 1575 (1971).

540. T. Alfrey, L. Arond, and C. G. Overberger, *J. Polymer Sci.*, **4**, 539 (1949).

541. R. E. Florin, *J. Am. Chem. Soc.*, **73**, 4468 (1951).

542. R. E. Florin, *J. Am. Chem. Soc.*, **71**, 1867 (1949).

543. M. Kato, *J. Polymer Sci., A-1*, **7**, 2405 (1969).

544. M. Kato and H. Kamogawa, *J. Polymer Sci., A-1,* 4, 1773 (1966).

545. H. Staudinger and E. Dreher, *Ann.,* 517, 73 (1935).

546. S. Matsushita, T. Higashimura, and S. Okamura, *Kobunshi Kagaku,* 17, 456 (1960).

547. P. M. Kamath and H. C. Haas, *J. Polymer Sci.,* 24, 143 (1957).

548. Y. Imanishi, S. Matsushita, T. Higashimura, and S. Okamura, *Makromol. Chem.,* 70, 68 (1964).

549. S. Okamura, N. Kanoh, and T. Higashimura, *Makromol. Chem.,* 47, 19 (1961).

550. G. Natta, G. Dall'Asta, G. Mazzanti, and A. Casale, *Makromol. Chem.,* 58, 217 (1962).

551. H. Kamogawa and H. C. Cassidy, *J. Polymer Sci., A,* 2, 2409 (1964).

552. J. P. Tortai, M. Mayen, and E. Maréchal, Abstracts, IUPAC International Symposium Macromolecules, Boston, 1971, Vol. 1, p. 172.

553. R. H. Wiley and N. R. Smith, *J. Polymer Sci.,* 3, 444 (1948).

554. A. B. Hershberger, J. C. Reid, and R. G. Heiligmann, *Ind. Eng. Chem.,* 37, 1073 (1945).

555. S. Bywater in *The Chemistry of Cationic Polymerization,* P. H. Plesch, Ed., Macmillan, New York, 1963. p. 313.

556. Picco Technical Bulletin PPN 131, 132 and 133, Pennsylvania Industrial Chemicals Co., Pennsylvania.

557. S. Bywater in *The Chemistry of Cationic Polymerization,* P. H. Plesch, Ed., Macmillan, New York, 1963, p. 305.

558. S. Brownstein, S. Bywater, and D. J. Worsfold, *Makromol. Chem.,* 48, 127 (1961).

559. Y. Matsuguma and T. Kunitake, *Polymer J.,* 2, 353 (1971).

560. Y. Sakurada, K. Imai, and M. Matsumoto, *Kobunshi Kagaku,* 20, 429 (1963).

561. Y. Sakurada, T. Mochizuki, and M. Matsumoto, *Kobunshi Kagaku,* 20, 436 (1963).

562. D. J. Worsfold and S. Bywater, *J. Am. Chem. Soc.,* 79, 4917 (1957).

563. S. Okamura, T. Higashimura, and Y. Imanishi, *Kobunshi Kagaku,* 16, 69 (1959).

564. S. Okamura, T. Higashimura, and Y. Imanishi, *Kobunshi Kagaku,* 16, 129 (1959).

565. S. Okamura, T. Higashimura, and Y. Imanishi, *Kobunshi Kagaku,* 16, 244 (1959).

566. S. Okamura, T. Higashimura, and Y. Imanishi, *J. Polymer Sci.,* 33, 491 (1958).

567. Y. Ohsumi, T. Higashimura, and S. Okamura, *J. Polymer Sci., A,* 3, 3729 (1965).

568. Y. Manishi, T. Higashimura, and S. Okamura, *Kobunshi Kagaku,* 17, 236 (1960).

569. Y. Sakurada, *J. Polymer Sci., A,* 1, 2407 (1963).

570. Y. Ohsuma, T. Higashimura, and S. Okamura, *J. Polymer Sci., A1,* 4, 923 (1966).

571. T. Kunitake and C. Aso, *J. Polymer Sci., A-1,* 8, 665 (1970).

572. H. Staudinger and F. Breusch, *Ber.,* 62, 452 (1929).

573. Y. Inaki, S. Nozakura, and S. Murahashi, *J. Macromol. Sci. Chem., A6,* 313 (1972).

574. R. Branchu, H. Cheradame, and P. Sigwalt, *Compt. Rend.,* 268, 1292 (1969).

575. D. Braun and G. Heufer, *Makromol. Chem.,* 79, 98 (1964).

576. D. O. Jordan and A. R. Mathieson, *J. Chem. Soc.,* 1952, 2354.

577. Y. Sakurada and M. Veda, *Kobunshi Kagaku,* 20, 417 (1963).

578. Y. Sakurada, K. Imai and M. Matsumoto, *Kobunshi Kagaku,* 20, 422 (1963).

579. Y. Sakurada, M. Matsumato, K. Imai, A. Nishioka, and Y. Kato, *Polymer Letters,* 1, 633 (1963).

580. F. S. Dainton, R. H. Tomlinson, and T. L. Batke in *Cationic Polymerisation and Related Complexes*, P. H. Plesch, Ed., W. Heffer and Sons, Cambridge, 1953, p. 80.

581. F. S. Dainton and R. H. Tomlinson, *J. Chem. Soc.*, 1953, 151.

582. D. O. Jordan and A. R. Mathieson, *J. Chem. Soc.*, 1957, 3507.

583. K. Hirota, G. Meshitsuka, F. Takemura, and T. Tanaka, *Bull. Chem. Soc. Japan*, 33, 1316 (1960).

584. D. Braun, G. Heufer, U. Johnsen, and K. Kolbe, *Ber. Bunsenges*, 68, 959 (1964).

585. K. C. Ramey, G. L. Statton, and W. C. Jankowski, *J. Polymer Sci.*, *B*, 7, 639 (1969).

585.a. M. Irie, S. Tomimoto, and K. Hayashi, *J. Polymer Sci.*, *Polymer Chem. Ed.*, 10, 3235 (1972).

586. K. C. Ramey and G. L. Statton, *Makromol. Chem.*, 85, 287 (1965).

587. T. Higashimura, T. Kodama, and S. Okamura, *Kobunshi Kagaku*, 17, 163 (1960).

588. T. Higashimura, T. Yonezawa, S. Okamura, and K. Fukui, *J. Polymer Sci.*, 39, 487 (1959).

589. G. Smets and L. DeHaes, *Bull. Soc. Chem. Belges*, 59, 29 (1950).

590. E. M. Banas and O. O. Juveland, *J. Polymer Sci.*, *A-1*, 5, 397 (1967).

591. J. M. Van der Zanden and T. R. Rix, *Rec. Trav. Chim. Pays-Bas*, 75, 1166 (1956).

592. H. Brunner, A. L. Pallvel, and D. J. Walbridge, *J. Polymer Sci.*, 28, 629 (1958).

593. A. A. D'Onofrio, *J. Appl. Polymer Sci.*, 8, 521 (1964).

594. V. V. Korshak and N. G. Matveeva, *Bull. Acad. Sci.*, *USSR*, *Div. Chem. Sci.*, 1953, 547.

595. V. V. Korshak and N. G. Matveeva, *Bull. Acad. Sci.*, *USSR*, *Div. Chem. Sci.*, 1955, 855.

596. B. Elliott, A. G. Evans, and E. D. Owen, *J. Chem. Soc.*, 1962, 689.

597. J. P. Kennedy, *J. Polymer Sci.*, *A*, 2, 5171 (1964).

598. A. Mizote, T. Tanaka, T. Higashimura, and S. Okamura, *J. Polymer Sci.*, 3, 2567 (1965).

599. S. Murahashi, S. Nozakura, K. Tsuboshima, and Y. Kotake, *Bull. Chem. Soc. Japan*, 37, 706 (1964).

600. S. Murahashi, S. Nozakura, K. Tsuboshima, and Y. Kotake, *Bull. Chem. Soc. Japan*, 38, 156 (1965).

601. A. Shimizu, T. Otsu, and M. Imoto, *Bull. Chem. Soc. Japan*, 41, 953 (1968).

602. A. Mizote, T. Higashimura, and S. Okamura, *Pure Appl. Chem.*, 16, 457 (1968).

603. T. Higashimura and M. Hoshino, *J. Polymer Sci.*, *A-1*, 10, 673 (1972).

604. J. Gerhardt, *J. Prakt. Chem.*, 36 (I) 270 (1845).

605. H. Staudinger and M. Brunner, *Helv. Chim. Acta*, 12, 972 (1929).

606. M. Secci and L. Mameli, *Ann. Chim.*, 47, 580 (1957).

607. P. Sigwalt, *Compt. Rend.*, 252, 3988 (1961).

608. P. Cerrai, F. Andruzzi, and P. Giusti, *Chim. Ind. (Milano)*, 51, 681 (1969).

609. P. Cerrai, F. Andruzzi, and P. Giusti, *Chim. Ind. (Milano)*, 51, 687 (1969).

610. T. Higashimura, K. Kawamura, and T. Masuda, *J. Polymer Sci.*, *A-1*, 10, 85 (1972).

611. G. S. Whitby and M. Katz, *J. Am. Chem. Soc.*, 50, 1160 (1928).

612. H. Staudinger, A. A. Ashdown, M. Brunner, H. A. Bruson, and S. Wehrli, *Helv. Chim. Acta*, 12, 934 (1929).

613.   N. D. Prishchepa, Y. Y. Goldfarb, and B. A. Krentsel, *Usp. Khim.*, **35**, 1986 (1966).

613.a. C. P. Brown and A. R. Mathieson, *J. Chem. Soc.*, **1957**, 3608.

614.   S. Murahashi, S. Nozakura, and Y. Kotake, Annual Meeting of the Society of Polymer Science, Japan, May 1-3, 1965, Tokyo, Paper 3C/04.

614.a. P. Sigwalt, *J. Polymer Sci.*, **52**, 15 (1961).

615.   D. C. Pepper, *Polymer Preprints*, 7, 426 (1966).

615.a. I. Bkouche-Waksman, and P. Sigwalt, *Compt. Rend.*, **255**, 680 (1962).

616.   E. Maréchal, *J. Polymer Sci., A-1*, 8, 2867 (1970).

617.   H. Cheradame, *Chim. Ind. (Paris), Genie Chim.*, **97**, 1500 (1967).

618.   H. Cheradame and P. Sigwalt, *Compt. Rend.*, **260**, 159 (1965).

619.   H. Cheradame, N. A. Hung, and P. Sigwalt, *Compt. Rend.*, **268**, 476 (1969).

620.   A. Polton, *Chim. Ind. (Paris), Genie Chem.*, **102**, 875 (1968).

621.   A. Polton and P. Sigwalt, *Compt. Rend.*, **268**, 1214 (1969).

622.   A. Polton and P. Sigwalt, *Bull. Soc. Chim. France*, **1**, 131 (1970).

623.   A. Polton and P. Sigwalt, *Compt. Rend.*, **265**, 1303 (1967).

624.   P. Sigwalt and E. Maréchal, *European Polymer J.*, **2**, 15 (1966).

625.   E. Maréchal and P. Evrard, *Bull. Soc. Chim. France*, **1969**, 2039.

626.   A. Miztoe, T. Tanaka, T. Higashimura, and S. Okamura, *J. Polymer Sci., A-1*, **4**, 869 (1966).

627.   A. D. Eckard, A. Ledwith, and D. C. Sherrington, *Polymer*, **12**, 444 (1971).

628.   O. C. Filho and A. S. Gomes, *Polymer Letters*, 9, 891 (1971).

629.   A. S. Gomes and O. C. Filho, *Polymer Letters*, 10, 725 (1972).

630.   E. Maréchal, J. J. Basselier, and P. Sigwalt, *Bull. Soc. Chim. France*, **8**, 1740 (1964).

631.   P. Caillaud, J. M. Huet, and E. Maréchal, *Bull. Soc. Chim. France*, **1970**, 1473.

632.   E. Maréchal, *Bull. Soc. Chim.*, **1969**, 1459.

633.   E. Maréchal, *Compt. Rend.*, **268C**, 1121 (1969).

634.   D. S. Brackman and P. H. Plesch, *Chem. Ind. (London)*, **1955**, 255.

635.   A. G. Heilbrunn and E. Maréchal, *Compt. Rend.*, **274**, 1149 (1972).

636.   E. Maréchal, P. Evrard, and P. Sigwalt, *Bull. Soc. Chim. France*, **1969**, 1981.

637.   A. Anton, J. Zwegers, and E. Maréchal, *Bull. Soc. Chim. France*, **1970**, 1466.

638.   A. Anton and E. Maréchal, *Bull. Soc. Chim. France*, **1971**, 2256.

639.   J. P. Quere and E. Maréchal, *Bull. Soc. Chim., France*, **1971**, 2983.

640.   E. Maréchal and B. Hamy, *Compt. Rend.*, **268**, 41 (1969).

641.   E. Maréchal and P. Sigwalt, *Bull. Soc. Chim. France*, **1966**, 1071.

642.   E. Maréchal and P. Sigwalt, *Bull. Soc. Chim. Frnace*, **1966**, 1075.

643.   E. Maréchal and P. Sigwalt, *Bull. Soc. Chim. France*, **1966**, 3847.

644.   E. Maréchal and A. Lepert, *Bull. Soc. Chim. France*, **1967**, 2954.

645.   J. P. Tortai and E. Maréchal, *Bull. Soc. Chim., France*, **1971**, 2673.

646.   A. W. Schmidt and V. Schöller, *Brennstoff-Chem.*, **23**, 247 (1942).

647.   G. F. D'Alelio, A. B. Finestone, L. Taft, and T. J. Miranda, *J. Polymer Sci.*, **45**, 83 (1960).

648.   E. B. Davidson, *Polymer Letters*, 4, 175 (1966).

649.   J. P. Kennedy, C. A. Cohen, and W. Naegele, *Polymer Letters*, 2, 1159 (1964).

650. J. M. Van der Zanden, and G. de Vries, *Rec. Trav. Chim. Pays-Bas,* 75, 1166 (1951).

651. J. M. Van der Zanden, and G. de Vries, *Rec. Trav. Chim. Pays-Bas,* 71, 879 (1952).

652. J. M. Van der Zanden, and G. de Vries, *Rec. Trav. Chim. Pays-Bas,* 74, 876 (1955).

653. K. Dziewonski and T. Stolyhow, *Ber.,* 57, 1531 (1924) and earlier references till 1915.

654. R. G. Flowers and H. F. Miller, *J. Am. Chem. Soc.,* 69, 1388 (1947).

655. I. Jones, *J. Appl. Chem.,* 1, 568 (1951).

656. M. Imoto and K. Takemoto, *J. Polymer Sci.,* 15, 271 (1955).

657. M. Imoto and I. Soematsu, *Bull. Chem. Soc. Japan,* 34, 26 (1961).

658. V. M. Story and G. Canty, *J. Res. Natl. Bur. Std. A,* 68, 165 (1964).

659. K. Saotome and M. Imoto, *Kobunshi Kagaku,* 41, 368 (1958).

660. K. Saotome and M. Imoto, *Kobunshi Kagaku,* 41, 373 (1958).

661. E. D. Bergman and D. Katz, *J. Chem. Soc.,* 1958, 3216.

662. R. H. Michel, *J. Polymer Sci., A,* 2, 2533 (1964).

663. R. G. Flowers and F. S. Nichols, *J. Am. Chem. Soc.,* 71, 3104 (1949).

664. H. J. Foster and G. Manecke, *Makromol. Chem.,* 133, 53 (1970).

665. C. Aso and R. Kita, *Kogyo Kagaku Zasshi,* 68, 707 (1965).

666. C. Aso, T. Kunitake and R. Kita, *Makromol. Chem.,* 97, 31 (1966).

667. C. Aso, T. Kunitake, Y. Matsuguma, and Y. Imaizumi, *J. Polymer Sci., A-1,* 6, 3049 (1968).

668. C. S. Marvel and E. J. Gall, *J. Org. Chem.,* 25, 1784 (1960).

669. N. D. Field, *J. Org. Chem.,* 25, 1784 (1960).

670. M. Kato and H. Kamogawa, *J. Polymer Sci., A-1,* 6, 2993 (1968).

671. E. Maréchal, P. Evrard, and P. Sigwalt, *Bull. Soc. Chim., France,* 1968, 2049.

672. J. P. Quere and E. Maréchal, *Bull. Soc. Chim., France,* 1971, 2227.

673. S. Kohjiya, K. Nakamura, and S. Yamashita, *Angew. Macromol. Chem.,* 27, 189 (1972).

674. M. Kucera, J. Svabik, and K. Majerova, *Coll. Czech. Chem. Commun.,* 32, 2708 (1972).

675. C. E. H. Bawn, C. Fitzsimmons, and A. Ledwith, *Proc. Chem. Soc.,* 1964, 391.

676. J. M. Barrales-Rienda and D. C. Pepper, *Polymer,* 8, 351 (1967).

677. J. M. Barrales-Rienda and D. C. Pepper, *Polymer,* 8, 337 (1967).

678. M. Imoto and K. Saotome, *J. Polymer Sci.,* 31, 208 (1958).

679. J. P. Kennedy in *Copolymerization,* G. E. Ham, Ed., Interscience, New York, 1964, chap. 5.

680. P. Belliard and E. Maréchal, *Bull. Soc. Chim. France,* 1972, 4255.

681. P. Cerrai, F. Andruzzi, and P. Giusti, *Makromol. Chem.,* 117, 128 (1968).

682. P. Giusti, P. Cerrai, F. Andruzzi, and P. L. Magagnini, *Kinetics Mech. Polyreactions,* 1, 385 (1969).

683. A. L. Müller, L. Toldy, and Z. Racz, *Ber.,* 77, 777 (1944).

684. G. Heublein and G. Agatha, *J. Pract. Chem.,* 312, 300 (1970).

685. C. A. Barson, J. R. Knight, and J. C. Robb, *Brit. Polymer J.,* 4, 427 (1972).

686. G. Heublein and H. Dawczynski, *J. Prakt. Chem.,* 314, 557 (1972).

687. A. Priola, S. Cesca, and G. Ferraris, *Markrolmol. Chem.,* 160, 41 (1972).

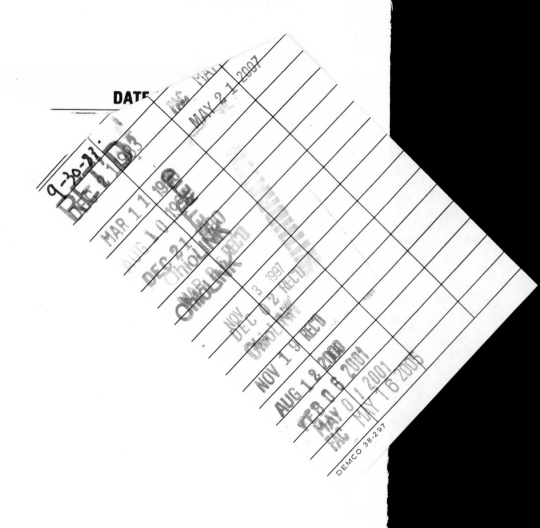